Protein Targeting

EDITED BY

Stella M. Hurtley

SCIENCE
Cambridge, UK

IRL PRESS
—at—
OXFORD UNIVERSITY PRESS
Oxford New York Tokyo

Oxford University Press, Walton Street, Oxford OX2 6DP

Oxford   New York
Athens   Auckland   Bangkok   Bombay
Calcutta   Cape Town   Dar es Salaam   Delhi
Florence   Hong Kong   Istanbul   Karachi
Kuala Lumpur   Madras   Madrid   Melbourne
Mexico City   Nairobi   Paris   Singapore
Taipei   Tokyo   Toronto
and associated companies in
Berlin   Ibadan

Oxford is a trade mark of Oxford University Press

Published in the United States
by Oxford University Press Inc., New York

A catalogue record for this book is available from the British Library

Library of Congress Cataloging in Publication Data
(Data available)
ISBN 0 19 963562 5 (Hbk)
ISBN 0 19 963561 7 (Pbk)

Typeset by Footnote Graphics, Warminster, Wilts
Printed in Great Britain by Bath Press Ltd, Bath, Avon

Protein Targeting

This i
or befo

Frontiers in Molecular Biology

SERIES EDITORS

B. D. Hames

Department of Biochemistry and Molecular Biology
University of Leeds, Leeds LS2 9JT, UK

AND

D. M. Glover

Cancer Research Campaign Laboratories
Department of Anatomy and Physiology
University of Dundee, Dundee DD1 4HN, UK

TITLES IN THE SERIES

Preface

Protein targeting — how proteins find the correct, functional location in the cell — is one of the fundamental processes of life. Within the pages of this book the molecular mechanisms of several protein targeting events are described. Because the majority of proteins are manufactured in the cytosol, most of the targeting processes outlined look at the process by which such proteins find their way to the correct intracellular destination from the cytosol. All of the chapters concentrate on protein targeting within eukaryotes; in order to keep the book to a manageable size, protein targeting within bacteria and virus assembly have been neglected. Also I must apologize to those of you whose first love is plant biology. While much has been learnt recently about protein targeting to the chloroplast, the general principles that are emerging are very similar to what has been learnt by looking at mitochondrial import (Chapter 1). Similarly, studies on targeting to other organelles within plant cells are yielding results similar to those found in other eukaryotic cells.

Throughout the pages of the book similar principles will emerge for protein targeting to distinct organelles. Newly synthesized proteins contain protein targeting signals which act as molecular address tags to allow the protein to engage with the protein targeting machinery on any particular organelle. Sometimes targeting actually occurs co-translationally (Chapter 4). Targeting signals may consist of linear sequences of amino acids, as found in signal sequences for targeting to the endoplasmic reticulum (Chapter 4) or to the mitochondria (Chapter 1). Alternatively, they may consist of more complicated motifs such as the tyrosine-containing β-turns that cause plasma membrane proteins to be targeted to coated vesicles during endocytosis (Chapter 7). Once in the correct organelle, proteins may undergo further trafficking that will require the sequential reading of multiple targeting signals, which may consist of sugars, the oligomeric state of a protein, or other signals, some of which have yet to be identified (Chapters 5–7). The various molecular machineries that recognize targeting signals also have similar basic characteristics. For proteins which must be translocated across one or more membranes to achieve their final, functional location, cytosolic proteins recognize and bind to the newly synthesized targeting signals and often act as molecular chaperones to maintain the protein in a transport-competent state. At the target organelle a receptor molecule recognizes and binds to the protein–chaperone complex and translocation will be initiated. The mechanism of translocation itself is generally thought to involve some sort of 'pore' in the target membrane to allow transport. Energy, in the form of ATP, will generally be required to drive translocation and to ensure that the transport reaction remains vectorial. Inside the target organelle the translocating molecules will often be met by further chaperone molecules which may play a role both in the folding and assembly of the newly translocated proteins and also in providing the energy-source for translocation itself. At the final destination targeting signals will often be destroyed by proteolysis or other post-translational modifications which may be involved both in making transport irreversible, and in producing a mature, functional protein.

The initial targeting of a newly synthesized protein will often not be the end of a protein's requirement for targeting. Nearly all of the proteins found in the Golgi apparatus, secretory vesicles and granules, plasma membrane, endosomes, and lysosomes start out by being co-translationally translocated into the endoplasmic reticulum (Chapter 4). They will then need further sorting signals to allow them to be transported by vesicle-mediated intracellular transport to the correct destination. Both signal-mediated transport and retention mechanisms are required to maintain the distinctive protein compositions of the various organelles, which continually exchange components by vesicular traffic (Chapters 5–7).

Import into the nucleus provides a variation on the general theme. The nucleus is surrounded by the nuclear envelope which is perforated by pores which could allow the free diffusion of molecules from the cytosol into the nucleus and vice versa. However, most nuclear proteins, are in fact, specifically targeted to the nucleus by a mechanism that involves the recognition of targeting signals in the protein by cytoplasmic receptor molecules, transport to the nucleus, and binding to and active transport across the nuclear pore (Chapter 2). Because the nucleus actually disassembles during cell division, releasing its contents to the cytosol, the signals on nuclear proteins are not destroyed and remain active for multiple rounds of nuclear import and release over the lifetime of the protein.

A general failure of protein targeting would be incompatible with life. However, defects in the targeting of specific proteins to specific destinations are known, and will generally lead to relatively severe disease. For example, lysosomal storage diseases are often associated with defects in the recognition of the mannose-6-phosphate signal found on lysosomal enzymes. Similarly the defect in some other disorders has been shown to involve protein import into peroxisomes (Chapter 3). The characterization of certain disorders has been one of the major keys in the elucidation of the mechanisms involved in protein targeting—for example, the characterization of the defect in the low-density lipoprotein receptor which can lead to familial hypercholesterolaemia was one of the key findings in the elucidation of the mechanism of protein targeting to clathrin-coated endocytic vesicles.

Yeast strains with general defects in protein targeting have also been isolated. As would be expected the defects are generally conditional lethal mutations, but characterization of such mutants has been invaluable in advancing our knowledge of the molecular mechanisms of protein targeting.

In addition to knowledge gained from studying cells with defects in protein targeting, much molecular insight into the steps involved in protein targeting has been gained from the reconstitution of protein targeting in cell-free systems. Many protein targeting steps have been reconstituted *in vivo* including protein translocation into the ER (Chapter 4), mitochondria (Chapter 1), chloroplasts and peroxisomes (Chapter 3); protein import into isolated nuclei (Chapter 2); and also multiple stages of intracellular transport (Chapters 5–7).

I hope that through reading this book you will be infected with the enthusiasm of all of the authors for getting to grips with the molecular details of protein targeting. While much has been learnt, the area is still a very active one for numerous molecular and cell biologists world-wide.

S. M. H.

Cambridge
June 1995

Acknowledgements

I would like to take the opportunity to thank all those who have contributed to the production of this book, including the authors, my friends and colleagues in Edinburgh and at SCIENCE, the Series Editors, and the team at Oxford University Press who have helped throughout the process.

Contents

Contributors

RUDOLF BAUERFEIND
Institute for Neurobiology, University of Heidelberg, Im Neuenheimer Feld 364, D-69120 Heidelberg, Germany.

ANN CORSI
Molecular and Cell Biology, 401 Barker Hall, University of California at Berkeley, Berkeley, CA 94720, USA.

CHRISTOPHER J. DANPURE
MRC Laboratory for Molecular Cell Biology and Department of Biology, University College London, Gower Street, London WC1E 6BT, UK.

STELLA M. HURTLEY
SCIENCE, 14 George IV Street, Cambridge CB2 1HH, UK.

WIELAND B. HUTTNER
Institute for Neurobiology, University of Heidelberg, Im Neuenheimer Feld 364, D-69120 Heidelberg, Germany.

DAVID A. JANS
Nuclear Signalling Laboratory, Division for Biochemistry and Molecular Biology, John Curtin School of Medical Research, Australian National University, PO Box 334, Canberra City, ACT 2601, Australia.

NICHOLAS T. KTISTAKIS
Department of Biochemistry, University of Texas Southwestern Medical Center at Dallas, 5323 Harry Hines Boulevard, Dallas, TX 75235-9038, USA.

TREVOR LITHGOW
School of Biochemistry, La Trobe University, Bundoora 3083, Australia.

CAROLYN E. MACHAMER
Department of Cell Biology and Anatomy, School of Medicine, Johns Hopkins University, 725 N. Wolfe Street, Baltimore, MD 21205-2105, USA.

MASATO OHASHI
Institute of Neurobiology, University of Heidelberg, Im Neuenheimer Feld 364, D-69120 Heidelberg, Germany.

KARIN RÖMISCH
MRC Institute of Molecular Cell Biology, University College, Gower Street, London WC1E 6BT, UK.

MICHAEL G. ROTH
Department of Biochemistry, University of Texas Southwestern Medical Center at Dallas, 5323 Harry Hines Boulevard, Dallas, TX 75235-9038, USA.

GOTTFRIED SCHATZ
Department of Biochemistry, Biozentrum, University of Basel, Klingelbergstrasse 70, Basel CH-4056, Switzerland.

Abbreviations

α2,6-ST	α2,6-sialyltransferase
ADHIII	alcohol dehydrogenase
AGT	alanine:glyoxylate aminotransferase
AP	associated protein (also known as adaptin)
ARF	ADP-ribosylation factor
BFA	brefeldin A
C6-NBD-SM	*N*-(*N*-[7a-nitro-2,1,3-benzoxadiazol-4-yl]-ε-aminohexanoyl)-sphingosylphosphorylcholine
CaM	calmodulin
CAT	chloramphenicol acetyltransferase
CGAT	now known as VMAT-1
CGN	*cis* Golgi network
CHO	Chinese hamster ovary
CIT1	peroxisomal citrate synthase
CIT2	mitochondrial citrate synthase
CKII	casein kinase II
CLSM	confocal laser scanning microscopy
COP	coat protein
COXIV	cytochrome oxidase subunit IV
COXVa	cytochrome oxidase subunit Va
CREB	cAMP-response element binding protein
CSV	constitutive secretory vesicle
DBP	DNA-binding protein
DHFR	dihydrofolate reductase
DPAP	dipeptidylaminopeptidase A
DPPIV	dipeptidylpeptidase IV
drNLS	developmentally regulated NLS
ECV	endosomal carrier vesicle
EGFR	epidermal growth factor receptor
EM	electron microscope
ER	endoplasmic reticulum
FFL	firefly luciferase
GAP	GTPase activating protein
GAPDH	glyceraldehyde-3-phosphate dehydrogenase
GPI	glycosylphosphatidylinositol
GR	glucocorticoid receptor
HA	haemagglutinin
HDAg	hepatitis delta antigen

hsp	heat-shock protein
ICP8	infected-cell protein 8
IMP	inner membrane protease
IP3	inositol tris-phosphate
IR	insulin receptor
IRD	infantile Refsum disease
ISG	immature secretory granule
LAP	lysosomal acid phosphatase
LDL	low-density lipoprotein
LDLR	low-density lipoprotein receptor
MAPK	mitogen-activated protein kinase
MAS	mitochondrial assembly
MDCK	Madin–Darby canine kidney
mft	mitochondrial fusion targeting
mhsp	mitochondrial heat-shock protein
MOM	mitochondrial outer membrane
MSF	mitochondrial import stimulating factor
MSG	mature secretory granule
MTS	mitochondrial targeting sequence
mUBF	mouse upstream binding factor
NAGT-1	N-acetylglucosaminyltransferase 1
NALD	neonatal adrenoleukodystrophy
NEM	N-ethylmaleimide
NF	nuclear factor
NF-AT	nuclear factor of activated T cells
NLS	nuclear localization sequence
NLSBP	NLS-binding protein
NMR	nuclear magnetic resonance
NOS	nucleolus-localization signal
NP	nucleoprotein
NPC	nuclear pore complex
NSF	N-ethylmaleimide-sensitive factor
OTC	ornithine transcarboxylase
PAM	peptidylglycine-α-amidating monooxygenase
PBF	presequence-binding factor
PC	phosphatidylcholine
PGK	phosphoglycerate kinase
PH1	primary hyperoxaluria type 1
PI	phosphatidylinositol
pIgR	polymeric immunoglobulin receptor
PK-A	cAMP-dependent protein kinase
PK-C	Ca^{2+}/phospholipid-dependent protein kinase
PMP	peroxisomal membrane protein
POMC	pro-opiomelanocortin

PTS	peroxisomal targeting sequence
RCDP	rhizomelic chondrodysplasia punctata
β1,4-GT	β1,4-galactosyltransferase
SCP	sterol-carrier protein
SLMV	synaptic-like microvesicle
SNAP	soluble NSF attachment protein
SNARE	SNAP receptor
SRP	signal recognition particle
SVAT	now known as VMAT-2
T-ag	tumour antigen
TF	transcription factor
TGN	*trans* Golgi network
TNF	tumour necrosis factor
TPR	tetratricopeptide repeat
TR	transferrin receptor
Ub	ubiquitin
VMAT	vesicular monoamine transporter
VSV	vesicular stomatitis virus
WGA	wheat-germ agglutinin
ZS	Zellweger syndrome

1 | Targeting precursor proteins to mitochondria

TREVOR LITHGOW and GOTTFRIED SCHATZ

1. Introduction

In order to grow and divide, eukaryotic cells need mitochondria. This statement is underscored most dramatically by the observation that deletion of any of the several genes essential for mitochondrial biogenesis kills yeast cells (1) and that failure to direct at least one mitochondrion into the daughter cell after mitosis prevents the final steps of yeast cell division (2).

Mitochondria cannot be made *de novo* (3). To assemble a new mitochondrion, a eukaryotic cell directs specific precursor proteins, synthesized in the cytosol, to the surface of its existing mitochondria, from where the precursors are directed into the organelle and sorted to one of the four submitochondrial compartments (Fig. 1). Assembly of new precursors into existing mitochondria then leads to the birth of new organelles through a poorly understood fission process. A long-standing question has been how eukaryotic cells target mitochondrial precursors specifically to that organelle.

Many mitochondrial precursors are synthesized with N-terminal presequences which are lost from the precursors upon import into the mitochondrion. Deletion of the presequence usually prevents import of the precursor; and fusion of a mitochondrial presequence to a foreign 'passenger' protein targets the resulting fusion protein to mitochondria.

The targeting sequences of many mitochondrial precursor proteins have been defined using the cytosolic enzyme dihydrofolate reductase (DHFR) as the passenger protein (see Table 1 for specific references). There is no simple consensus in the primary structure of these targeting sequences. The common property appears to be that a sequence with a propensity to form a basic amphipathic helix, exposed at the N-terminus of DHFR, can direct a fusion protein into the mitochondrial matrix.

In addition to defining the amino acid sequences that can target a passenger protein to the mitochondria, these early studies supplied some basic information about the requirements for translocation of a precursor protein across the mitochondrial membranes. A tightly folded DHFR domain prevents translocation, although partially folded domains might still be translocation-competent. Import of these fusion proteins is independent of cytosolic factors and independent of ATP outside

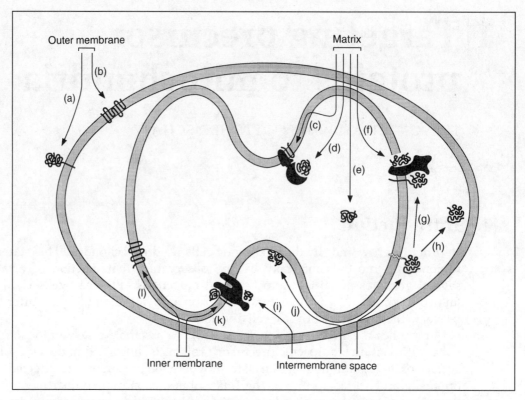

Fig. 1 Mitochondrial protein import. These various precursors are targeted to the mitochondria and sorted into one of the four mitochondrial compartments. Outer membrane, (a) Mas70p; (b) porin. Matrix, (c), subunit 9 of the F_1F_0-ATPase; (d), the β-subunit of the F_1-ATPase; (e), alcohol dehydrogenase III; (f), the iron–sulfur protein of complex III of the respiratory chain. Intermembrane space, (g), cytochrome c_1; (h), cytochrome b_2; (i), cytochrome c; (j), cytochrome c haem lyase. Inner membrane, (k), subunit Va of cytochrome oxidase; (l), ATP/ADP carrier. This figure is redrawn from ref. 4 with permission.

the mitochondria, but absolutely requires ATP in the mitochondrial matrix and an electrochemical potential across the mitochondrial inner membrane.

Over the past ten years several mitochondrial precursors have been discovered which do not conform to all of the rules for import suggested from these early studies. Some precursors have no apparent presequence and some, although tightly folded, can be actively unfolded during translocation across the mitochondrial membranes. Import of some precursors depends on cytosolic factors (i.e. molecular chaperones) and on ATP outside the mitochondria. Some precursors can be imported independently of ATP inside the mitochondria, so long as they do not need to be translocated completely across the inner membrane. The diversity of requirements for import is shedding new detail on to the mechanisms employed by cells to target precursor proteins into mitochondria. In this chapter we will review what is known about the targeting mechanisms for each of these apparent 'classes' of precursors and provide a general model to explain the targeting of precursors into the mitochondria.

Table 1 Targeting of dihydrofolate reductase (DHFR) into the mitochondria

Fusion protein	Length of presequence	Fusion protein targeted to	Reference
COXIV–DHFR	25	Matrix	5
COXIV–DHFR	12	Matrix	5
COXIV–DHFR	7	Cytosol	5
COXIV–DHFR*	7	Matrix	6
Cyt.c_1-DHFR	64	Intermembrane space	7
Cyt.c_1-DHFR	31	Matrix	7
ADHIII–DHFR	28	Matrix	7
OTC–DHFR	36	Matrix	8
Mas70–DHFR	29	Outer membrane	9
Bcl-2–DHFR	22	Outer membrane	10
F_1F_0 Su9-DHFR	69	Matrix	11

The short presequences indicated here were fused to DHFR and were able to redirect this normally cytosolic protein into the specified submitochondrial compartment. DHFR* refers to a mutant (Cys_{157}, Cys_{183}), destabilized DHFR protein

2. Targeting sequences

The information for targeting to the mitochondria and sorting within the mitochondria is usually encoded within an N-terminal sequence which is often referred to as a presequence. Table 1 illustrates that fusing the presequence from matrix-targeted precursors like yeast alcohol dehydrogenase (ADHIII) or rat ornithine transcarbamylase (OTC) to DHFR not only sends the passenger to the mitochondria, but also directs it into the matrix. Similarly, the presequence of the intermembrane space protein cytochrome c_1 (cyt c_1) directs DHFR into the intermembrane space, and the N-terminal region of the outer membrane import receptor Mas70p anchors DHFR in the outer mitochondrial membrane. Thus these N-terminal presequences contain both mitochondrial targeting and intramitochondrial sorting information (4). At the extreme N-terminus of the precursor is a short region, typically 9–12 amino acids long, which contains the matrix-targeting signal. Regions of the presequence directly behind this matrix-targeting information can specify a stop to the transfer of the precursor across either the outer (e.g. Mas70p; 9, 12, 13) or the inner (e.g. cytochrome b_2 and cytochrome c_1; 7, 14, 15) membrane. This point has been elegantly demonstrated with the matrix-targeted precursor to OTC. Inserting a transmembrane helix at one of two distinct points in the OTC precursor stopped the transfer of the fusion protein across either the outer or the inner mitochondrial membrane (16). In some cases the targeting information can be considered bipartite: the cytochrome c_1 presequence can be thought of as a 12 amino acid matrix-targeting sequence followed by a distinct intermembrane space-targeting, stop-transfer sequence: a truncated form of the c_1 presequence, missing some of the stop-transfer signal, directs DHFR into the matrix (4, 7). The sequences that mediate mitochondrial targeting and stop-transfer for the outer membrane in Mas70p overlap (12) and the composite signal is referred to as a signal-anchor sequence (9): the

Fig. 2 What is a basic, amphipathic helix? The COXIV$_{(1-12)}$DHFR fusion protein. This representation is based on limited proteolysis of the fusion protein that suggests that the presequence is accessible to proteases, and NMR analysis of synthetic presequences that suggests the residues 3–11 can assume a helical conformation. References to this work are discussed in the text. The basic residues in the presequence, Arginine$_5$, Arginine$_9$, and Lysine$_{12}$, are coloured black. With a pitch to the helix of 100°, these residues are aligned to form a discrete, positively-charged surface.

transmembrane signal-anchor alone, with the N-terminal 12 amino acids eliminated, can target DHFR into the mitochondrial outer membrane (9). Several other outer membrane proteins, including Mas20p (17), OM45 (18), and porin (19), appear to have similar signal-anchor domains, but a consensus signal-anchor targeting sequence for the outer membrane remains to be demonstrated.

In some cases the presequence also contains processing information. In the case of cytochrome c_1–DHFR, the stop-transfer sequence specifies the location of the fusion protein in the inner membrane, and processing signals direct the IMP protease to release the DHFR passenger from the inner membrane into the intermembrane space (7). In most cases, where assembly of the precursor into a multicomponent complex in the inner membrane requires regions of the precursor distinct from the presequence, the presequence alone will direct the DHFR passenger only into the matrix (11). Translocation of precursors into and across the inner membrane (20, 21) and targeting to the various mitochondrial subcompartments (4) have been recently reviewed.

2.1 N-terminal targeting sequences

There are no obvious similarities in the primary structure of N-terminal mitochondrial targeting signals. Experiments with presequence peptides (22, 23) and randomly generated peptide libraries (24), as well as computer-aided sequence analysis (25, 26) have shown that the mitochondrial targeting region of these N-terminal sequences can form a basic, amphipathic helix. Several studies, comprehensively reviewed by Roise (27), were made to determine if these targeting sequences actually do fold into a basic amphipathic helix. Circular dichroism and NMR measurements on synthetic peptides representing the presequences of COXIV, OTC and F_1-ATPase β-subunit have shown that these peptides do not assume a helical conformation in aqueous solution (22, 23, 28), but do so in the presence of negatively charged detergents or phospholipids. NMR analysis of the 25 amino acid COXIV presequence in the presence of phospholipid micelles enabled Endo et al. (29) to demonstrate that only residues 3–11 formed an α-helix. Such a helix would indeed be basic and amphipathic (Fig. 2) and represents most of the minimal COXIV presequence that can target DHFR into isolated mitochondria (5). Thus, negatively charged phospholipid surfaces can stabilize a common secondary structure in the N-terminus of many mitochondrial precursors.

The concept that targeting factors (which might be cytosolic chaperones or import receptors) can stabilize a general feature of the targeting sequence, in this case the amphipathic helix, has already been proposed as a mechanism for targeting precursors to the endoplasmic reticulum (30, see Chapter 4). A model that integrates the involvement of phospholipid surfaces and targeting factors as they might function in protein targeting to mitochondria is provided in Fig. 3. A nascent precursor could interact reversibly with the negatively charged phospholipids on the mitochondrial surface which stabilize the helical structure of the presequence. A putative cytosolic chaperone might also bind to the presequence to stabilize the targeting information in the presequence. In either of these cases, which are not mutually exclusive, the precursor would arrive at the import receptors in a conformation that could be recognized, resulting in the import of the precursor.

A search has begun for the proteins that can bind basic, amphipathic presequences. Two proteins, MSF and PBF, have been purified from the cytosol of rat liver cells by chromatography using immobilized precursors. These factors appear to function by binding authentic mitochondrial precursors (33, 34). The possible role of cytosolic factors in mitochondrial import will be discussed further in Section 4. There is also tentative evidence that presequences interact with proteins on the mitochondrial surface. Synthetic presequence peptides are imported by isolated mitochondria, and import is reduced if proteins are first shaved from the mitochondrial surface with proteases (35). While it is not known which, if any, of the known components of the import machinery bind the presequences directly, radiolabelled COXIV presequence peptide incubated with mitochondria can be cross-linked to three known components of the import machinery, Mas70p, Isp42p, and Isp45p (36; M. Cumsky, personal communication). The role of Mas70p as a precursor receptor is discussed in Section 5.1. Isp42p and Isp45p are probably central components of the import sites in the outer and inner membranes, respectively (37). In addition, presequence peptides are specifically bound by the matrix-located processing peptidase (38). Thus, the N-terminal presequences contain information for targeting, for translocation, and for the processing of mitochondrial precursors.

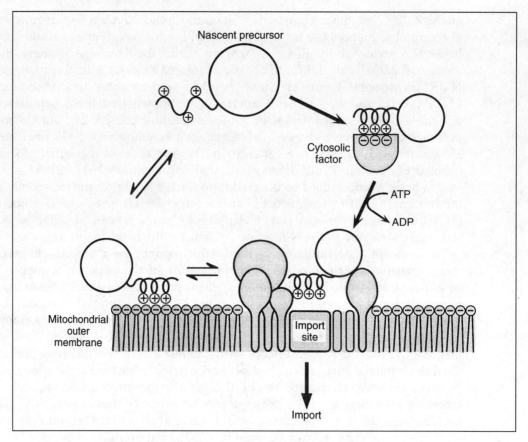

Fig. 3 A model of targeting. NMR analysis of synthetic presequences suggests that the amphipathic helical structure might only form if it is stabilized by an appropriate negatively charged surface. As discussed in the text, the phospholipid surface of the mitochondria and some of the putative cytosolic chaperones might allow such surfaces to increase the binding of the precursor to the import machinery in the mitochondrial outer membrane. This model is based on suggestions from refs 31 and 32.

2.2   C-terminal targeting sequences

In at least one case a discrete C-terminal sequence has been shown to be both necessary and sufficient to mediate mitochondrial targeting. The product of the mammalian proto-oncogene *bcl-2* is located in the perinuclear endoplasmic reticulum, the nuclear envelope, and the mitochondrial outer membrane (39, 40). The extreme C-terminal 22 amino acids of Bcl-2 include a transmembrane span that anchors Bcl-2 on the mitochondrial surface; deletion of only these 22 amino acids converts Bcl-2 into a soluble, cytosolic protein (41) and fusion of these 22 amino acids to the C-terminus of DHFR is sufficient to target the fusion protein to the mitochondrial outer membrane (10). Since this same region of the protein is necessary for the insertion of Bcl-2 into the endoplasmic reticulum (42), the targeting mechanism may well be peculiar to the class of tail-anchored proteins (43), most of which are associated with endoplasmic reticulum-derived membranes. It is not known how these tail-anchored proteins are targeted to their final destinations. They might insert into any membrane *in vivo*, as some have been shown to do *in vitro* (43), but might only be stably maintained in some of those membranes. Since Bcl-2 can form heterodimers with the Bax protein (41), the availability of a partner protein might determine its relative distribution into the endoplasmic reticulum (and nuclear envelope), or the mitochondrial outer membrane. Alternatively, the cytosolic domain of Bcl-2 might fold, leaving the C-terminal tail exposed. Since this tail is superficially similar to the signal-anchor domain that targets Mas70p into the mitochondrial outer membrane, cytosolic factors and/or import receptors might then direct the insertion of Bcl-2 into the mitochondrial outer membrane (10).

2.3 Internal sequences

Several mitochondrial precursors do not have cleavable, N-terminal targeting sequences. Some, like rhodanese and the matrix-located molecular chaperone cpn10, have an N-terminus that resembles a matrix-targeting signal, but the signal is not proteolytically removed after import (44–46). For several other imported proteins, the region responsible for mitochondrial targeting has not been identified. Most proteins of this type, including porin (19, 47), monoamine oxidase (48), Isp42p (49, 50), benzodiazepine receptor (51), and MOM22 (52), are found on the mitochondrial surface. Very little is known about the signals that target these proteins to the mitochondria and lead to their assembly on the mitochondrial surface. In the well-studied case of the ATP/ADP carrier, the precursor lacks a cleavable presequence (53), but is translocated across the mitochondrial outer membrane by the same import apparatus used by precursors that have cleavable N-terminal presequences.

Depletion of extramitochondrial ATP inhibits import of the ATP/ADP carrier into isolated mitochondria, suggesting that the protein binds to ATP-dependent cytosolic chaperones (54). By manipulating the temperature and the electrochemical potential across the inner membrane, the ATP/ADP carrier can be arrested at distinct stages of import. At low temperature and in the absence of a membrane potential, the protein binds to the surface of mitochondria; by increasing the temperature and restoring the membrane potential the protein can be chased into the mitochondria, where it gradually assembles into the functional nucleotide carrier (55). Arrested on the mitochondrial surface, the ATP/ADP carrier can be cross-linked to the import receptors MOM19 and MOM72 (56, see also Fig. 5) and its binding to Mas70p/MOM72 is a rate-limiting step in the import pathway *in vitro* (57–59). The results suggest that the ATP/ADP carrier is bound by Mas70p/MOM72 and Mas20p/MOM19 even though its N-terminus is apparently unable to form a basic, amphipathic helix.

While the N-terminal 72 residues of the ATP/ADP carrier, fused to DHFR, are unable to function as a mitochondrial targeting signal, the first 111 amino acids do direct the attached DHFR to the mitochondria. Interestingly, residues 71–97 constitute a putative transmembrane helix, preceded by a basic residue and followed by a basic, amphipathic region that extends to position 111 (60). It has not been demonstrated that the stretch between residues 71 and 111 alone can function as a mitochondrial targeting signal, but the structure of this region is at least super- ficially similar to the signal-anchor domain responsible for targeting Mas70p to the mitochondria (9), followed by an internal, basic, amphipathic helix. Either one or both of these elements might be the signal that targets the carrier to the mito- chondrial surface.

For some time, mitochondrial precursors like porin and the ATP/ADP carrier have waited in the 'too hard' basket. These proteins clearly contain information for mitochondrial targeting, but this information is probably not encoded at their extreme N-termini. A promising approach for defining the internal targeting sequences of these proteins would be through the identification of the components that bind these proteins on their way into the mitochondria. For some precursors, these components might be found in the cytosol as well as on the mitochondrial surface.

3. Precursor subclasses

3.1 Some precursors seem to have exposed presequences

In vitro-translated COXIV–DHFR forms a compact structure in which the pre- sequence (and a small C-terminal peptide) is readily accessible to externally added proteases (61–63). Limited proteolysis of several other precursors, including those of hsp60 and cytochrome b_2, suggests that these also contain relatively tight domains and that their presequences are exposed (T. Lithgow, unpublished data). Complete folding of the DHFR moiety, stabilized by the ligand methotrexate, prevents translocation of artificial mitochondrial precursors containing DHFR as the passenger and only the presequence of the fusion protein can cross the mito- chondrial membranes (61, 64, 65). However, when bound to the mitochondria in the absence of this ligand, the compact DHFR domain unfolds to a protease-sensitive conformer, which can then be translocated across the membranes (63). Association with the mitochondrial surface is also sufficient to mediate the partial unfolding of several other precursors, including aspartate aminotransferase and malate dehydro- genase (66). However, some precursor domains remain folded even during associa- tion with the mitochondrial surface and can therefore prevent complete trans- location across the outer membrane. The haem-binding domain of cytochrome b_2 is a case in point. The ATP-dependent action of the mitochondrial hsp70 (mhsp70) inside the mitochondrion is required to pull the cytochrome b_2 haem-binding domain apart and allow the complete translocation of the precursor across the outer membrane and into the intermembrane space (67, 68).

The only difference in the import of compact and loosely folded precursors appears to be the rate at which they enter the mitochondria. Loosely folded pre- cursors, such as COXVa (69) and urea-denatured COXIV–DHFR (64, 70) are imported by isolated mitochondria up to 100 times faster than partially folded precursors. In both cases the import machinery appears to interact readily with the exposed presequence. This is not so for another class of precursors whose targeting sequences appear to be protected by cytosolic components of the protein import machinery.

3.2 Some precursors need ATP in the cytosol to enter the mitochondria

A recent study on the energetics of mitochondrial protein import (54) was undertaken with an *in vitro* system in which the levels of ATP inside or outside the mitochondria could be selectively depleted. This allowed, for the first time, the localized ATP requirements of import for several mitochondrial precursors to be studied side by side. With this system, it has become clear that precursors such as DHFR fusion proteins, cytochrome b_2 and hsp60 are imported into mitochondria without any requirement for ATP outside the mitochondria. However, other precursors, including those of alcohol dehydrogenase III, cytochrome c_1, F_1-ATPase β-subunit, and ATP/ADP carrier require extramitochondrial ATP for optimal import rates. In most cases, the requirement for external ATP can be overcome by presenting the precursors to mitochondria in a urea-denatured state. It is likely that all mitochondrial precursors bind to cytosolic chaperones transiently after translation (Fig. 4). Some precursors probably remain bound to cytosolic chaperones after translation and require ATP in the cytosol for import, whereas others are rapidly released from the chaperones without losing their import competence. Indeed, purified artificial precursors containing DHFR as the passenger protein can be imported efficiently in the absence of cytosolic proteins (61, 71).

Precursors that only remain import-competent while bound to cytosolic chaperones are probably only released immediately before or during translocation across the mitochondrial membranes, in a reaction requiring extramitochondrial ATP. Alcohol dehydrogenase III and cytochrome c_1 are examples of this class of precursor. Their presequences are completely protected against levels of protease that can remove the presequences from hsp60, cytochrome b_2, and DHFR fusion proteins (T. Lithgow, unpublished), suggesting that these precursors are bound to, and protected by, cytosolic chaperones.

Precursors whose import requires external ATP might provide a novel insight into the mitochondrial targeting machinery. Some of these precursors have N-terminal presequences (F_1-ATPase β-subunit, alcohol dehydrogenase III, cytochrome c_1) and some do not (porin, ATP/ADP carrier, monoamine oxidase). Several important questions are raised by these findings.

Firstly, do chaperones only recognize the targeting sequences, or do they also bind to the mature part of the precursor? Second, how is the targeting information decoded while it is bound by the chaperones? One possibility is that a mitochondrial receptor recognizes the chaperone–precursor complex (72), similar to the SRP-mediated targeting to the endoplasmic reticulum (see Chapter 4). In this case, the cytosolic chaperones would function as targeting factors. Alternatively, the cytoso-

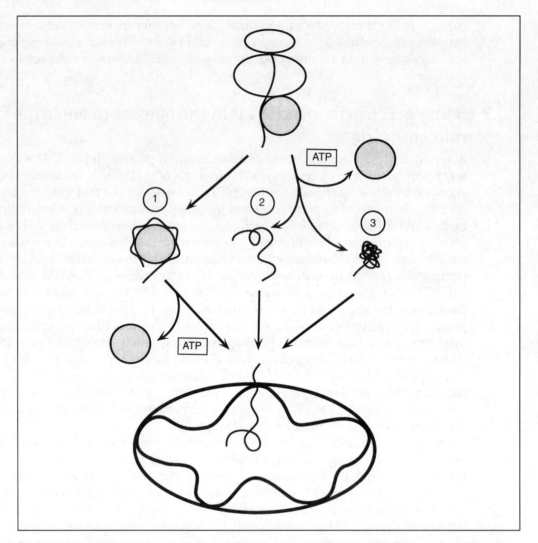

Fig. 4 Precursors might approach the mitochondria in one of three different states. Three classes of precursors can be defined by *in vitro* import assays. Group 1, including the alcohol dehydrogenase III and ATP/ADP carrier, remain bound to cytosolic chaperones and require ATP outside the mitochondria as a prerequisite for efficient import. Group 2 and Group 3 precursors are released rapidly from cytosolic chaperones, and might remain loosely folded (Group 2) or assume a more compact conformation (Group 3). In an *in vitro* import assay, there is no extramitochondrial ATP requirement for the import of the Group 2 and 3 precursors. Reproduced from ref. 54 with permission.

lic chaperones could release and rebind the precursors until the transiently exposed targeting sequences can initiate import. In that case, the cytosolic chaperones would have no role in targeting, but would promote efficient protein import by preventing precursor conformations that would decrease import efficiency.

The third and most important question is the identity of the cytosolic factors that are bound to these precursors. Several candidates have already been identified.

4. Cytosolic factors

4.1 Hsp70

The first defined cytosolic chaperone implicated in the assistance of mitochondrial protein import was the heat-shock protein, hsp70. Yeast cells have six genes encoding cytosolic isoforms of hsp70. Two of these (*SSB1* and *SSB2*) encode proteins that are largely associated with 80S ribosomes where they appear to function in mediating the smooth passage of nascent polypeptides through the ribosome tunnel (73). While expression of the four other hsp70 genes (*SSA1*, *SSA2*, *SSA3*, and *SSA4*) are differentially regulated, the four protein products are functionally redundant (74). Deshaies *et al.* (75) engineered a yeast strain which was crippled for expression of three of the four chromosomal *SSA* genes, but which had a plasmid-borne copy of *SSA1* under the control of the repressible GAL promoter. When the *SSA1*-encoded hsp70 protein was depleted from the cytosol of these cells, the precursor form of the F_1-ATPase β-subunit accumulated (75). Independently, it was shown that addition of purified Ssa1p stimulated the import of the F_1-ATPase β-subunit precursor *in vitro* (76). F_1-ATPase β-subunit is one of the precursors whose import *in vitro* requires extramitochondrial ATP, and probably an ATP-dependent cytosolic chaperone (54). As the release of bound polypeptides requires ATP hydrolysis by hsp70, cytosolic hsp70 might be at least partially responsible for the external ATP requirement.

4.2 Cytosolic DnaJ homologues

Purified cytosolic hsp70 has a weak ATPase activity which is stimulated ten-fold by the cytosolic protein Ydj1p, whose sequence resembles that of bacterial DnaJ. Ydj1p regulates the affinity of the interaction between Ssa1p and a loosely folded polypeptide substrate (77): the hsp70–Ydj1p complex releases a bound polypeptide more efficiently than Ssa1p alone. It has also been shown that bacterial DnaJ can prevent the processing of the precursor of the F_1-ATPase β-subunit by crude yeast extracts (78), although it is unclear if the chaperone protects the precursor from cytoplasmic proteases or prevents import of the precursor into mitochondria that should also be present in the crude extract. By either interpretation, purified DnaJ probably associates directly with the precursor of the F_1-ATPase β-subunit. The *YDJ1* gene was independently identified as the *MAS5* gene (79). A temperature-sensitive *mas5* mutant accumulates the precursor form of F_1-ATPase β-subunit at the non-permissive temperature (79, 80). A small fraction of cytosolic hsp70 is actually associated with the mitochondrial surface (81), where it would be perfectly positioned to stimulate release of Ydj1p from mitochondrial precursors.

Hsp70 and Ydj1p probably bind not only mitochondrial precursors but also most of the polypeptides destined for other intracellular locations. This makes it unlikely that these chaperones function in the actual targeting process; instead, they probably maintain precursors in an import-competent conformation.

4.3 Chaperones specific for mitochondrial precursors

Two other cytosolic proteins appear to discriminate mitochondrial precursors from polypeptides not destined for the mitochondria. These proteins might play a direct role in the targeting of precursors to the mitochondria. MSF (mitochondrial import stimulating factor) is a heterodimer of approximately 30 kDa subunits (34). PBF (presequence-binding factor) is probably a monomer of 50 kDa (33). Both proteins can be purified from the cytosol of rat liver cells or rabbit reticulocytes by chromatography on immobilized mitochondrial precursor proteins. Neither chaperone binds to the corresponding mature proteins. A stable complex between the precursor of OTC and PBF has been demonstrated. Whereas urea-denatured OTC precursor rapidly aggregates after dilution from 8 M urea, addition of purified PBF maintains the precursor in a soluble complex that contains approximately equimolar amounts of PBF and the precursor. This complex surrenders the precursor to isolated mitochondria and the precursor is imported and processed to the mature form. Under the same assay conditions, no import of the OTC precursor is observed in the absence of PBF (33). Purified PBF stimulates the import of some, but not all precursors (82).

Whereas PBF activity is apparently independent of ATP, MSF stimulates the import of the adrenodoxin precursor by an ATP-dependent mechanism. Purified PBF and MSF can function as molecular chaperones for some mitochondrial precursors, but it is not clear to what extent these cytosolic chaperones can influence the targeting of the precursors to which they bind. Such a function would imply a direct interaction between the chaperone and a component of the mitochondrial surface (72).

Genetic screens have been developed to identify genes that effect the targeting of the precursor form of the F_1-ATPase β-subunit *in vivo*. In one approach, the presequence of F_1-ATPase β-subunit was fused to the passenger protein β-galactosidase. The fusion protein is targeted to the mitochondria, where it disrupts respiratory function (83). Respiration-competent yeast mutants were identified which still expressed the fusion protein, but were unable to target it to the mitochondria (84). A plasmid carrying two yeast genes was cloned which complemented the *mft1* (mitochondrial fusion targeting) mutation. Initially the complementing activity was assigned to the wrong gene on the plasmid (84), but subsequent work assigned the complementing activity to the other gene, *orf2* (85, J. Garrett, personal communication). No work has been described with the purified Mft1p, but as a candidate targeting factor for mitochondrial precursors the predicted sequence of Mft1p is intriguing: the C-terminal third of the protein is 40% aspartic and glutamic residues. How could the basic, amphipathic targeting sequence of the F_1-ATPase β-subunit precursor resist such a protein!

While PBF, MSF, and Mft1p appear to act selectively on some mitochondrial precursor proteins, it has still to be shown that they mediate the targeting of these precursors to the mitochondrial surface. Indeed, at least some mitochondrial precursors are recognized by components of the import machinery on the mitochondrial surface in the absence of cytosolic chaperones.

5. The import machinery on the mitochondrial surface decodes targeting signals

The protein import apparatus in the mitochondrial outer membrane is composed of several protein components (Fig. 5). The identification of the various components has been reviewed recently (37).

Two basic strategies have been employed to determine which of the components of the import apparatus are involved in targeting and translocation of precursors into mitochondria. Antibodies monospecific for a distinct protein component of the outer membrane have been tested for their ability to inhibit protein import into isolated mitochondria. Also, the genes encoding several of the putative components have been disrupted. With the exception of *ISP42* (49), disruption of genes encoding these components does not kill cells; import of precursors can thus be measured both *in vivo* and *in vitro* in the absence of a single component of the import machinery. Taken together, the various data have identified import receptors which bind mitochondrial precursors and deliver them to the putative protein transport channel across the outer membrane.

5.1 Mitochondrial protein import receptors

There are at least three proteins in the outer membrane that can function as receptors for mitochondrial protein import. By definition, the receptors are those com-

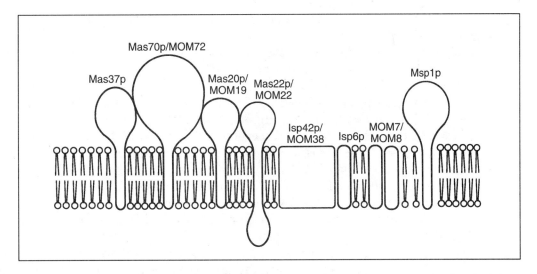

Fig. 5 The protein import machinery in the mitochondrial outer membrane. The involvement of each of these proteins as components of the import machinery has been described as follows: Mas37p (100), Mas70p/MOM72 (57–59), Mas20p/MOM19 (17, 86), Mas22p/MOM22 (52, 56), Isp42p/MOM38 (49, 50), Isp6p (88), MOM7/MOM8 (56), Msp1p (89). The architecture of the import machinery is not yet known.

ponents on the mitochondrial surface that recognize the targeting information of a mitochondrial precursor and deliver the precursor to the import site for translocation across the outer membrane. Two of these receptors have been described in detail. They have protein subunits of approximately 20 and 70 kDa. In yeast these two receptors are called Mas20p and Mas70p (for mitochondrial assembly subunits) and in *Neurospora* MOM19 and MOM72 (for mitochondrial outer membrane subunits). Each receptor exposes a large domain to the cytosol; removal of this domain with low levels of protease inhibits protein import into isolated mitochondria.

The import receptor Mas70p/MOM72 promotes the efficient import of a distinct set of precursors including those of the F_1-ATPase β-subunit, cytochrome c_1 (57, 91), and the ATP/ADP carrier (57–59). Each of these precursors requires ATP outside the mitochondria (Section 3.2), suggesting that these precursors are delivered to the mitochondria attached to cytosolic chaperones. Mas70p does not promote the import of DHFR-containing fusion proteins or of cytochrome b_2 (57) which are imported independently of extramitochondrial ATP. On the other hand, Mas70p also promotes import of urea-denatured alcohol dehydrogenase III (91) which does not require extramitochondrial ATP (54). Since inhibition of Mas70p does not accelerate the import of the undenatured alcohol dehydrogenase III precursor (which is slower and requires extramitochondrial ATP), the rate-limiting step in the import of the undenatured precursor appears to be the release of the precursor from cytosolic chaperones. By presenting the precursor to mitochondria in a urea-denatured form, this release reaction is bypassed and the import-promoting function of Mas70p is revealed. The Mas70p receptor is important to these precursors after their release from the cytosolic chaperones. It is not yet clear which targeting signals in the ATP/ADP carrier and the alcohol dehydrogenase III precursor are recognized by the Mas70p receptor.

The import receptor MOM19 was initially believed to be a 'master receptor' that was required for import of all precursors with cleavable presequences, and thought not to be involved in the import of precursors without cleavable presequences (86, 87). However, arrest of the ATP/ADP carrier precursor, which does not have a presequence, at an early stage of import revealed that it was in contact with both MOM19 and MOM72 (56), and dissection of the import of this precursor has shown Mas20p/MOM19 and Mas70p/MOM72 to be functionally redundant (17, 57, 59, 92). Inactivation of the *MAS20* gene (17, 90) or the *MOM19* gene (92) is not lethal to yeast or *Neurospora* cells (Section 5.2), but impairs the import of a subset of mitochondrial precursors, both *in vitro* and *in vivo*. While most of these precursors have cleavable presequences, yeast mitochondria lacking Mas20p are not perturbed in the import of alcohol dehydrogenase III, another precursor with a cleavable presequence (93). Mas20p may well recognize precursors that do not remain associated with cytosolic chaperones, such as the presequence–DHFR fusion proteins, cytochrome b_2, and hsp60. Our working model is that Mas20p binds directly to the exposed targeting sequence, and that this interaction is the rate-limiting step in the import of these precursors.

5.2 Do Mas70p and Mas20p function as hetero-oligomers?

A structural analysis of the mitochondrial import receptors is required if we are to understand how targeting sequences function to direct precursors to the mitochondria. No three-dimensional structure of an import receptor has been solved, but there is some indirect evidence to suggest that the functional receptor might be a hetero-oligomer.

Both Mas20p (17, 90) and Mas70p (57) contain TPR motifs: 34 amino acid helices proposed to mediate protein–protein interactions (94). These helices are thought to allow tight packing via conserved tyrosine/phenylalanine residues (whose side groups would project out of the helices) and glycine/alanine residues (which leave pockets in a helix). Thus, two opposed helices could pack together. Mas20p and Mas70p could interact with each other via these TPR helices. In addition, the trans-membrane domain of Mas70p can mediate the dimerization of a passenger protein (95). One surface of the transmembrane helix, characterized by a predominance of alanine residues, is also represented in the transmembrane domain of Mas20p, which might allow Mas70p to form either homo- or hetero-dimers *in vivo*. These speculative ideas are represented in Fig. 6.

Initially, loss of the *MAS20/MOM19* gene has serious consequences for cells (17, 90, 92). Yeast cells without functional Mas20p import many mitochondrial pre-cursors poorly, grow slowly with glucose as a carbon source, and are respiration-deficient. In this condition, disruption of the *MAS70* gene is lethal to the cells (17). However, yeast cells adapt to the loss of Mas20p, and after several days these Mas20p-deficient cells can respire and import precursors into their mitochondria at

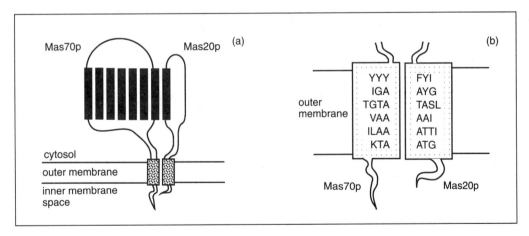

Fig. 6 Mas20p and Mas70p: TPRs and transmembrane alanines. The cartoon sketches the structures in the import receptors Mas70p and Mas20p that might allow interaction with each other and/or other components of the import machinery. The cytosolic domain of Mas70p contains seven copies of the 34 amino acid TPR helices (coloured black), whereas Mas20p has only one. The transmembrane helices pack against each other. The alanine-rich surfaces formed on the transmembrane helices of Mas70p and Mas20p might allow the for-mation of homodimers or heterodimers of the receptor subunits.

the same rate as wild-type cells. These adapted, Mas20p-deficient cells can tolerate loss of Mas70p: cells without either Mas20p or Mas70p are viable, able to respire, and protein import into the isolated mitochondria still requires proteins exposed on the mitochondrial surface (93). There are two proteins which could function as import receptors in the absence of Mas20p and Mas70p: Mas37p and Mas22p. These two proteins might actually represent partners of Mas70p and Mas20p in wild-type mitochondria.

The *MAS37* gene encodes a 37 kDa protein in the mitochondrial outer membrane. Loss of Mas37p renders cells temperature-sensitive for respiration (100). Mitochondria lacking Mas37p import the urea-denatured alcohol dehydrogenase III precursor only poorly, but import of DHFR-fusion proteins and several other precursors is apparently not affected. In this respect, loss of Mas37p has similar consequences to loss of Mas70p (57, 91). Overexpression of Mas37p appears to stabilize overexpressed Mas70p and, conversely, overexpression of Mas70p stabilizes overexpressed Mas37p (100). Although yeast tolerate deletion of either the *MAS70* or *MAS37* gene, deletion of both *MAS70* and *MAS37* genes is lethal (100). This synthetic lethality, and the observations mentioned above, suggest that Mas37p forms a functional complex with Mas70p.

The *Neurospora* protein MOM22 (52) and the corresponding yeast protein Mas22p (101), are outer membrane proteins with a very acidic domain exposed to the cytosol. Down-regulation of MOM19 expression leads to a substantial decrease in the level of MOM22 in the outer membrane, and overexpression of MOM19 leads to an increase in the level of MOM22 (92). Thus, the two proteins might form a dimer, or higher-ordered structure, in the outer membrane. Antibodies against MOM22 inhibit the import of numerous mitochondrial precursors into mitochondria containing both MOM19 and MOM22 (52), but do not inhibit protein import into MOM19-deficient mitochondria (92). Thus, the activity of MOM22 appears to require the presence of MOM19.

The models proposed by Hines *et al.* and Söllner *et al.* (57, 58) are still useful to explain the recognition of precursors by the mitochondrial receptors. It was proposed that Mas70p might perform its receptor function as a complex with another protein and that additional, distinct receptor subunits might exist in the outer membrane (Fig. 7).

6. Future directions

6.1 Dissection of the targeting process

Targeting of precursors to the mitochondria appears to involve at least two recognition steps. It should be possible to develop assays to study each of these. In the first step, cytosolic chaperones recognize nascent polypeptides, and some chaperones might specifically recognize nascent mitochondrial precursors. In the second step, receptors on the mitochondrial surface recognize and bind mitochondrial precursors and deliver them to the import site. Components of the import site

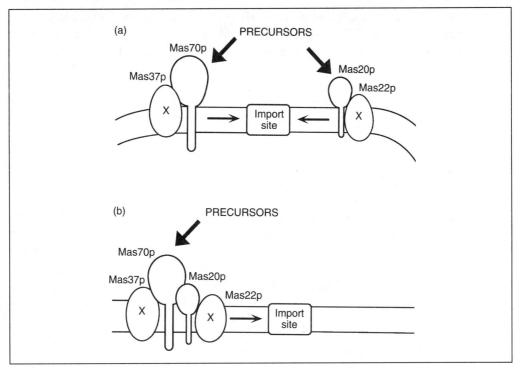

Fig. 7 Hypothesis for receptor function. Distinct partner proteins (X) might exist that function together with Mas70p and Mas20p. (a) If the import receptors mediate multiple convergent import pathways, precursors such as ATP/ADP carrier and alcohol dehydrogenase III are predominantly imported via the Mas70p receptor, and other precursors such as hsp60 and the presequence-DHFR fusion proteins largely via the Mas20p receptor. This model implies that Mas20p and Mas70p do not need to interact with each other. (b) With a single common import pathway, Mas70p and Mas20p, together with other putative receptor subunits (X), form a single oligomeric receptor which mediates the import of all classes of precursor. Even in this model, the activity of the Mas70p subunit might be more important (i.e. rate-limiting) to some precursors than to other precursors. The cartoon is redrawn from ref. 57.

might then mediate additional recognition processes, for example in discriminating between precursors that should be inserted into the outer membrane or delivered into one of the three other mitochondrial compartments.

In one case it has been demonstrated that the presence of isolated mitochondria is sufficient to destabilize the chaperone–precursor complex (33). If cytosolic chaperones can play a role in precursor targeting then, given the possibility to trap mitochondrial precursors on cytosolic chaperones, it should be possible to develop an assay for the hypothetical receptor that recognizes the chaperone–precursor complex.

An assay has been developed to measure precursor translocation into purified vesicles derived from the outer mitochondrial membrane. The assay has been used to measure the integration of precursors normally targeted into the outer membrane (such as porin and MOM22) or across the membrane into the intermembrane space (cytochrome *c* haem lyase) (96). These vesicles can also specifically bind the precursor of the ATP/ADP carrier, but can not mediate translocation of this precursor. If the outer membrane vesicle assay can be scaled-up successfully, it should be possible to solubilize and purify from the vesicles the components of the outer membrane that form the ATP/ADP carrier receptor, to address whether either Mas20p/MOM19 or Mas70p/MOM72 alone can act as receptors, or whether receptor function requires an oligomeric receptor composed of these, and perhaps other, subunits (see Section 4.2).

6.2 A full circle: what is sufficient to target a protein to the mitochondria?

In the vast majority of cases studied to date, an N-terminal sequence with a propensity to form two or three turns of a basic, amphipathic helix is responsible for targeting precursor proteins to mitochondria. Perhaps because these presequences would interfere with the activity of mature mitochondrial proteins, or be disruptive to the mitochondrial membranes, the targeting sequence is usually removed from the precursor subsequent to import and could then be degraded. During the course of the evolution of mitochondria from endosymbiotic bacteria (97), most mitochondrial targeting sequences probably arose through rearrangements at the 5'-end of some nuclear genes, resulting in the addition of N-terminal presequences to the proteins they encode (98, 99).

Some membrane proteins, such as the ATP/ADP carrier and Mas70p, appear to have targeting signals that are not simply a basic amphipathic helix. We do not know if these proteins share a common pathway, or whether each of these proteins is imported by a distinct pathway. Most of these precursors appear to share at least part of the import machinery with precursors targeted via basic, amphipathic presequences. Most likely, the different targeting pathways deliver the precursors to the same import channel across the outer membrane. A detailed comparison of the different targeting mechanisms should lead us to a deeper understanding of precisely how the targeting information of mitochondrial precursors is decoded.

Acknowledgements

Thanks to Margrit Jäggi for help with the artwork, to Michael Cumsky, Jinnie Garret, Roland Lill, Klaus Pfanner, Gordon Shore, Sepp Kohlwein, and Sabine Grazer for discussing their work prior to publication, and to Sabine Rospert and Kevin Hannavy for critical reading of the manuscript. T. L. was supported by a long-term fellowship from the Human Frontiers Scientific Program.

References

1. Baker, K. P. and Schatz, G. (1991) Mitochondrial proteins essential for viability mediate protein import into yeast mitochondria. *Nature* , **349**, 205.

2. McDonnell, S. J., Stewart, L. C., Talin, A., and Yaffe, M. P. (1990) Temperature-sensitive yeast mutants defective in mitochondrial inheritance. *J. Cell Biol.*, **111**, 967.

3. Schatz, G. (1993). From granules to organelles: how yeast mitochondria became respectable. In *The early days of yeast genetics*. Hall, M. N. and Linder, P. (ed.). Cold Spring Harbor Laboratory Press, Cold Spring Harbor, NY, p. 241.

4. Glick, B. S., Beasley, E. M., and Schatz, G. (1992) Protein sorting in mitochondria. *Trends Biochem. Sci.*, **17**, 453.

5. Hurt, E. C., Müller, U., and Schatz, G. (1985) The first twelve amino acids of a yeast mitochondrial outer membrane protein can direct a nuclear-encoded cytochrome oxidase subunit to the mitochondrial outer membrane. *EMBO J.*, **4**, 3509.

6. Verner, K. and Lemire, B. D. (1989) Tight folding of a passenger protein can interfere with the targeting function of a mitochondrial presequence. *EMBO J.*, **8**, 1491.

7. van Loon, A. G. P. M., Brändli, A. W., and Schatz, G. (1986) The presequences of two imported mitochondrial proteins contain information for intracellular and intramitochondrial sorting. *Cell*, **44**, 801.

8. Sheffield, W. P., Nguyen, M., and Shore, G. C. (1990) Effects of 70 kDa heat shock protein on polypeptide folding, aggregation and import competence. *J. Biol. Chem.*, **265**, 11069.

9. McBride, H. M., Millar, D. G., Li, J. M., and Shore, G. C. (1992) A signal-anchor sequence selective for the mitochondrial outer membrane. *J. Cell Biol.*, **119**, 1451.

10. Nguyen, M., Millar, D. G., Yong, V. W., Korsmeyer, S. J., and Shore, G. C. (1993) Targeting of Bcl-2 to the mitochondrial outer membrane by a COOH-terminal signal anchor sequence. *J. Biol. Chem.*, **268**, 25265.

11. Pfanner, N., Müller, H. K., Harmey, M. A., and Neupert, W. (1987) Mitochondrial protein import: involvement of the mature part of a cleavable precursor protein in the binding to receptor sites. *EMBO J.*, **6**, 3449.

12. Hase, T., Müller, U., Riezman, H., and Schatz, G. (1984) A 70 kDa protein of the yeast mitochondrial outer membrane is targeted and anchored by its extreme amino terminus. *EMBO J.*, **3**, 3157.

13. Nakai, M., Hase, T., and Matsubara, H. (1989) Precise determination of the mitochondrial import signal contained in a 70 kDa protein of the yeast mitochondrial outer membrane. *J. Biochem.*, **105**, 513.

14. van Loon, A. P. G. M. and Schatz, G. (1987) Transport of proteins to the mitochondrial intermembrane space: the sorting domain of the cytochrome c_1 presequence is a stop-transfer sequence specific for the mitochondrial inner membrane. *EMBO J.*, **6**, 2441.

15. Beasley, E. M., Müller, S., and Schatz, G. (1993) The signal that sorts yeast cytochrome b_2 to the mitochondrial intermembrane space contains three distinct functional regions. *EMBO J.*, **12**, 2303.

16. Nguyen, M., Bell, A. W., and Shore, G. C. (1988) Protein sorting between mitochondrial membranes specified by position of the stop-transfer domain. *J. Cell Biol.*, **106**, 1499.

17. Ramage, L., Junne, T., Hahne, K., Lithgow, T., and Schatz, G. (1993) Functional cooperation of mitochondrial protein import receptors in yeast. *EMBO J.*, **12**, 4115.

18. Yaffe, M. P., Jensen, R. E., and Guido, E. C. (1989) The major 45 kDa protein of the yeast mitochondrial outer membrane is not essential for cell growth or mitochondrial function. *J. Biol. Chem.*, **264**, 21091.

19. Mihara, M. and Sato, R. (1985) Molecular cloning and sequencing of cDNA for yeast porin, an outer mitochondrial membrane protein: a search for a targeting signal in the primary structure. *EMBO J.*, **4**, 769.

20. Poyton, R. O., Duhl, D. M. J., and Clarkson, G. H. D. (1992) Protein export from the mitochondrial matrix. *Trends Cell Biol.*, **2**, 369.
21. Horst, M., Kronidou, N. G., and Schatz, G. (1993) Through the mitochondrial inner membrane. *Curr. Biol.*, **3**, 175.
22. Roise, D., Horvath, S. J., Tomich, J. M., Richards, J. H., and Schatz, G. (1986) A chemically synthesized presequence of an imported mitochondrial protein can form an amphiphillic helix and perturb natural and artificial phospholipid bilayers. *EMBO J.*, **5**, 1327.
23. Epand, R. M., Hui, S. W., Argan, C., Gillespie, L. L., and Shore, G. C. (1986) Structural analysis and amphiphilic properties of a chemically synthesized mitochondrial signal peptide. *J. Biol. Chem.*, **261**, 10017.
24. Lemire, B. D., Fankhauser, C., Baker, A., and Schatz, G. (1989) The mitochondrial targeting function of randomly generated peptide sequences correlates with predicted helical amphiphilicity. *J. Biol. Chem.*, **264**, 20206.
25. von Heijne, G. (1986) Mitochondrial targeting sequences may form amphiphilic helices. *EMBO J.*, **5**, 1335.
26. von Heijne, G., Steppuhn, J., and Hermann, R. G. (1989) Domain structure of mitochondrial and chloroplast targeting peptides. *Eur. J. Biochem.*, **180**, 535.
27. Roise, D. (1993) The amphipathic helix in mitochondrial targeting sequences. In *The amphipathic helix.* Epand, R. M. (ed.). CRC Press, p. 257.
28. Hoyt, D. W., Cyr, D. M., Gierasch, L. M., and Douglas, M. G. (1991) Interaction of peptides corresponding to mitochondrial presequences with membranes. *J. Biol. Chem.*, **266**, 21693.
29. Endo, T., Shimada, L., Roise, D., and Inagaki, F. (1989) N-terminal half of a mitochondrial presequence peptide takes a helical conformation when bound to dodecylphosphocholine micelles: a proton nuclear magnetic resonance study. *J. Biochem.*, **106**, 396.
30. Rothman, J. E. (1989) GTP and methionine bristles. *Nature*, **340**, 433.
31. Maduke, M. and Roise, D. (1993) Import of a mitochondrial presequence into protein-free phospholipid vesicles. *Science*, **260**, 364.
32. Lithgow, T., Timms, M., Høj, P. B., and Hoogenraad, N. J. (1991) Identification of a GTP-binding protein in the contact sites between inner and outer mitochondrial membranes. *Biochem. Biophys. Res. Commun.*, **180**, 1453.
33. Murakami, K. and Mori, M. (1990) Purified presequence binding factor (PBF) forms an import-competent complex with a purified mitochondrial precursor protein. *EMBO J.*, **9**, 3201.
34. Hachiya, N., Alam, R., Sakasegawa, Y., Sakaguchi, M., Mihara, K., and Omura, T. (1993) A mitochondrial import factor purified from rat liver cytosol is an ATP-dependent conformational modulator for precursor proteins. *EMBO J.*, **12**, 1579.
35. Glaser, S. M. and Cumsky, M. G. (1990) A synthetic presequence reversibly inhibits protein import into yeast mitochondria. *J. Biol. Chem.*, **265**, 8808.
36. Gaikwad, A. S. and Cumsky, M. G. (1994) The use of chemical cross-linking to identify proteins that interact with a mitochondrial presequence. *J. Biol. Chem.*, **269**, 6437.
37. Hannavy, K., Rospert, S., and Schatz, G. (1993) Protein import into mitochondria: a paradigm for the translocation of polypeptides across membranes. *Curr. Opin. Cell Biol.*, **5**, 694.
38. Yang, M., Geli, V., Oppliger, W., Suda, K., James, P., and Schatz, G. (1991) The *MAS*-encoded processing protease of yeast mitochondria. *J. Biol. Chem.*, **266**, 6416.

39. Krajewski, S., Tanaka, S., Takayama, S., Schibler, M. J., Fenton, W., and Reed, J. C. (1993) Investigations of the subcellular distribution of the Bcl-2 oncoprotein: residence in the nuclear envelope, endoplasmic reticulum and mitochondrial outer membranes. *Cancer Res.*, **53**, 4701.

40. Lithgow, T., van Driel, R., Bertram, J. F., Mason, D. Y., and Strasser, A. (1994) The protein product of the oncogene bcl-2 is a component of the nuclear envelope, the endoplasmic reticulum and the outer mitochondrial membrane. *Cell Growth Differ.*, **5**, 411.

41. Oltvai, Z. N., Milliman, C. L., and Korsmeyer, S. J. (1993) Bcl-2 heterodimerizes *in vivo* with a conserved homolog, Bax, that accelerates programmed cell death. *Cell*, **74**, 609.

42. Chen-Leavy, Z. and Cleary, M. L. (1990) Membrane topology of the Bcl-2 protooncogene protein demonstrated *in vitro*. *J. Biol. Chem.*, **265**, 4929.

43. Kutay, U., Hartmann, E., and Rapoport, T. A. (1993) A class of membrane proteins with a C-terminal anchor. *Trends Cell Biol.*, **3**, 72.

44. Miller, D. M., Delgado, R., Chirgwin, J. M., Hardies, S. C., and Horowitz, P. M. (1991) Expression of cloned bovine adrenal rhodanese. *J. Biol. Chem.*, **266**, 4686.

45. Rospert, S., Junne, T., Glick, B. S., and Schatz, G. (1993) Cloning and disruption of the gene encoding yeast mitochondrial chaperonin 10, the homolog of *E. coli* groES. *FEBS Lett.*, **335**, 358.

46. Ryan, M., Hoogenraad, N. J., and Høj, P. B. (1994) Isolation of a cDNA clone specifying rat chaperonin 10, a stress-inducible mitochondrial matrix protein synthesized without a cleavable presequence. *FEBS Lett.*, **337**, 152.

47. Pfaller, R. and Neupert, W. (1987) High-affinity binding sites involved in the import of porin into mitochondria. *EMBO J.*, **6**, 2635.

48. Mitoma, J. and Ito, A. (1992) Mitochondrial targeting signal of rat liver monoamine oxidase B is located at its carboxy terminus. *J. Biochem.*, **111**, 20.

49. Baker, K. P., Schaniel, A., Vestweber, D., and Schatz, G. (1990) A yeast mitochondrial outer membrane protein essential for protein import and cell viability. *Nature*, **348**, 605.

50. Kiebler, M., Pfaller, R., Söllner, T., Griffiths, G., Horstmann, H., Pfanner, N., and Neupert, W. (1990) Identification of a mitochondrial receptor complex required for recognition and membrane insertion of precursor proteins. *Nature*, **348**, 610.

51. Anholt, R. R. H., Pedersen, P. L., de Souza, E. B., and Snyder, S. H. (1986) The peripheral-type benzodiazepine receptor. *J. Biol. Chem.*, **261**, 576.

52. Kiebler, M., Keil, P., Schneider, H., van der Keli, I. J., Pfanner, N., and Neupert, W. (1993) The mitochondrial receptor complex: a central role of MOM22 in mediating preprotein transfer from receptors to the general insertion pore. *Cell*, **74**, 483.

53. Adrian, G. S., McCammon, M. T., Montgomery, D. L., and Douglas, M. G. (1986) Sequences required for the delivery and localization of the ADP/ATP translocator to the mitochondrial inner membrane. *Mol. Cell. Biol.*, **6**, 626.

54. Wachter, C., Schatz, G., and Glick, B. S. (1994) Protein import into mitochondria: the requirement for external ATP is precursor-specific whereas intramitochondrial ATP is universally needed for translocation into the matrix. *Mol. Biol. Cell*, **5**, 465.

55. Pfanner, N. and Neupert, W. (1987) Distinct steps in the import of the ADP/ATP carrier into mitochondria. *J. Biol. Chem.*, **262**, 7528.

56. Söllner, T., Rassow, J., Wiedman, J., Schlossman, J., Keil, P., Neupert, W., and Pfanner, N. (1992) Mapping of the protein import machinery in the mitochondrial outer membrane by crosslinking of translocation intermediates. *Nature*, **355**, 84.

57. Hines, V., Brandt, A., Griffiths, G., Horstmann, H., Brütsch, H., and Schatz, G. (1990) Protein import into yeast mitochondria is accelerated by the outer membrane protein MAS70. *EMBO J.*, **9**, 3191.

58. Söllner, T., Pfaller, R., Griffiths, G., Pfanner, N., and Neupert, W. (1990) A mitochondrial import receptor for the ADP/ATP carrier. *Cell*, **62**, 107.
59. Steger, H. F., Söllner, T., Kiebler, M., Dietmeier, K. A., Pfaller, R., Trülzsch, K. S., Tropschug, M., Neupert, W., and Pfanner, N. (1990) Import of ADP/ATP carrier into mitochondria: two receptors act in parallel. *J. Cell Biol.*, **111**, 2353.
60. Smagula, C. and Douglas, M. G. (1988) Mitochondrial import of the ADP/ATP carrier protein in *Saccharomyces cerevisiae*. *J. Biol. Chem.*, **263**, 6783.
61. Eilers, M. and Schatz, G. (1986) Binding of a specific ligand inhibits import of a purified protein into mitochondria. *Nature*, **322**, 228.
62. Verner, K. and Schatz, G. (1987) Import of an incompletely folded precursor protein into isolated mitochondria requires an energized inner membrane, but no added ATP. *EMBO J.*, **6**, 2449.
63. Endo, T., Eilers, M., and Schatz, G. (1989) Binding of a tightly folded artificial mitochondrial precursor protein to the mitochondrial outer membrane involves a lipid-mediated conformational change. *J. Biol. Chem.*, **264**, 2951.
64. Eilers, M., Hwang, S., and Schatz, G. (1988) Unfolding and refolding of a purified precursor protein during import into isolated mitochondria. *EMBO J.*, **7**, 1139.
65. Rassow, J., Guiard, B., Wienhues, U., Herzog, V., Hartl, F. U., and Neupert, W. (1989) Translocation arrest by reversible folding of a precursor protein imported into mitochondria. A means to quantitate translocation contact sites. *J. Cell Biol.*, **109**, 1421.
66. Hartmann, C. M., Gehring, H., and Christen, P. (1993) The mature form of imported mitochondrial proteins undergoes conformational changes upon binding to isolated mitochondria. *Eur. J. Biochem.*, **218**, 905.
67. Glick, B. S., Wachter, C., Reid, G. A., and Schatz, G. (1993) Import of cytochrome b_2 to the mitochondrial intermembrane space: the tightly folded heme-binding domain makes import dependent upon matrix ATP. *Protein Sci.*, **2**, 1901.
68. Gambill, B. D., Voos, W., Kang, P. J., Miao, B., Langer, T., Craig, E. A., and Pfanner, N. (1993) A dual role for the mitochondrial heat shock protein 70 in membrane translocation of preproteins. *J. Cell Biol.*, **123**, 109.
69. Miller, B. R. and Cumsky, M. G. (1991) An unusual mitochondrial import pathway for the precursor to yeast cytochrome c oxidase subunit Va. *J. Cell Biol.*, **112**, 833.
70. Vestweber, D. and Schatz, G. (1988) Point mutations destabilizing a precursor protein enhance its post-translational import into mitochondria. *EMBO J.*, **7**, 1147.
71. Becker, K., Guiard, B., Rassow, J., Söllner, T., and Pfanner, N. (1992) Targeting of a chemically pure preprotein to mitochondria does not require the addition of a cytosolic recognition factor. *J. Biol. Chem.*, **267**, 5637.
72. Lithgow, T., Høj, P. B., and Hoogenraad, N. J. (1993) Do cytosolic factors prevent promiscuity at the membrane surface? *FEBS Lett.*, **329**, 1.
73. Nelson, R. J., Ziegelhoffer, T., Nicolet, C., Werner-Washburne, M. and Craig, E. A. (1992) The translocation machinery and 70 kDa heat shock protein cooperate in protein synthesis. *Cell*, **71**, 97.
74. Craig, E. A. (1990) Regulation and function of the HSP70 multigene family of *S. cerevisiae*. In *Stress proteins in biology and medicine*. Morimoto, R. I., Tissieres, A., and Georgopolous, C. (ed.). Cold Spring Harbor Laboratory Press, Cold Spring Harbor, NY, p. 301.
75. Deshaies, R. J., Koch, B. D., Werner-Washburne, M., Craig, E. A., and Schekman, R. (1988) A subfamily of stress proteins facilitates translocation of secretory and mitochondrial precursor polypeptides. *Nature*, **332**, 800.

76. Chirico, W. J., Waters, M. G., and Blobel, G. (1988) 70K heat shock related proteins stimulate protein translocation into microsomes. *Nature*, **332**, 805.

77. Cyr, D. M., Lu, X., and Douglas, M. G. (1992) Regulation of HSP70 function by a eukaryotic DnaJ homolog. *J. Biol. Chem.*, **267**, 20927.

78. Hendrick, J. P., Langer, T., Davis, T. A., Hartl, F. U., and Weidmann, M. (1993) Control of folding and membrane translocation by binding of the chaperone DnaJ to nascent polypeptides. *Proc. Natl Acad. Sci.*, **90**, 10216.

79. Atencio, D. P. and Yaffe, M. P. (1992) MAS5, a yeast homolog of DnaJ involved in mitochondrial protein import. *Mol. Cell. Biol.*, **12**, 283.

80. Caplan, A. J., Cyr, D. M., and Douglas, M. G. (1992) YDJ1p facilitates polypeptide translocation across different intracellular membranes by a conserved mechanism. *Cell*, **71**, 1143.

81. Lithgow, T., Ryan, M., Anderson, R., Høj, P., and Hoogenraad, N. (1993) Identification of the constitutive form of heat-shock protein 70 as a component of the outer membrane of mitochondria from rat liver. *FEBS Lett.*, **332**, 277.

82. Murakami, K., Tannase, S., Morino, Y., and Mori, M. (1992) Presequence binding factor-dependent and independent import of proteins into mitochondria. *J. Biol. Chem.*, **267**, 13199.

83. Douglas, M. G., Geller, B. L., and Emr, S. D. (1984) Intracellular targeting and import of an F_1 ATPase β-subunit–β-galactosidase hybrid protein into yeast mitochondria. *Proc. Natl Acad. Sci.*, **81**, 3983.

84. Garrett, J. M., Singh, K. K., von der Haar, R. A., and Emr, S. (1991) Mitochondrial protein import: isolation and characterization of the *Saccharomyces cerevisiae MFT1* gene. *Mol. Gen. Genet.*, **225**, 483.

85. Ito, M., Yasui, A., and Komamine, A. (1993) Precise mapping and molecular characterization of the *MFT1* gene involved in the import of a fusion protein into mitochondria in *Saccharomyces cerevisiae. FEBS Lett.*, **320**, 125.

86. Söllner, T., Griffiths, G., Pfaller, R., Pfanner, N., and Neupert, W. (1989) MOM19, an import receptor for mitochondrial precursor proteins. *Cell*, **59**, 1061.

87. Schneider, H., Söllner, T., Dietmeier, K., Eckershorn, C., Lottspeich, K., Trülzsch, B., Neupert, W., and Pfanner, N. (1991) Targeting of the master receptor MOM19 to mitochondria. *Science*, **254**, 1659.

88. Kassenbrock, C. K., Cao, W., and Douglas, M. G. (1993) Genetic and biochemical characterization of ISP6, a small mitochondrial outer membrane protein associated with the protein translocation complex. *EMBO J.*, **12**, 3023.

89. Nakai, M., Endo, T., Hase, T., and Matsubara, H. (1993) Isolation and characterization of the yeast Msp1 gene which belongs to a novel family of putative ATPases. *J. Biol. Chem.*, **268**, 24262.

90. Moczko, M., Ehmann, B., Gartner, F., Honlinger, A., Schafer, E., and Pfanner, N. (1994) Deletion of the receptor MOM19 strongly impairs import of cleavable preproteins into *Saccharomyces cerevisiae* mitochondria. *J. Biol. Chem.*, **269**, 9045.

91. Hines, V. and Schatz, G. (1993) Precursor binding to yeast mitochondria. A general role for the outer membrane protein Mas70p. *J. Biol. Chem.*, **268**, 449.

92. Harkness, T. A. A., Nargang, F. E., van der Keli, I., Neupert, W., and Lill, R. (1994) A crucial role of the mitochondrial protein import receptor MOM19 for the biogenesis of mitochondria. *J. Cell Biol.*, **124**, 637.

93. Lithgow, T., Junne, T., Wachter, C., and Schatz, G. (1994) Yeast mitochondria lacking the two import receptors Mas20p and Mas70p can efficiently and specifically import precursor proteins. *J. Biol. Chem.*, **269**, 15325.

94. Goebl, M. and Yanagida, M. (1991) The TPR snap helix: a novel protein repeat motif from mitosis to transcription. *Trends Biochem. Sci.*, **16,** 173.
95. Millar, D. G. and Shore, G. C. (1993) The signal anchor sequence of mitochondrial Mas70p contains an oligomerization domain. *J. Biol. Chem.*, **268,** 18403.
96. Mayer, A., Lill, R., and Neupert, W. (1993) Translocation and insertion of precursor proteins into isolated outer membranes of mitochondria. *J. Cell Biol.*, **121,** 1233.
97. Gray, M. W. (1989). The evolutionary origins of organelles. *Trends Genet.*, **5,** 294.
98. Vassarotti, A., Stroud, R., and Douglas, M. G. (1987) Independent mutations at the amino terminus of a protein act as surrogate signals for mitochondrial protein import. *EMBO J.*, **6,** 705.
99. Bibus, C. R., Lemire, B. D., Suda, K., and Schatz, G. (1988) Mutations restoring import of a yeast mitochondrial protein with a non-functional presequence. *J. Biol. Chem.*, **263,** 13097.
100. Gratzer, S., Lithgow, T., Bauer, R. E., Lamping, E., Paltauf, F., Kohlwein, S. D., Hauke, V., Junne, T., Schatz, G., and Horst, M. (1995) Mas37p, a novel receptor subunit for protein import into mitochondria. *J. Cell Biol.*, **129,** 25.
101. Lithgow, T., Junne, T., Suda, K., Gratzer, S., and Schatz, G. (1994) The mitochondrial outer membrane protein Mas22p is essential for protein import and viability of yeast. *Proc. Natl Acad. Sci.*, **91,** 11973.

2 | Protein transport to the nucleus and its regulation

DAVID A. JANS

1. Introduction: nuclear–cytoplasmic processes in the eukaryotic cell

Eukaryotic cells by definition possess a nucleus (karyon) and the existence of a nuclear subcellular compartment means that the genetic information, the DNA, is separated from the site of protein synthesis, i.e. the cytoplasm. Separation is effected by a double membrane structure, the nuclear envelope, which is both contiguous with the endoplasmic reticulum (ER) and largely ER-like in terms of lipid and enzyme composition. Since gene transcription and translation take place in separate compartments, specific transport events must occur: first, mRNA must make its way from the nucleus into the cytoplasm in order to be translated into protein; and secondly, the proteins that are required in the nucleus need to be specifically transported from their site of synthesis in the cytoplasm into the nucleus. Nucleocytoplasmic transport processes in both directions are precisely controlled, whereby the regulation of the subcellular distribution of proteins which regulate gene transcription, RNA processing, or other nuclear events is integral to many cellular processes (1). Since nuclear localization of transcription factors (TFs) (2–5), morphogens (6, 7), or oncogene products (8–11) appears to precisely accompany changes in the differentiation or metabolic state of eukaryotic cells, it is clear that nuclear protein import is a key control point in regulating gene expression and signal transduction (1).

This chapter will seek to describe what is known about protein transport to the nucleus, concentrating on the components of the transport apparatus that have been characterized, including those of the nuclear pore complex (Section 2), nuclear localization sequences (Section 3) and their binding proteins, GTP-binding proteins, and molecular chaperones, as well as the mechanisms regulating signal-dependent nuclear localization (Sections 4 and 5). Section 7 will introduce *in vivo* and *in vitro* systems to analyse nuclear protein transport kinetics quantitatively, and present two examples of proteins whose import kinetics and regulation by phosphorylation have been analysed in detail employing such systems.

2. The nucleus

2.1 The nuclear pore complex as a molecular sieve

All passive and active transport into and out of the nucleus occurs through the nuclear pore complex (NPC) (12–14) present in the nuclear envelope. NPC structure and function are conserved in all eukaryotes indicating the NPCs central role in eukaryote cell function. It is composed of at least 30 distinct protein components in varying stoichiometries and has a near organelle-like mass of about 10^5 kDa. Between 10^2 and 5×10^7 NPCs are present per nucleus depending on the cellular metabolic and differentiation state. This corresponds to a range of about 3 NPCs/μm^2 in the nuclear envelope of an inactive oocyte, to about 15–20/μm^2 in that of normal metabolically active differentiated somatic cells. As its name suggests, the NPC has a pore-like, molecular sieve function, whereby molecules smaller than 40–45 kDa can diffuse freely in and out of the nucleus independent of temperature and energy. Proteins larger than 45 kDa require a nuclear localization sequence (NLS) (15–17) in order to be targeted specifically to the nucleus. NLS-dependent protein transport is ATP- and temperature-dependent (18, 19), and can be inhibited by the lectin wheat-germ agglutinin (WGA) (20, 21) which binds to NPC proteins. Antibodies to NPC components block transport (22–25), reinforcing the idea that the NPC is the sole path of protein entry into the nucleus.

2.2 Nuclear pore complex structure

A simplistic representation of the NPC including its relationship to the nuclear lamina (see Section 2.4) is shown in Fig. 1. A number of models of NPC structure have been proposed based largely on electron microscopic (EM) techniques (e.g. 26, 27). All have to account for the molecular exclusion properties of the NPC, as well as its ability to transport much larger molecules (e.g. 970 kDa IgM molecules) in an NLS-dependent fashion. Most models suggest an octagonal rotational symmetry for the approximately 100 nm diameter complex, with two concentric ring structures probably on the cytoplasmic and nucleoplasmic sides of the nuclear envelope, and spoke structures surrounding a central pore of about 90 Å in diameter through which active nucleocytoplasmic transport takes place (26, 27). The eight circular granules which line the cytoplasmic ring have been hypothesized to be the remnants of fibrils compacted during sample preparation (see Fig. 1 and references in 14). EM data alone give no hint as to the mechanism by which the small central channel 'opens' to allow the passage of molecules up to 40 nm in diameter. Even though molecular chaperonin proteins have been implicated in nuclear protein transport (28–30, see Section 5.2), unfolding of the transport substrate is clearly not the mechanism for this, since very large (280 Å diameter) colloidal gold molecules can be transported into the nucleus if covalently labelled with NLS-bearing molecules (31, 32). A double iris/camera-shutter-like mechanism (12, 26) has been proposed, whereby ATP-dependent stretching and contraction of NPC components

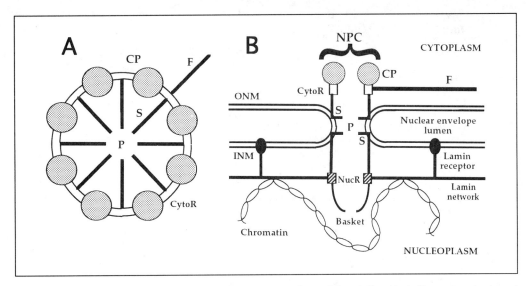

Fig. 1 Model, adapted from ref. 14, of the nuclear pore complex and its relationship to the nuclear lamin and chromatin (A) from above (cytoplasmic side) and (B) in cross-section. CP, cytoplasmic particle (compacted fibril?); CytoR, cytoplasmic ring; F, fibril (pore connecting?); INM, inner nuclear membrane; NucR, nuclear ring; ONM, outer nuclear membrane; P, central pore; S, spoke.

in an actin–myosin-like fashion may constitute the mechanistic basis of the opening of the pore during active nuclear transport (27). Indeed, several nuclear envelope-associated ATPases have been identified (33–35) including a 40–46 kDa ATPase (35) which is important for RNA export from the nucleus.

An alternative model for transport through the NPC relies, in part, on the refinement of both sample preparation and EM image-averaging techniques. This has elaborated the model of NPC structure to include basket-like structures on the nucleoplasmic side of the pore, and fibrillar structures attached to the cytoplasmic ring as shown in Fig. 1 (36). This has encouraged support for solid-phase models of nuclear transport (see ref. 14), where transport is purported to be directed along a network of filaments mediated by NLS-binding proteins (NLSBPs—see Section 5.1). Movement of transport protein substrates along the filaments may be through molecules analogous to kinesin molecules which are involved in the directed ATP-dependent movement of organelles and vesicles in neuronal cells (e.g. ref. 38). Immuno-EM data for the NLSBP Nopp140, which shuttles between nucleus and cytoplasm, has been interpreted as representing direct evidence for linear tracks of protein movement through the NPC (37).

2.3 Components of the nuclear pore complex

Relatively few of the multiple protein components constituting the NPC have been characterized at the molecular level (see Table 1). These include the nuclear envelope-anchored concanavalin A-binding high-mannose Gp210 protein (25, 39),

Table 1 Components of the nuclear pore complex

Proteins	Origin	Lectin binding	Role in nuclear transport	References
Transmembrane/nuclear envelope associated:				
Gp210	Rat	Con A	Protein import[a]	25, 39, 41
POM121[b]	Rat	WGA	?	40
Soluble nucleoporins (F–X–F–G repeated motif):				
Gp62[c]	Vertebrate	WGA	Protein import[a]	22, 42, 43
Nsp1p[d]	Yeast	WGA?	?	43, 44
Nup1p	Yeast	WGA?	?	45
Nup2p	Yeast	WGA?	?	46
Nup153p[e]	Rat	WGA	RNA export?	see 14, 47
Soluble nucleporins (G–L–F–G repeated motif):				
Nup145p[f]	Yeast	WGA	RNA export Protein import	48
Nup100p[f]	Yeast	WGA	RNA export? Protein export?	48–50
Nup/Nsp116p	Yeast	WGA	RNA export Protein import?	48–51
Nup/Nsp49p	Yeast	WGA	?	48–50

[a]Monoclonal antibodies specific to the respective proteins block protein nuclear import.
[b]POM121 contains an F–X–F–G repeat motif similar to those of the nucleoporins.
[c]Other nucleoporins (Gp58 and Gp54) complex to Gp62 to form oligomers (52).
[d]Nsp1p complexes to the nucleoporin NIC96 (53).
[e]Nup153p also contains four zinc finger motifs typical of DNA-binding proteins, but also suggestive of RNA-binding capability (see 14).
[f]Nup145p contains the octapeptide R–G–Y–G–C–I–T–F, which is homologous to a single-stranded DNA/RNA-binding site (48). Nup100p and Nup/Nsp116p contain similar motifs.
Abbreviations: ConA, concanavalin A; WGA, wheat-germ agglutinin.

most of whose mass is located within the nuclear envelope lumen adjacent to the NPC. Its function may be to anchor the NPC in the nuclear envelope, with a putative role in fusing the inner and outer nuclear membranes during NPC assembly (25, 39). The more recently described 121 kDa WGA binding protein POM121 also appears to be localized in the pore membrane region (40). The signal for specific sorting of Gp210 to the NPC membrane has been shown to be contained within the single transmembrane segment of the molecule (41).

Of the NPC mass 5–10% is made up of a class of peripheral, non-membrane-anchored NPC protein components called nucleoporins. They possess O-linked glycosidic N-acetylglucosamine moieties, which are the targets of WGA-mediated inhibition of nuclear protein transport, and appear to complex with Gp210. NPCs depleted of WGA-binding proteins are inactive in terms of nuclear protein import (54); as mentioned above, monoclonal antibodies to nucleoporins reversibly block nuclear protein import (22–25).

Nucleoporins identified by molecular cloning appear to form two classes (see Table 1) based on the presence of repeat units (F-X-F-G and G-L-F-G respectively, where the single letter amino acid code is used and X is any amino acid). The Gp62 protein (42), which is related to both Nsp1p and Nup1p (44, 45), has a distinctive primary structure including a region of heptad hydrophobic repeats similar to coiled-coil proteins such as myosin, which may have functional significance in the context of the models of transport through the NPC. In a similar fashion to Nsp1p and Nup1p (50, 53), Gp62 with other nucleoporins forms a stable hetero-oligomeric complex that functions in transport through the NPC (14, 52). The C-termini of both Nup145p and Nsp1p are capable of targeting normally cytosolic carrier proteins to the nuclear periphery, implying that they contain specific signals for localization to the NPC (48, 55). In contrast, deletion analysis indicates that the N-terminus of Nup1p contains its NPC targeting signal (56). The zinc finger-containing Nup153p appears to be localized exclusively on the nucleoplasmic side of the NPC, possibly associated with the nucleoplasmic basket structures (14, 47). Although the mechanistic role of nucleoporins in protein nuclear import is not clear; their distinctive structure has encouraged speculation that the coiled-coil C-terminal domains of Gp62 or Nsp1p (together with complexing proteins—50, 52, 53) may constitute part of the fibrillar structures on the cytoplasmic side of the NPC, whilst the central parts of the molecule may function as hinge domains together with the glycosylated N-termini to 'open' and 'close' the central pore (see ref. 14).

2.4 The nuclear lamina

Structural elements of the nucleus in addition to those described above include the nuclear lamina, a protein meshwork of intermediate filament-type proteins called lamins (57) which can be regarded as the nucleoskeleton. Lamins possess a central α-helical rod domain flanked by N- and C-terminal non α-helical end domains. In addition to an NLS, lamins contain a C-terminal tetrapeptide motif known as the CaaX box (Cys–aliphatic amino acid–aliphatic amino acid–any amino acid) which is subject to the post-translational modifications (farnesylation, proteolytic trimming, and carboxyl-methylation) necessary for the association of lamins with the nuclear membrane which is also mediated by the lamin receptor (57, 58). A- and B-type lamins are distinguished according to structural and biochemical properties, the A-type lamins being developmentally controlled (57). When lamins depolymerize at the onset of mitosis, A-type lamins are solubilized, but B-types remain membrane-associated (57, 59). Mitotic disassembly of the nuclear lamina is proposed to result from lamin hyperphosphorylation, probably by the cyclin-dependent kinase (cdk) cdc2 (cell division cycle control kinase 2) (59).

The nuclear lamin network has sites at which the chromosomes are attached (see Fig. 1). This attachment serves to protect the fragile interphase chromatin from mechanical stress, but its principal role may be to localize the chromatin close to the NPC. This means that TFs entering the nucleus through the NPC are in close proximity to their specific target genes and promoters.

2.5 Other nuclear processes

A body of evidence indicates that Ca^{2+} does not simply diffuse passively between nucleus and cytoplasm (60). This may seem in contradiction to the proposed molecular sieve properties of the NPC discussed in Section 2.1, but can be attributed in part to the ER-type Ca^{2+}-ATPase and inositol tris-phosphate (IP3)-dependent Ca^{2+} channels present in the nuclear envelope (61). Regulation of intranuclear Ca^{2+} levels provides the possibility of exerting control over nuclear enzymatic activities and ultimately gene expression, through:

- The action of Ca^{2+}-binding proteins such as calmodulin (CaM), which is variously located in nucleoplasm, chromatin, nucleolus, and nuclear matrix, and whose nuclear concentration appears to be hormonally regulated (see ref. 60).
- CaM-binding proteins such as calcineurin, CaM-dependent protein kinase II (CaMPK II), and myosin light-chain kinase, whose activities are modulated by CaM (see ref. 60).
- Ca^{2+}/IP3-regulated phosphorylation on the part of the Ca^{2+}/phospholipid-dependent protein kinase (PK-C) (see ref. 60).

A role has been proposed for Ca^{2+}/CaM in regulating nuclear processes such as nuclear protein phosphorylation, and thereby DNA repair and replication and gene expression (see ref. 60).

Many nuclear proteins, especially transcription activators such as c-myc, c-fos, the yeast MATα2, and the tumour suppressing p53, have very short half-lives (see 62, 63). This process in the nucleus can be mediated by the 76 amino acid proteolysis co-factor ubiquitin (Ub), whose covalent attachment to a protein labels it for destruction via a proteolytic cascade (63). Such processes are integral to cellular function, since they enable the cell to alter the nuclear levels of proteins regulating gene expression very rapidly and precisely. An example is that of the DNA-binding c-myc proto-oncogene product, whose nuclear localization has a decisive role in regulating development in *Xenopus*. The c-myc protein is stockpiled in the cytoplasm and is unusually stable in the early oocyte (8, 64); at fertilization it is rapidly transported into the nucleus of the dividing embryo, accompanied by a drastic reduction in its half-life (8, 64) which may be Ub-mediated (63). The combination of precisely timed nuclear entry and regulated nuclear degradation of c-myc enables its activity to be controlled very precisely.

The cell-cycle regulated degradation of cyclins may also be controlled in a similar fashion, since they appear to be synthesized and degraded at different phases of the cell cycle, thus influencing the function of cdks (65). Whilst Ub may be involved in this process, there is evidence that the Pro/Glu/Ser/Thr-rich (PEST) sequences within the cyclin C-terminus are also involved (65, 66). Clearly, nuclear protein degradation is an important means of regulating the levels of nuclear proteins, and thereby nuclear and cellular function.

3. Signals for nuclear targeting

3.1 Nuclear localization sequences

NLSs (15–17) are the short peptide sequences which are necessary and sufficient for the nuclear localization of their respective proteins. These sequences have been identified based on one or more of three basic criteria:

- Mutation or deletion of the NLS leads to cytoplasmic localization of the protein in question.
- The NLS is active in nuclear targeting as a peptide covalently coupled to a normally cytoplasmic localized protein.
- The NLS is able to target a normally cytoplasmic localized protein to the nucleus when encoded in a fusion protein.

NLSs are regarded as functioning via recognition/ligand–receptor-like interactions (e.g. with NLSBPs—see Section 5.1) and not through DNA or histone binding (e.g. refs 67–70). That the NLS is an entry rather than a retention signal has been shown by the fact that NLS-deficient carrier proteins microinjected into the nucleus remain nuclear (68). Photobleaching experiments show that NLS-containing proteins which have been accumulated in the nucleus are highly laterally mobile in the nucleus (71), implying that binding in the nucleus to chromatin or other structures such as the nucleoskeleton, is unlikely to be the mechanism by which NLSs function, at least in the case of the SV40 large tumour antigen (T-ag) NLS (12, 13, see however ref. 72 and references in 14). NLSs are fundamentally different from other peptide signal or targeting sequences in that they are not cleaved during transport. This is because, in contrast to ER (see Chapter 4 of this volume) and other targeting signals, NLSs are required to function many times, through a number of cell divisions, which in the case of higher eukaryotes, involve dissolution of the nuclear envelope (12–14).

Despite the fact that proteins which are below the NPC mol. wt cut-off point do not require an NLS in order to enter the nucleus, many, including polypeptide ligands, accumulate in the nucleus either constitutively or in response to stimuli (see ref. 73). Histones are probably the best characterized example of such proteins; although their nuclear accumulation is probably largely through complexation in chromatin, they possess defined NLSs which are functional in targeting large carrier proteins such as β-galactosidase to the nucleus (74). Breeuwer and Goldfarb (75) showed that the addition of an NLS to a small protein causes its transport to the nucleus to be temperature- and ATP-dependent, properties of NLS-dependent transport. This implies that an NLS confers upon the protein carrying it the specific regulatory properties associated with active rather than passive transport.

As can be seen in Table 2, there is no general consensus sequence for NLSs (76–78). The best characterized NLS is that of the T-ag ($PKKKRKV^{132}$), whereby a single amino acid substitution of N or T for the critical K^{128} residue of the NLS abolishes its function and results in complete cytoplasmic localization. Although not all NLSs resemble that of T-ag (e.g. those of MATα2 and influenza virus nucleoprotein (NP)) many of the NLSs listed in Table 2 have been identified on the basis of homology to the T-ag NLS, but in some cases not formally tested using all of the criteria above.

In the absence of a consensus NLS, several secondary structure prediction analyses have been performed (see refs 76 and 16, 110, 111). Apart from the fact that all known NLSs are very hydrophilic (and basic, with the few exceptions already mentioned),

Table 2 Nuclear localization sequences in nuclear proteins

Protein	NLS (References)	Critical residue[a]	Carrier proteins targeted[b]
SV40 T-ag	PKKKRKV[132] (15, 79, 80)	K[128]	PK/BSA/IgG/βG/CSA
Polyoma T	VSRKRPRP[196] (79, 81)		PK/CSA
	PPKKARED[286] (79, 81)		(not PK or CSA)
Nucleoplasmin	KRPAATKKAGQAKKKKLD[172] (79, 80, 82)		IgG/PK/βG/CSA
Xenopus laevis N1/N2	VRKKRKT[537] (69)	K[534]	
	AKKSKQEP[554] (69)	K[549]	
Lamin L_I	VRTTKGKRKRIDV[240] (79, 83)		CSA
Lamin A/C	SVTKKRKLE[422] (84)	K[417]	
Cofilin	PEEVKKRKKAV[36] (85)		BSA
Human-c-myc	PAAKRVKLD[328] (79, 86)		PK/HSA/CSA
	RQRRNELKRSF[374] (79, 86)		PK
Ad7 E1a	KRPRP[289] (68, 79, 80)		IgG/GK/CSA
SV40 VP1	GAPTKRKGS[8] (79, 87)		POL (not PK or CSA)
SV40 VP2/3	PNKKKRK[323] (88, 89)	K[320]	POL/PK/BSA/βG/IgG
Human p53	PQPKKKPL[323] (79, 90, 91)	K[320]	PK/βG (not CSA)
NF-κB p50	QRKRQK[372] (92, 93)	K[369]R[370]	
NR-κB p65	EEKRKR[286] (94)		
Chicken v-rel	KSKKQK[295] (95)	K[292]K[293]	βG
Mouse c-abl IV	SALIKKKKKMAP[631] (9, 144)		
Influenza virus NS1	DRLRR[38] (96)		
	PKQKKRK[221] (96)		
Hepatitis virus delta antigen	RKLKKKIKKL[44] (97)		(not βG)
	PRKRP[89] (97)		(not βG)
Chicken c-ets-1	GKRKNKPK[383] (98)		
v-jun/c-jun	KSRKRKL[253] (99)		IgG
Ribosomal protein L29	KTRKHRG[12] (100)	R[8]	βG
	KHRKHPG[29] (100)	R[25]	βG
Yeast MATα2	NKIPIKDLLNPQ[13] (17, 70, 79, 101)		βG (not IgG or CSA)
	VRILESWFAKNIENPYLDT[159] (70)		βG
Influenza virus nucleoprotein	AAFEDLRVLS[345] (102)		α_1G
Hepatitis B virus core protein	SKCLGWLWG[29] (103)		
Yeast histone 2B	GKKRSKAK[36] (74)	K[31]	βG
Monkey v-sis	RVTIRTVRVRRPPKGKHRK[255] (104)[c]		PK
Human PDGF A (longer form)[d]	RESGKKRKRKRLKPT (106)		PK, CAT, DHFR
Mouse Mx1	REKKKFLKRR[615] (107)	R[614]	
Prothymosin α	TKKQKT[107] (108)		βG
VirD2 protein (octopine – A. tumefaciens)	EYLSRKGKLEL[38] (109)[e]		βG

[a]Residue shown by mutation analysis to be essential for NLS function.
[b]Carrier proteins that have been shown to be able to be targeted to the nucleus by the NLS sequence either as a fusion protein derivative or peptide covalently coupled to the carrier (see 12, 76, 77). α_1G, α_1-globin; βG, β-galactosidase; BSA, bovine serum albumin; CAT, chloramphenicol acetyltransferase; CSA, chicken serum albumin; DHFR dihydrofolate reductase; GK, galactokinase; GLUC, β-glucuronidase; HSA, human serum albumin; IgG, immunoglobulin; PK, pyruvate kinase; POL, polio virus VP1.
[c]Two copies are necessary to target PK to the nucleus. The v-sis NLS is also present in the platelet-derived growth factor (PDGF)-B related c-sis proto-oncogene (104, 105).
[d]Alternatively spliced longer form of PDGF A (106).
[e]Bacterial NLS capable of nuclear targeting in plant and yeast cells.

generally preceded by a β-turn/random coil region, and highly antigenic surface structures (see ref. 111), no definitive consensus structural motif has been identified. Secondary structure and conformation, however, play a role in the nuclear localization of many NLS-bearing proteins such as p53 (112) and gal4 (113), presumably through influencing the accessibility of the NLS (see Section 4).

3.2 Multiple and bipartite nuclear localization sequences

A number of proteins possess two or more NLSs (e.g. polyoma large T-antigen, c-myc, N1/N2, influenza virus NS1, MATα2, and yeast ribosomal protein L29—see Table 2) which are required in concert to achieve 'complete' nuclear localization (see also 91, 114–116 for p53, influenza virus PBP1 and 2, and adenovirus DNA binding protein (DBP), respectively). Multiple copies of an NLS appear to be more efficient than only one copy, especially in the case of a 'weak' NLS (32, 101, 110, 111, 117).

Possibly a variant of multiple NLSs are bipartite NLSs (78) which consist of two series of basic residues separated by a 10–12 amino acid spacer (118, 119). Whilst varying the length of this spacer using an S-P-G-G insert had no effect on nuclear targeting efficiency in the case of the nucleoplasmin NLS, a more hydrophobic, bulky Q-P-W-L spacer markedly reduced its targeting efficiency (118). This implies that conformation and/or hydrophobicity may be important, perhaps for co-recognition of the two arms of the NLS (see also Section 5.2). Again, few of the bipartite NLS sequences shown in Table 3 have been formally tested using the NLS criteria listed in the previous section, but have been identified rather through their homology to the nucleoplasmin NLS (78, 118).

Longer NLS-functioning sequences ('non-peptide' NLSs) have been described for the yeast transcription factor gal4 (the N-terminal 74 amino acids—113) and the snRNP U1A protein (amino acids 94–204; 133). Multiple mutation sites/deletions affect the nuclear localization of these proteins, meaning that it has not (as yet) been possible to narrow down the NLS to a more definitive shorter sequence. It is also possible that as above, a secondary structure may play a role by influencing NLS accessibility. The size of the carrier used to test nuclear targeting sufficiency of an NLS can be important, as indicated by studies with the gal4 NLS in which amino acids 1–74 were shown to be necessary to target a β-galactosidase fusion protein (approximately 480 kDa) to the nucleus, whereas only residues 1–29 are needed to target invertase (about 120 kDa) (134). Similar results have been reported by Yoneda *et al.* (135) for T-ag (94 kDa) where the NLS alone (amino acids 126–132) is sufficient to target bovine serum albumin (67 kDa) to the nucleus, but not to target IgM (about 970 kDa). Larger carrier proteins presumably have more stringent targeting requirements than smaller proteins and may more realistically approximate to the *in vivo* situation.

3.3 Signals for localization to 'subnuclear compartments'

3.3.1 Nucleolar localization signals

There is some experimental evidence for sequence-dependent targeting to subnuclear compartments such as the nucleolus. Nucleolus-localization signals (NOSs) appear to be related to NLSs in that they also comprise short sequences of essentially basic amino acids, which may relate to the fact that entry into the nucleus is initially required before a subnuclear compartment is encountered. A well-characterized

Table 3 Bipartite nuclear localization sequences in nuclear proteins

Protein	NLS (References)	Carrier proteins targeted[a]
Xenopus		
NO38	**KR**AAPNAASKVPL**KKTR**[153] (118, 120)	
Nucleoplasmin	**KR**PAATKKAGQA**KKKKL**[171] (79, 82, 117, 118)	PK/βG/IgG/CSA
N1/N2	**RKKRK**TEEESPLKDKDA**KKSK**QEP[554] (69)	βG
S. cerevisiae		
SWI5	**KK**YENVVIKRSP**RKRG**RP**RK**[655] (4)	βG
GCN4	**KR**ARNTEAA**RRS**RA**RK**[245]	
Human		
poly(ADP-ribose) polymerase	**KRK**GDEVDGVDEVA**KKK**S**KK**[226] (119)	βG
DNA polymerase	**KK**S**KK**GRQEALERL**KKAK**[41]	
	KKQFFTPKVLQDY**RKLK**[1434]	
Topoisomerase I	**RK**EEKVRASGDA**KIKKE**[105]	
	KKIKTEDT**KK**E**KKRK**LEEEEDG**KLKKPK**[178]	
c-fos	**KRRIRRE**RNKMAAAKCRN**KRRRL**[161] (121)	PK
p110[Rb]	**KR**SAEFFNPPKPL**KKLR**[869] (122)	
Mouse		
Nucleolin	**KK**EMTKQKEAPEA**KKQK**[298]	
Steroid hormone receptors (human)		
Glucocorticoid	**RK**CLQAGMNLEA**RKTKK**[495] (123, 124)[b]	βG
Progesterone	**RK**CCQAGMVLGG**RKFKK**[495]	
Androgen	**RK**CYEAGMTLGA**RKLKK**[448]	
Oestrogen	**RK**CYEVGMMKGGI**RKDR**[495] (127)	
erb-A	**RK**LAKRKLIEEN**REKRR**[196]	
Thyroid β	**KR**LAKRKLIEEN**REKRR**[195]	
Viral		
Herpes ICP-8	**RKR**AFHGDDPFGEGPPD**KK**[1188]	PK
Plants		
TGA-1A (tobacco)	**KK**LAQNREAA**RK**S**RL**R**KK**[92] (129)	GLUC
TGA-1B (tobacco)	**KKK**ARLVRNRESAQLS**RQRKK**[364] (129)	GLUC
Opaque-2 (maize)	**RKRK**ESNRESA**RRS**RY**RK**[247] (130)	
A. tumefaciens[c]		
VirD2 protein (octopine)	**KR**PRDRHDGELGG**RKRAR**[413] (109, 131)	βG
VirD2 protein (nopaline)	**KR**PREDDDGEPSD**RKRER**[434] (109, 131)	GLUC
VirE2 protein	**KLR**PEDRYIQTEKYG**RREIQKR**[249] (132)	GLUC
	RAIKTKYGSDTEI**KLKSK**[309] (132)	GLUC
'Consensus'	**K/R–K/R**–10–12 amino acids–**K/R–K/R–K/R** (118)	

[a]See legend to Table 2. βG, β-galactosidase; CSA, chicken serum albumin; GLUC, β-glucuronidase; IgG, immunoglobulin; PK, pyruvate kinase. The two arms of basic residues of the bipartite NLS are shown in bold type.

[b]Steroid hormone receptors contain in addition to the above bipartite NLS conferring constitutive nuclear localization, a second dominant ligand-binding-dependent NLS (e.g. wild-type GR is cytoplasmic in the absence of ligand; 123, 124), which has also been shown for the progesterone (125, 126) and oestrogen (127) receptors. Ylikomi et al. (128) report that the situation with respect to the oestrogen and progesterone receptors may be more complicated, whereby three NLS-functioning sequences contribute to nuclear localization, in addition to the ligand-binding-dependent NLS.

[c]Bacterial NLS sequences capable of nuclear targeting in plant (and yeast) cells.

example is that of the nucleolar protein NO38 which has distinct NLS (see Table 3) and NOS (present in the C-terminal 108 amino acids) sequences (120). Similar results have been reported for chicken nucleolin (136). Dang and Lee (90) have suggested a dual NLS structure for the consensus NOS, which comprises regions homologous to both the T-ag and polyoma T-ag NLSs, separated by a glutamine residue (see Table 4), but this is clearly not consistent with the requirement

Table 4 Nucleolar localization sequences in nuclear proteins

Protein	Nucleolar localization sequence (NOS) (References)	Carrier proteins targeted[a]
HTLV-1 (p27x-111) Rex	MP**K**T**RRR**P**RRR**SQ**RKR**PPTP[20] (137, 138)	βG
HIV Tat	G**RKKRR**Q**RRR**AP[59] (90)	PK
c-myc–HIV Tat (fusion)	PAA**KR**V**K**LDQ**RRR**AP (90)	PK
HSP70	F**KRK**H**KK**DISQN**KR**AV**RR**[267] (90)	PK
'Consensus'	T-ag NLS–Q–polyoma T NLS **PKKKRK**V–Q–VS**RKR**P**R**P	

[a]Carrier proteins that have been shown to be able to be targeted to the nucleolus by the NOS sequence as a fusion protein derivative. βG, β-galactosidase; PK, pyruvate kinase. Basic residues are highlighted in bold type.

for other sequences of proteins such as the mouse upstream binding factor (mUBF) and nucleolin explained below.

Since no membrane structure defines the nucleolus, it is likely that binding in the nucleolus is the major mechanism of localization. This contention is supported by results for the nucleolar-localized rRNA gene transcription factor mUBF (139), which possesses an NLS (amino acids 449–480), including the KKKAK[456] sequence) capable of targeting β-galactosidase to the nucleus, but no identifiable NOS (139). Nucleolar localization requires sequences other than the NLS, including the first of six HMG-boxes (about 80–90 amino acid sequences homologous to repeat sequences of members of the HMG—high-mobility group—class of nuclear non-histone proteins) and the acidic C-terminal region (139). It is suggested that the mUBF NLS targets the protein to the nucleus, followed by sequestration in the nucleolus through binding to other proteins at the rRNA promoter mediated by the mUBF HMG-boxes and acidic tail (139). CKII-specific phosphorylation within the mUBF C-terminus activates promoter binding and gene transcription possibly through increasing the affinity of these interactions (140). The NLSBPs Nopp140 (37) and Nsr1 (141) both contain multiple putative CKII sites in addition to a number of candidate NLSs, and may localize in the nucleolus by a similar mechanism (142). Nucleolar localization of chicken nucleolin, which is dependent on both RNA-binding motifs and a glycine/arginine-rich C-terminal domain, has also been interpreted as being due to binding to specific nucleolar components (72).

3.3.2 Signals for targeting to other subnuclear compartments

The existence of other discrete subnuclear compartments has been proposed (see 13) which include those defined on the basis of 'specific' subnuclear localization of proteins involved in RNA splicing ('nuclear speckles'; 138) and snRNP-protein components ('foci'; 143). The functionally interchangeable Arg/Ser-rich regions of the nuclear speckle-localizing *Drosophila melanogaster* RNA processing regulators, *Tra* and *Su(w^a)*, can target β-galactosidase to nuclear speckles (138), implying that localization may be signal sequence-specific. Such sequences appear to be unique to RNA processing regulators such as SC35 and SF2/ASF (143).

4. Modulators of NLS function

4.1 Protein context and competing signals

A number of studies have shown that the possession of an NLS alone may not be sufficient to result in nuclear localization of a particular protein (see refs 76, 77). One factor which plays a role is that of protein context; the position of the NLS within a protein (111). The general consensus is that an NLS has to be on the surface of the protein in order to be accessible for recognition (111, 145). Moreland *et al.* (145), for example, showed that the N-terminal NLS (amino acids 1–21) of the yeast ribosomal L3 protein was only functional in targeting a β-galactosidase fusion protein to the nucleus when made accessible via inclusion of an eight amino acid Gly–Pro spacer. There are several examples of proteins containing NLSs which are functional only in certain protein contexts. The best characterized are the NF-κB p50/110 proteins, whereby p50 comprises the N-terminal NLS-containing portion of the p110 protein. Whereas the p50 protein is predominantly nuclear, the p110 precursor is exclusively cytoplasmic (92, 93) through intramolecular masking of the NLS (see Section 4.3.2). This conclusion is supported by the observation that monoclonal antibodies to the NLS recognize native p50 but fail to recognize p110 unless it has been denatured (93).

A further example of NLS-containing proteins which are cytoplasmic is the onco-genic p160^{gag}/$^{v-abl}$ variant of mouse p150 type IV c-abl (9). Both proteins contain a functional T-ag-like NLS (KKKKKMA^{632}), but only c-abl is nuclear (9, 144). Nuclear localization of c-abl appears to be mediated, in part, by c-abl residues 72–126, which are lacking from p160^{gag}/$^{v-abl}$, since deletion of these residues results in pre-dominantly cytoplasmic localization of c-abl. These residues contain no identifiable NLS, but instead constitute part of the SH3 (*src* homology-3) domain, which is responsible for association with specific protein factors such as the guanosine triphosphatase-activating protein (GAP)-rho-like 3BP-1. A role for binding to other proteins, in concert with the c-abl-NLS, is implicated in effecting c-abl's nuclear localization (see also Section 4.2) (9, 144).

Signals for localization to subcellular compartments other than the nucleus can override NLSs (see ref. 77). An example is the fact that the T-ag NLS inserted into the sequence of the normally plasma membrane-associated polyoma middle T-antigen (111) is not functional in nuclear targeting. Comparable effects are seen for a SV40 T-ag frameshift mutant in which an altered C-terminus resulting in an additional largely hydrophobic sequence of about 50 amino acids overrides T-ag NLS-mediated nuclear localization (146). Analogous results have been obtained for the herpes simplex virus major DNA-binding protein (infected-cell protein 8—ICP8), whose C-terminal NLS is capable of targeting a heterologous protein to the nucleus (see Table 3). Deletions in multiple domains of ICP8 inhibit its nuclear localization (144), implying that there are distinct conformational constraints limiting the function of the NLS (147). Similar results have been reported for pyruvate kinase fusion proteins carrying short sequences for the c-*fos* gene, including the bipartite NLS and 'cytoplasmic localization' determinants (121) and for the snRNP U1A protein (133). The hepatitis delta antigen (HDAg) provides a further example, whereby two NLSs (RKLKKKIKKL[44] and PRKRP[89]) are separated by a hydrophobic domain (amino acids 50–65) which impairs the function of one of the NLSs (97).

The question of dominance of signals for targeting to different subcellular compartments, and in particular of the competition between NLSs and mitochondrial localization and secretory signals, has been addressed in several studies (see ref. 77). It appears that the most N-terminal signal dominates (148, see also 149). The yeast TRM1 protein, required in both nucleus and mitochondria for tRNA modification, interestingly appears to carry functional signals for both mitochondrial (N-terminal) and nuclear (amino acids 70–213) localization, as shown by the targeting of β-galactosidase fusion proteins (150).

4.2 'Piggy back' and cell-type-specific effects

A number of examples exist of proteins greater than 45 kDa which, although lacking a functional NLS, are predominantly nuclear, whereby the mechanism appears to be via association with NLS-bearing proteins and co-transport into the nucleus ('piggy back'). The non-histone protein HMG1 is able to co-transport a specifically reacting 170 kDa monoclonal antibody to the nucleus (151). Nuclear localization of NLS-defective p53 observed in T-ag-expressing COS cells (91) is presumed to be through specific complex formation with T-ag and co-transport into the nucleus. Similar observations have been made for the adenovirus DNA polymerase, which is localized to the nucleus through association with the preterminal protein pTP1 (NLS RLPVRRRRRRVP[373]) (152), and for the two subunits of the mammalian pancreas-specific TF PTF1 which require a third 75 kDa glycoprotein for nuclear localization (153). Nuclear localization of NLS-deleted c-fos fusion protein variants is similarly thought to be through co-transport with c-fos's NLS-containing AP-1 transcription complex partner, c-jun (see below and 121). Predominant localization of the NLS-deficient cdk cdc2 kinase in the nucleus is believed to be through complex formation with cyclin B, which possesses three putative NLSs: PKKRHA[61], SKKRRQP[117], and PKKLKKD[160] (154, 155). The ability of cdc2 and other cdks to form complexes with many other nuclear factors, including p107[Rb], cyclin A, E2F, etc. (e.g. ref. 156), provides a number of possibilities both for their nuclear transport and for the co-transport of proteins and the TFs that interact with them. The reverse process—that of retention of an NLS-bearing protein in the cytoplasm via association with a cytoplasmically localized interacting protein—has been demonstrated in the case of p53 localized in the cytoplasm either by NLS-defective T-ag (157) or by non-nuclear localized mutant p53 in inactive oligomers (158).

Cell-type-specific differences in terms of NLS effectivity have been observed for T-ag (110), adenovirus DBP (116), and p53 (91). cAMP-dependent protein kinase

(PK-A) regulation of nuclear localization of rNFIL-6 (rat nuclear factor induced by interleukin-6) appears to be cell-type specific, since it localizes to the nucleus upon elevation of intracellular cAMP levels in rat PC12 phaeochromocytoma cells, but does not do so in HeLa cells (see ref. 159). The reason for such cell-type variability may relate, in part, to interactions with other proteins such as those mentioned above (and see also below with respect to the regulation of nuclear protein transport).

4.3 Regulated (conditional) nuclear localization

TFs regulating gene expression in the nucleus are no different from other proteins in that they are synthesized in the cytoplasm and therefore subject to the mechanisms regulating nuclear protein import. Whilst some proteins appear to be constitutively targeted to the nucleus, such as histones, others are only conditionally targeted to the nucleus, often being preferentially in the cytoplasm (see ref. 1). Advantages of cytoplasmic localization for a TF include the potential to control its nuclear activity through regulating its nuclear uptake and its direct accessibility to cytoplasmic signal-transducing systems. TFs able to undergo inducible nuclear import include the glucocorticoid receptor (GR; 121), the α-interferon-regulated factor ISGF3 (160), the nuclear v-jun oncogenic counterpart of the AP-1 transcription complex member c-jun (99), the yeast TF SWI5 (4), the *D. melanogaster* morphogen dorsal (6, 7), and the nuclear factors NF-κB (2, 3) and NF-AT (nuclear factor of activated T cells, 5).

Pathways of signal transduction from extracellular signal to nucleus leading to the ultimate regulation of gene expression (see ref. 73) include the hormone-stimulated response to elevated cAMP levels. This results in translocation of the PK-A C-subunit to the nucleus (73, 161, 162) where it phosphorylates and thereby activates nuclear TFs such as the cAMP-response element binding protein (CREB). In a similar fashion, mitogen-activated protein kinases (MAPKs) move to the nucleus in response to cellular stimulation by mitogens, whilst the PK-C α-subunit translocates to the nuclear envelope upon phorbol ester treatment (see refs 1, 73).

4.3.1 Cytoplasmic retention factors

One mechanism of regulating nuclear protein import is that of cytoplasmic retention, whereby a cytoplasmically localized 'anchor' protein or specific retention factor binds an NLS-containing protein and prevents it from migrating to the nucleus. Cytoplasmic retention has been described for c-fos (10) and also for the GR, where the HSP90 heat-shock protein complexes with and retains the GR in the cytoplasm in the absence of glucocorticoid hormone (123). Hormone binding by the GR dissociates the complex to allow NLS-dependent nuclear translocation of the receptor (123), recent results suggesting that cell-cycle-dependent dephosphorylation of the GR may also be involved (see refs 1, 163). Cytoplasmic retention has also been described for the B-type cyclins, which translocate to the nucleus as cyclin B–cdk complexes at the onset of mitosis (164). Cyclin B1 residues 109–160 can induce cytoplasmic location of the normally nuclear human cyclin A, indicating that these residues are responsible for cyclin B1 cytoplasmic retention; cytoplasmic retention of cyclin B1 can be overridden by the nucleoplasmin NLS. Phosphorylation has been implicated in regulating the release of cyclin B1 from cytoplasmic retention (164).

The PK-A C-subunit moves to the nucleus upon dissociation from an inactive PK-A holoenzyme complex with the PK-A regulatory (R-) subunit subsequent to cAMP binding by the R-subunit upon hormonal stimulation (see previous section; 161, 162). The PK-A R-subunit can be regarded as playing a cytoplasmic anchor role similar to that of HSP90 and the *rel/dorsal* family I-κB proteins (see below), since it functions to retain the C-subunit in the cytoplasm in the vicinity of the perinuclear Golgi in the case of the type II R-subunit and PK-A holoenzyme (162) in the absence of cAMP-mediated stimulation.

A cytoplasmic anchor function has been described for the NF-κB-binding inhibitor protein I-κB, also known as MAD-3, in retaining the NLS-carrying NF-κB p65 subunit in an inactive complex in the cytoplasm. Phorbol ester, or other treatment, induces the release of p65 and its migration to the nucleus (2, 3)—*in vitro* experiments imply that this can also be effected by phosphorylation either of I-κB by PK-C or of p65 by PK-A (3, and see ref. 165). Roles similar to that of I-κB have been proposed for the sequence-related molecules cactus, which negatively regulates nuclear localization of the *D. melanogaster* morphogen dorsal (6, 7); and p40^{rel}, which may retain the c-*rel* proto-oncogene product in the cytoplasm in analogous fashion. Phosphorylation of dorsal has been shown to effect its release from cactus and nuclear translocation (166), dorsal being constitutively nuclear in the absence of cactus (6, 7). Overexpression of dorsal in the presence of cactus can overcome cytoplasmic retention implying that it is titratable (see ref. 165).

Other I-κB family members include I-κBγ, a discrete gene product identical to the 70 kDa C-terminus of the NF-κB p50 precursor p105/p110 (167), and the proto-oncogene product bcl-3 (168), both of which can bind the mature NF-κB p50 subunit *in vitro* and may function to retain it in the cytoplasm *in vivo*. All of the I-κB/cactus family members contain five to seven ankyrin repeats, structural elements involved in protein–protein interactions. Some of these appear to be directly involved in cytoplasmic retention since deletion of ankyrin repeat 7 together with part of repeat 6 inactivates bcl-3 binding of NF-κB p50 (168). The mechanism of cytoplasmic retention in the case of the I-κBs appears to be through NLS-masking (see next section).

Phosphorylation by cdc2 at Thr^{124} adjacent to the NLS inhibits nuclear transport of T-ag fusion proteins, drastically reducing the level of maximal nuclear accumulation (see Section 7.2.1) (169). Cytoplasmic anchoring appears to be the basis of this inhibition, since maximal inhibition of transport of tetrameric T-ag–β-galactosidase fusion proteins (containing four copies of each of the NLS and the cdc2 site) is effected by a stoichiometry of phosphorylation of only one at the cdc2 site (169), which indicates that one phosphorylated cdc2 site is sufficient to retain the protein in the cytoplasm even in the presence of three non-cdc2-phosphorylated sites. Phosphorylation presumably increases the affinity of the specific interaction between T-ag and a putative cytoplasmic retention factor. The inhibitory effect of cdc2 phosphorylation of Thr^{124} can be overcome by increasing the concentration of the cytosolic T-ag fusion protein (D. A. Jans, unpublished), implying that there may be a finite, titratable level of this cellular factor.

4.3.2 NLS masking

A number of proteins possessing apparently functional NLSs are predominantly cytoplasmic due to the inaccessibility or masking of their NLSs. This may be effected through interaction with another protein (e.g. a factor binding to the NLS itself) or conformational effects whereby the NLS is masked by other parts of the molecule (1). Phosphorylation is an efficient and potentially rapidly responsive means of modulating NLS accessibility (1).

The active (nuclear) form of NF-κB is composed of the p50 and p65 (relA) protein components, both of which are homologous within an approximately 300 amino acid NLS-containing sequence (the *rel* homology region) required for DNA binding and dimerization. This domain is shared with members of the *rel* oncogene TF family including the *D. melanogaster* developmental control gene *dorsal*. It has recently become clear that masking of the NLSs of both of the NF-κB components is the mechanism by which they are retained in the cytoplasm (93, 94, 170). The C-terminus of the p105 precursor of NF-κB p50 (or the I-κBγ molecule—see previous section) appears to retain the NF-κB p50 subunit in the cytoplasm through intramolecular masking of its NLS. Antibodies specific to the NLS recognize p50 but not p105, implying that the NLS is inaccessible in the larger precursor (93). One mechanism of unmasking the NLS appears to be through proteolysis of the p105 C-terminus (92, 171). Intermolecular masking of its NLS by I-κB is the mechanism of NF-κB p65's cytoplasmic retention (94), whereby deletion or mutation of the p65 NLS eliminates binding to I-κB (170). NLS unmasking appears to be brought about through I-κB degradation (172) induced by PK-C phosphorylation (probably by the ζ isotype, which is known to be activated by tumour necrosis factor α, one of NF-κB's physiological stimuli), which may be proteasome-mediated (173). PK-A-phosphorylation of p65 (3, and see ref. 165) can also effect the dissociation of p65 from I-κB.

Nuclear localization of the NF-κB TF is thus dually regulated by specific intra- and intermolecular masking of the NLSs of the two NF-κB subunits (see ref. 1). Signal transduction-triggered phosphorylation regulates the masking events precisely to enable rapid response in terms of nuclear translocation and gene induction.

NLS masking can also be directly effected by phosphorylation close to an NLS which masks or inactivates it through charge or conformational effects (1), as exemplified by cell cycle-dependent nuclear exclusion of the *Saccharomyces cerevisiae* TF SWI5, which is involved in mating type switching. SWI5 nuclear exclusion is effected by phosphorylation by the cdk *CDC28* (4), the yeast *cdc2* homologue. Three *CDC28* sites, one of which is within the spacer of the SWI5 bipartite NLS (see Table 3), inhibit nuclear localization by inactivating or masking the function of the SWI5 NLS (4). At anaphase, *CDC28* activity falls and SWI5 is dephosphorylated to effect its nuclear entry and the subsequent activation of transcription; removal by mutation of the *CDC28* sites results in constitutive nuclear localization (4, 174—see

Section 7.2.2). Cell cycle-dependent phosphorylation (probably by *cdc2*) of p110^{Rb}, the product of the retinoblastoma-susceptibility factor tumour-suppressor gene, also appears to regulate the tightness of its nuclear association ('nuclear tethering'), whereby binding is reduced in the hyperphosphorylated state (175–177).

NLS-dependent nuclear transport of lamin B2 is inhibited by phosphorylation at two PK-C sites (serines 410 and 411) N-terminally adjacent to the NLS (RS^{410}S^{411} RGKRRRIE—NLS underlined) both *in vivo* and *in vitro* in response to phorbol ester stimulation (178). Negative charge through phosphorylation close to the NLS is presumed to inactivate NLS function. In analogous fashion, nuclear translocation of the actin-binding protein cofilin upon heat shock treatment is accompanied by dephosphorylation at a consensus multifunctional CaMPK site (serine 24) adjacent to the cofilin NLS (VRKSS^{24}TPEEVKKRKKA; 85, 179). Phosphorylation at this site is proposed to mask the function of the NLS (179), in similar fashion to *CDC28*-phosphorylation-mediated inactivation of the SWI5- NLS (4).

NLS-dependent nuclear translocation of the nuclear oncogene product v-jun (in contrast to that of its cellular protooncogene counterpart, c-jun) has been shown to be dependent on the stage of the cell cycle (99). Interestingly, v-jun possesses a Cys to Ser mutation at position 248, exactly adjacent to its NLS (ASKS^{248}RKRKL), implying that cell-cycle-dependent phosphorylation at this site may regulate NLS function in similar fashion to SWI5, cofilin, and lamin B2 above.

4.3.3 Phosphorylation enhancing NLS-dependent nuclear transport

In addition to an inhibitory cdk/cdc2 site (see Section 4.3.1 above), a site for casein kinase II (CKII—the serine at position 112) close to the T-ag NLS also regulates the kinetics of T-ag nuclear transport, increasing the rate of transport so that maximal accumulation within the nucleus occurs within 15–20 min. This compares to the 10 hours it takes when CKII-phosphorylation is prevented through deletion or muta-tion of the CKII site (71, 180, 181—see Section 7.2.1 for details of the kinetics). Aspartic acid at position 112 can simulate phosphorylation at the CKII site in terms of accelerating the rate of nuclear import (181), implying that negative charge at the CKII site is mechanistically important in this process. The phosphorylated CKII site presumably represents a signal in addition to, or recognized in concert with the NLS, which is specifically recognized by the cellular nuclear transport apparatus to enable accelerated transport.

PK-A is directly implicated in enhancing nuclear localization of the product of the c-*rel* protooncogene (95, 121, 182). In comparable fashion to the CKII site and T-ag, alanine at the PK-A site (Ser^{266}) 22 amino acids N-terminal of the NLS (chicken c-rel: RRPS^{266}–22 amino acid spacer–KAKRQR) abolishes c-rel nuclear localization, whilst aspartic acid at the site simulates PK-A phosphorylation in inducing nuclear translocation (95). Significantly, this PK-A site, together with the NLS, is conserved in all members of the *rel*/*dorsal* family within the *rel* homology region, which has been shown to be the domain of dorsal to which cactus binds (183). Experiments in which the cDNAs encoding dorsal and the PK-A C-subunit have been co-transfected into Schneider cells clearly implicate PK-A and this conserved site (Ser^{312} in dorsal-RRPS^{312}–22 amino acid spacer–<u>RRKRQR</u>, in regulating dorsal nuclear localization in response to the activated toll receptor pathway (165). Co-expression of PK-A enhances *dorsal* nuclear localization, which can be blocked by specific PK-A peptide inhibitors. Similar to the results for c-rel, mutation of Ser^{312} to glutamine significantly reduces *toll*-induced nuclear localization, whilst aspartic acid in place of Ser^{312} can partially induce nuclear transport, indicating that a negative charge at the site is mechanistically important (165). PK-A phosphorylation of *c-rel*, *dorsal*, and perhaps also of the NF-κB subunits, thus appears to enhance their nuclear transport, possibly through modulating their interactions with cytoplasmic retention factors.

4.3.4 Developmentally regulated NLS

A 'developmentally regulated' NLS-functioning sequence (drNLS) has been reported for a 45 amino acid region of the adenovirus E1a protein (the sequence: FV–X$_{7-20}$–MXSLXYM–X$_4$–MF^{182}, where X is any amino acid; ref. 184), which is distinct from the constitutive E1a NLS shown in Table 2. This sequence is functional in targeting a carrier protein to the nucleus up to the early neurula stage of *Xenopus* embryonic development, when it is inactivated in a hierarchical fashion among the embryonic germ layers (184). Significantly, a sequence homologous to the drNLS of E1a (AV–X$_{17}$–MXILXYSXMF^{612}) has been identified in the rat (as well as mouse and human) GR hormone binding domain known to contain the ligand binding-dependent NLS (123).

5. Cellular components of the nuclear transport system

5.1 NLS-binding proteins

NLS-specific receptors or NLSBPs have been defined and postulated to play a direct role in active nuclear transport through recognition and direct binding of NLSs (185–191). A list of NLSBPs that have been mostly identified by cross-linking and/or ligand-blot experiments using labelled NLS peptides is presented in Table 5. A mechanistic role for NLSBPs in mediating docking at the nuclear envelope has been implied by the fact that antibodies to the NLSBP NBP70 inhibit binding at the NPC (195), as well as in reconstitution experiments (199—see also Sections 5.2 and 5.3).

cDNAs for two NLSBPs, yeast p67 Nsr1 (141) and rat Nopp140 (37), have been cloned and found to have very distinctive primary sequences. The nucleolarly localized Nsr1 is distinguished by an acidic N-terminus containing a series of serine clusters (including 21 consensus CKII sites) a basic C-terminus containing Arg–Gly repeats, and two RNA recognition motifs. It has recently been shown to be involved in ribosome biogenesis and pre-rRNA processing (142). Nopp140 is a very highly phosphorylated protein containing 10 approximately 46 amino acid repeat sequences consisting of an acidic-serine (CKII-site consensus) cluster adjacent to a cluster of basic amino acids (37). Like Nsr1, it is largely nucleolarly localized, which

Table 5 Nuclear localization sequence-binding proteins (NLSBPs)

Protein (kDa) (Reference)	Origin	Identification	NLS
67 (HSC/HSP70) (28, 185, 192)	Rat nucleus	Affinity chromatography using anti-D-D-D-E-D antibodies	T-ag
76/67/59 (187)	Rat nuclear envelope	Photoaffinity labelling	T-ag
66 (188, 193)	HeLa nucleus	Label transfer	T-ag
67 (Nsr1) (141, 142, 189)	Yeast nucleolus	Ligand blotting/ affinity chromatography	Histone H2B
70 (NBP70) (190, 194, 195)	Yeast, *Drosophila*, Hela, *Zea mays* nucleus/cytosol	Ligand blotting/ affinity chromatography	T-ag/histone H2B/nucleoplasmin
60/70 (196)	Rat nucleus/cytosol	Chemical cross-linking *In vitro* reconstitution	T-ag
54\56 (197)	Bovine erythrocytes	Chemical cross-linking	T-ag
140 (Nopp140) (37, 191)	Rat nucleolus	Ligand blot	T-ag

is somewhat difficult to reconcile with the fact that T-ag (and the T-ag NLS) is well known to be completely excluded from the nucleolus, and that Nopp140 contains no identifiable NOS (37). A cytoplasm–nucleolus shuttling NLS receptor role is proposed for the protein, supported by targeting of antibodies to Nopp140 to the nucleolus, apparently by a 'piggy back' mechanism (37). Using antibodies to a different NLSBP in immunocytochemical studies, Li *et al.* (193) found largely cytoplasmic staining, particularly peripheral to the nucleus, also interpreted to indicate NLSBP shuttling, in this case between the cytoplasm and nucleus. Phosphorylation, possibly by CKII, may be essential for NLS-binding activity by both Nsr1 and Nopp140 NLSBPs (37, 141), as it is for NBP70 from yeast (194, 195), implying that hormonal signals may regulate NLSBP activities and thereby modulate nuclear protein transport.

5.2 Molecular chaperones

In an attempt to identify NLSBPs, Yoneda *et al.* (185) raised antibodies to an 'anti-T-ag–NLS peptide' (D-D-D-E-D) and showed that they bound specifically to nuclear envelopes and could block the nuclear uptake of nucleoplasmin or T-ag–NLS–BSA conjugates. The antibodies recognized proteins of 69 and 59 kDa in Western blots and also appeared to be able to recognize proteins binding to NLSs other than that of T-ag (192). Later experiments showed that the antibodies were specific for the stress protein/chaperone HSP70 and its cytoplasmic correlate HSC70, and that other antibodies specific for HSC/HSP70 could also block T-ag–NLS–BSA conjugate transport to the nucleus (28). Direct evidence has also been obtained that HSC/HSP70 is necessary for NLS-dependent nuclear transport in reconstituted systems (29; see, however, ref. 30); whilst HSC/HSP70 activity in terms of *in vitro* nuclear protein import could be substituted by bacterially expressed HSP70 or HSC70, it could not be reconstituted by the mitochondrial chaperone cpn60 (29). HSC/HSP70 has been shown to bind NLSs directly (28) and hence can function as an NLSBP.

Dingwall and Laskey (198) have speculated that HSC/HSP70's role as an NLSBP is to bind bipartite NLSs, and bring together the two arms of basic residues to present them to the nuclear transport apparatus in a configuration similar to the more classical NLSs such as that of T-ag. Consistent with this, linker insertion analysis by Robbins *et al.* (118) showed that the sequence of the 10–12 amino acid spacer between the two arms of basic residues is important for function. However, whilst HSC/HSP70 is essential for nucleoplasmin nuclear transport *in vitro*; it appears not to be required for transport of the bipartite NLS-containing GR (30), and yet is apparently required for transport of T-ag which does not have a bipartite NLS. HSC/HSP70 appears to bind the T-ag NLS specifically (28), which would also seem to be at odds with the above hypothesis (198). The exact role of HSC/HSP70 in nuclear transport, especially with respect to an NLSBP function, would appear to remain unclear at this stage, and it is still possible that HSC/HSP70's chaperone role is indirect, i.e. through stabilization of factors such as other NLSBPs or other proteins (see next section), rather than through direct NLS binding.

A genetic approach has succeeded in implicating a role in nuclear protein transport for another molecular chaperone, the ER chaperone SEC63 (NPL1; 148), which, like HSP70, exhibits homology to the *E. coli* heat-shock protein DNAJ. The exact role of this protein in nuclear transport can only be speculated upon since SEC63 appears to be primarily necessary for the transport of proteins into the ER, where it is exclusively localized. Of significance in this regard may be the fact that the NPC component Gp210 has a consensus ER-localization signal (38), implying that NPC proteins may be assembled via the ER. Whilst there are clearly similarities between the transport of proteins into the ER (see Chapter 4 of this volume) mitochondrion (see Chapter 1 of this volume), and nucleus (e.g. the apparent role of molecular chaperones and low mol. wt monomeric GTP-binding proteins—see next section, etc.), nuclear protein transport is fundamentally different in that, as mentioned in Section 3.1, the signal sequences (NLSs) are not cleaved from the proteins during targeting.

5.3 Other components

The establishment of *in vitro* or reconstituted nuclear import assay systems (29, 148, 169, 194, 196, 197, 199) has greatly contributed to our understanding of nuclear protein import, enabling the identification of the roles of NLSBPs and HSC/HSP70 as discussed in the previous two sections (28–30). NLS-dependent nuclear transport can be shown to involve at least two stages (18, 19), both of which are dependent on cytosolic factors (169): an ATP-independent 'docking' or binding stage at the nuclear envelope/NPC; and an ATP-dependent active transport step into the nucleus. Roles have recently been established for a 97 kDa protein together with an NLSBP in the first docking step at the nuclear envelope (199); and for the 25 kDa

GTP-binding protein Ran/TC4 (and possibly the interacting components RCC1 and RanBP-1) in active import into the nucleus (200, 201). Transport can be inhibited by the non-hydrolysable GTP analogue GTPγS, as well as by the alkylating agent N-ethylmaleimide (NEM) (202).

Other proteins believed to be involved in the nuclear transport process include the NEM-sensitive *Xenopus* cytosolic factors NIF-1 and -2, identified via reconstitution of activity in an *in vitro* nuclear transport assay (202). Although not NLSBPs (see ref. 13), they appear to be necessary for transport substrate binding to the NPC of isolated rat nuclei (202). Their relationship to the factors described above is unclear.

Yeast systems appear to be somewhat different from those of higher eukaryotes—although being ATP- and NLS-dependent, nuclear transport is apparently not dependent on cytosolic factors and is not inhibited by WGA (13). This has been suggested to be due to differences in the yeast *in vitro* systems where cellular cytosolic factors may remain nuclear-associated, in contrast to those from higher eukaryotes (13).

6. Export of proteins from the nucleus and shuttling proteins

The nuclear envelope not only represents a barrier to protein entry into the nucleus, but also restricts the exit of macromolecules, including proteins, from the nucleus. Nuclear protein export appears to be an active, receptor-mediated process, which is saturable, temperature- and energy-dependent, and WGA-inhibitable as shown in the case of ribosomal subunit nuclear export (203).

Several proteins have been characterized which appear to 'shuttle' between nucleolus and cytoplasm, including the nucleolar proteins B23 (NO38) (204) and nucleolin (72, 204), HSC70s (205), various hnRNP protein components (206), and the NLSBP Nopp140 (37). These proteins appear to have roles in RNA transport processes, with the exception of HSC70 and Nopp140 which have roles in nuclear protein transport. Shuttling in these studies has been shown by monitoring the migration of proteins from nucleus to nucleus in interspecies hybrids (204), or by showing nuclear/nucleolar localization of antibodies microinjected into the cytoplasm via a 'piggy back' mechanism in the presence of inhibitors of *de novo* protein synthesis (37, 204, 206).

Bidirectional movement between nucleus and cytoplasm has been reported for the influenza virus ribonucleoprotein M1 (207) and for the yeast RNA-binding protein Npl13 (208). Energy-dependent shuttling between nucleus and cytoplasm of the progesterone, oestrogen, and even glucocorticoid hormone receptors has also been proposed (125–127, 163), recent experiments even suggesting that T-ag may be able to shuttle (126) albeit over a longer time-frame (see also ref. 208). Functional NLSs seem to be required for shuttling (126, 208), implying that NLSBPs, such as Nopp140 (37), HSC70 (205) and the p67 of Li *et al.* (193) which themselves appear to shuttle between nucleolus and cytoplasm, could be involved. Flach *et al.* (208), have shown that whilst T-ag- and histone H2B NLS-fusion proteins are strictly nuclear over 3 hours, Np13 can shuttle through the action of specific, but as yet unidentified, sequences distinct from its NLS-functioning C-terminus which is also required. Based on analysis of three protein constructs including nucleolin, Schmidt-Zachmann *et al.* (72) propose that nucleo-cytoplasmic shuttling is inversely related to the affinity of binding in the nucleus ('nuclear retention'), so that proteins which bind tightly in the nucleus or nucleolus are less likely to shuttle. In the case of most nuclear import events in response to hormonal or other stimuli, the rate of protein export is negligible in comparison with the rapid rate of entry. Although no rigorous kinetic analyses have been done to date, subsequent export/shuttling would appear to be a much slower process relating to long-term response.

7. Nuclear protein transport kinetics

7.1 Systems to analyse nuclear protein transport

A limitation of many studies relating to the identification of NLSs is the nature of the approach used, whereby eukaryotic cells are generally transfected with a protein-expressing plasmid construct, and then scored for subcellular localization 24–48 h later using indirect immunofluorescence in fixed cells. Apart from potential difficulties resulting from fixation-induced artefacts, this approach gives no information with respect to the rates of nuclear import since analysis is performed on cells at steady state, and the observed subcellular localization is a function not only of nuclear transport properties, but also of the level of expression of the transfected DNA, etc. In addition, the high levels of protein expression in the cells analysed are often not physiological, especially in the case of TFs which are normally present in very low amounts.

Some of the above problems can be avoided by examining the transport of carrier proteins containing covalently attached NLS–peptides microinjected into living cells. The constructs used, however, have generally been far from physiological due to very high NLS to protein mol. wt ratios, a factor which has been shown to be important in several studies (e.g. refs 134, 135). Alternative methods, such as subcellular fractionation, can provide quantitative information, but suffer from the caveat that protein redistribution may occur during the fractionation process.

Since we had previously used confocal laser scanning microscopy (CLSM) to measure the relative concentrations of fluorescently labelled proteins in different subcellular compartments (see references in 209), we set out to apply this technique to characterize quantitatively the kinetics of nuclear protein transport (169, 174, 180, 181). The use of microinjection in conjunction with CLSM has made it possible to analyse nuclear import kinetics in single living cells of the rat hepatoma HTC or African green monkey kidney Vero cell lines (169, 174, 180, 181). We have also established an *in vitro* nuclear transport assay system based on CLSM analysis, using HTC or other cell types which have been mechanically perforated (169). Consistent with the findings from other *in vitro* systems (see above), NLS-dependent nuclear protein import is dependent on ATP and cytosolic factors (169) and inhibited by WGA and GTPγS.

Our analysis of the rates and maximal extent of nuclear transport *in vivo* (169, 174, 181) and *in vitro* (169) has shown that nuclear transport is not an all-or-none phenomenon exclusively determined by the NLS—rather, other sequences function to modulate the kinetics of NLS-dependent nuclear transport very precisely. T-ag fusion protein nuclear transport both *in vivo* and *in vitro* is NLS-dependent, inhibited by cdc2-phosphorylation, and accelerated by CKII phosphorylation (see Figs 2–3; 169, 174, 181, and see below).

7.2 The kinetics of nuclear protein transport—the CcN motif as a discrete signal element regulating nuclear protein transport

7.2.1 SV40 large T-antigen

It has been known for a long time that nuclear localization of T-ag is dependent on amino acids 126–132. Our analysis of the kinetics of nuclear transport at the single cell level indicates that nuclear transport of T-ag is also regulated by phosphorylation (see Sections 4.3.1 and 4.3.3). Initial experiments to measure the kinetics of T-ag–β-galactosidase fusion proteins *in vivo* demonstrated that the T-ag NLS alone is not sufficient to target β-galactosidase fusion proteins to the nucleus within 1–2 h, although nuclear localization is complete within about 10 h (71). Only when the N-terminal flanking sequences (amino acids 111–125) are present together with the NLS is nuclear import rapid (steady state achieved within about 15–20 min) and extensive (about 15-fold higher concentrations in the nucleus, as opposed to in the cytoplasm) (see Fig. 2). This transport is NLS dependent since an NLS mutated derivative (Thr^{128}) of the T-ag amino acid 111–135 protein remains cytoplasmic even 10 h after microinjection (71, 180). Amino acids 111–125 include a cluster of five serine/threonine residues that are phosphorylated in SV40 infected cells (210), mutational analysis demonstrating that the CKII site (Ser^{112}) is responsible for this effect on the transport rate (180). When CKII-phosphorylation is prevented through deletion or point mutation of the CKII site, the rate of nuclear transport is reduced by about 40-fold (180, 181).

In contrast to the enhancing effect of the CKII site, phosphorylation at the cdc2 site at Thr^{124} inhibits nuclear transport (169). Whilst not affecting the transport rate significantly, it markedly reduces the level of maximal nuclear accumulation (169) (see Fig. 2). We have named the regulatory module for T-ag nuclear transport, comprising CKII ('C') and cdk/cdc2 ('c') sites and the NLS ('N'), the 'CcN motif' (169) (see Fig. 3). The effects on nuclear transport of phosphorylation at either the CKII or cdc2 site can be simulated by replacing the phosphorylated residues with aspartic acid residues (169, 181). This indicates that a negative charge at these sites, normally provided by specific phosphorylation, is what is functionally important in terms of

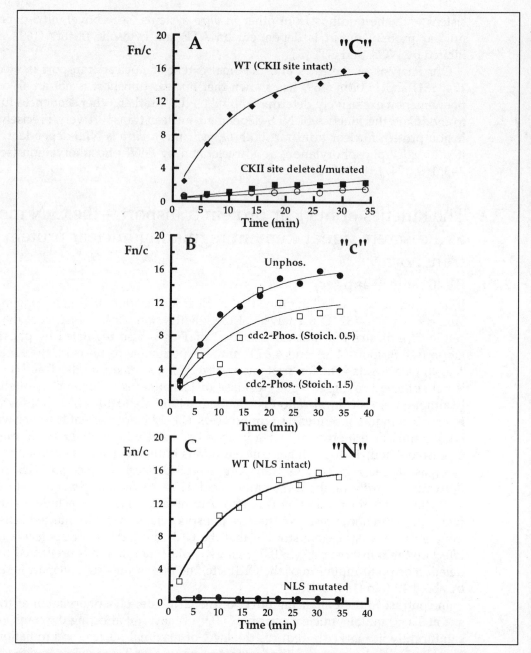

Fig. 2 Regulation of nuclear transport of T-ag by the CcN motif. Nuclear transport kinetics of fluorescently labelled T-ag[amino acids 111–135]-β-galactosidase[amino acids 9–1023] fusion protein derivatives in microinjected HTC cells were measured by quantitative CLSM. Fn/c represents fluorescence quantified in the nucleus relative to that in the cytoplasm (i.e. *n*-fold accumulation in the nucleus). Regulation/dependence of nuclear protein transport by/on the CcN motif component CKII site (A), cdc2 site (B), and NLS (C) is shown. The various CKII site deletion/point mutant derivatives (180, 181), *in vitro* cdc2-prephosphorylated wild-type (WT) fusion proteins (169) and NLS-deficient derivative (71) were as described.

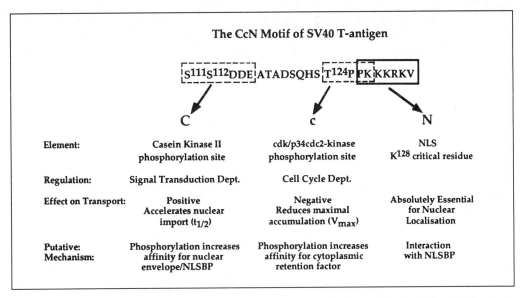

The CcN Motif of SV40 T-antigen

Element:	Casein Kinase II phosphorylation site	cdk/p34cdc2-kinase phosphorylation site	NLS K^{128} critical residue
Regulation:	Signal Transduction Dept.	Cell Cycle Dept.	
Effect on Transport:	Positive Accelerates nuclear import ($t_{1/2}$)	Negative Reduces maximal accumulation (V_{max})	Absolutely Essential for Nuclear Localisation
Putative: Mechanism:	Phosphorylation increases affinity for nuclear envelope/NLSBP	Phosphorylation increases affinity for cytoplasmic retention factor	Interaction with NLSBP

Fig. 3 Summary of the regulation of SV40 large T-antigen nuclear protein transport by the CcN motif. The single letter amino acid code is used, with phosphorylated residues numbered, the CKII and cdk sites with dashed lines, and the NLS with solid lines. Although mechanistically different, the yeast TF SWI5 has been confirmed to show identical regulation by the cdk sites of its CcN motif in terms of V_{max} (174).

regulating nuclear import. This implies that the phosphorylated sites may represent signals in themselves which are recognized by components of the cellular nuclear transport machinery. The results with Asp-substituted proteins also indicate that phosphatase activity at either of the CcN motif kinase sites can have no direct role in the nuclear transport process; i.e. nuclear dephosphorylation at either the CKII or cdc2 sites is not involved in T-ag retention in the nucleus (169, 181). Similarly, phosphorylation in the nucleus at either kinase site to effect nuclear retention is clearly not the mechanism of nuclear accumulation, as also shown by mutational analyses (180, 181). Both CKII and cdk site phosphorylations almost certainly occur in the cytoplasm *in vivo* (see 181), which implies that the initial regulatory events determining nuclear import kinetics are cytoplasmic, consistent with the cytosolic requirement for NLS-dependent nuclear transport *in vitro* (see refs 1, 169, 196, 199, 200, 202).

The existence of a complex regulatory system for SV40 T-ag nuclear localization involving two different kinases demonstrates the existence of specific mechanisms regulating nuclear entry. The NLS is clearly not the sole determinant of nuclear localization; rather, the kinetics of NLS-dependent nuclear import are regulated by phosphorylation in the vicinity of the NLS. Through a potentially signal transduction responsive CKII site (see refs 1, 181) positively regulating the rate of nuclear transport, and a cell-cycle-dependent cdk/cdc2 site negatively regulating the maximal extent of accumulation, the level of T-ag present in the nucleus can be precisely regulated as required with respect to the eukaryotic cell cycle and stages of the viral lytic cycle.

Significantly, a variety of proteins other than T-ag possess putative CcN motifs (169, see ref. 1), implying a general role for the CcN motif in regulating nuclear protein transport. These include oncogene products, viral proteins, and TFs (169), as well as the *D. melanogaster* 'Notch group' of genes and human homologues, the yeast TF SNF2 and human and *Drosophila* homologues, the family of interferon-induced TFs including IFi204 and IFi16, the Swi4 family of mismatch repair enzymes, various DNA repair and processing enzymes, the protein tyrosine phosphatase PEP and various other proteins including kinases and TFs (see references in 1). Putative CcN motifs are also found in the sequences of proteins localizing in the plant cell nucleus (e.g. see references in 1).

7.2.2 Cell-cycle-dependent nuclear localization of SWI5

As mentioned in Section 4.3.2, the yeast TF SWI5 is excluded from the nucleus in a cell-cycle-dependent fashion, mediated by phosphorylation by the cdk CDC28 (4). Nuclear entry occurs in G_1. We have examined the nuclear import kinetics of β-galactosidase fusion proteins carrying SWI5 amino acids 633–682, including the NLS in HTC and Vero cells using microinjection and quantitative CLSM. The SWI5 fusion proteins were transported to the nucleus in an NLS-dependent fashion, whilst the mutation to Ala of the cdk-site serines (Ser[646] and Ser[664]) increased the maximal level of nuclear accumulation from below 1 to about 8-fold (174). The rate of transport of this protein derivative was comparable to that of wild-type T-ag, whilst the maximal level of accumulation was about half of that of T-ag (174). Although restricted to only a few examples, our analysis implies that particular NLSs, together with their regulatory phosphorylation sites, intrinsically determine nuclear import parameters such as the maximal level of nuclear accumulation and the rate of transport.

Phosphorylation at the cdk sites thus appears to inhibit nuclear transport of SWI5, consistent with our observations for the inhibition of T-ag nuclear transport by cdc2-phosphorylation (169). The results show that a yeast bipartite NLS and, fascinatingly, its regulatory mechanisms, is functional in higher eukaryotes, implying the universal nature of regulatory signals for protein transport to the nucleus. The *X. laevis* N1N2 bipartite NLS is similarly functional in nuclear targeting in mammalian cells (unpublished), supporting this conclusion. NLSs from plants, yeast, and higher mammals have been shown by others to be functionally interchangeable, e.g. the T-ag NLS is functional in yeast and plant cells, etc. (211 and see 212), and the plant-cell NLS of *Agrobacterium tumefaciens* VirD2 protein in yeast cells (109).

8. Summary

Although the exact mechanism of nuclear protein import, its essential cellular components (Sections 2 and 5), and precise signal sequence/phosphorylation site requirements for nuclear localization (Sections 3 and 4), are not fully understood, a general picture is emerging. Multiple factors at all levels are necessary for energy and NLS-dependent nuclear protein import, including nucleoporins, NLSBPs (and other NEM-sensitive and NEM-insensitive cytosolic factors such as p97), the molecular chaperone HSC/HSP70, the monomeric GTP-binding protein Ran/TC4 and specifically interacting proteins, and, depending on the particular protein or TF to be transported, the protein factors specific to regulating nuclear import, such as cytoplasmic retention factors, kinases/phosphatases, etc. (Section 4.3). This complex array of factors is responsible for the high specificity of the cellular nuclear transport machinery, which is so essential to eukaryotic cell function. This is epitomized by the intricate complexity of structure of the NPC (Fig. 1), which enables the NPC to transport both small molecules passively, and larger molecules in active NLS-dependent fashion.

The amount of a particular protein in the nucleus and also the time and the rate at which nuclear entry occurs, according to the stage of the cell cycle or in response to hormonal or growth factor signals, can be precisely regulated (Section 7). The mechanisms determining this, such as CcN motifs, cytoplasmic retention factors, intra- and intermolecular NLS masking, the masking of NLSs by phosphorylation, etc., are highly specific and appear to be conserved across eukaryotes, implying that the signals regulating nuclear protein transport are as universal as the NLSs themselves (Section 7.2; refs 1, 174). This conservation at the level of NLS and NLS-modifying sequences clearly reflects the conservation of structure and function of the NPC and the rest of the cellular nuclear transport apparatus. A fuller understanding of the precise details of nuclear protein transport at all levels should help to establish the central importance of nuclear protein transport in the process of gene regulation, which is so integral to all aspects of eukaryotic cell function.

Acknowledgements

The Clive and Vera Ramaciotti Foundation is thanked for its support of our work on the T-ag CcN motif.

References

1. Jans, D. A. (1995) Phosphorylation-mediated regulation of signal-dependent nuclear protein transport—the 'CcN motif'. In *Membrane protein transport*. Rothman, S. (ed.). JAI Press Inc., (Greenwich) **2**, 161.
2. Shirakawa, F. and Mizel, S. B. (1989) *In vitro* activation and nuclear translocation of NK-KB catalyzed by cyclic AMP-dependent protein kinase and protein kinase C. *Mol. Cell. biol.*, **9**, 2424.
3. Ghosh, S. and Baltimore, D. (1990) Activation *in vitro* of NK-kB by phosphorylation of its inhibitor IkB. *Nature*, **344**, 678.
4. Moll, T., Tebb, G., Surana, U., Robitsch, H., and Nasmyth, K. (1991) The role of phosphorylation and the CDC28 protein kinase in cell cycle-regulated nuclear import of the *S. cerevisiae* transcription factor SWI5. *Cell*, **66**, 1.
5. Liu, J. (1993) FK506 and cyclosporin: molecular probes for studying intracellular signal transduction. *Trends Pharmacol. Sci.*, **14**, 182.

6. Steward, R. (1989) Relocalization of the *dorsal* protein from the cytoplasm to the nucleus correlates with its function. *Cell*, **59**, 1179.

7. Roth, S., Stein, D., and Nüsslein-Volhard, C. (1989) A gradient of nuclear localization of the *dorsal* protein determines dorsoventral pattern in the *Drosophila* embryo. *Cell*, **59**, 1189.

8. Gusse, M., Ghysdael, J., Evan, G., Soussi, T., and Mechali, M. (1989) Translocation of a store of maternal cytoplasmic c-*myc* protein into nuclei during early development. *Mol. Cell. Biol.*, **9**, 5395.

9. Van Etten, R. A., Jackson, P., and Baltimore, D. (1989) The mouse type IV c-*abl* gene product is a nuclear protein, and activation of transforming ability is associated with cytoplasmic localization. *Cell*, **58**, 669.

10. Roux, P., Blanchard, J.-M., Fernandez, A., Lamb, N., Jeanteur, Ph., and Piechaczyk, M. (1990) Nuclear localization of c-fos, but not v-fos proteins, is controlled by extracellular signals. *Cell*, **63**, 341.

11. Capobianco, A. J., Simmons, D. L., and Gilmore, T. D. (1990) Cloning and expression of a chicken c-*rel* cDNA: unlike $p59^{v-rel}$, $p68^{c-rel}$ is a cytoplasmic protein in chicken embryo fibroblasts. *Oncogene*, **5**, 584.

12. Hanover, J. A. (1992) The nuclear pore: at the crossroads. *FASEB J.*, **6**, 2288.

13. Stochaj, U. and Silver, P. A. (1993) Nucleocytoplasmic traffic of proteins. *Eur. J. Cell Biol.*, **59**, 1.

14. Agutter, P. S. and Prochnow, D. (1994) Nucleocytoplasmic transport. *Biochem. J.*, **300**, 609.

15. Kalderon, D., Roberts, B. L., Richardson, W. D., and Smith, A. E. (1984) A short amino acid sequence able to specify nuclear location. *Cell*, **39**, 499.

16. Lanford, R. E. and Butel, J. S. (1984) Construction and characterization of an SV40 mutant defective in nuclear transport of T-antigen. *Cell*, **37**, 801.

17. Hall, N. M., Hereford, L., and Herskowitz, I. (1984) Targeting of *E. coli* beta-galactosidase to the nucleus in yeast. *Cell*, **36**, 1057.

18. Richardson, W. D., Mills, A. D., Dilworth, S. M., Laskey, R. A., and Dingwall, C. (1988) Nuclear protein migration involves two steps: rapid binding at the nuclear envelope followed by slower translocation through nuclear pores. *Cell*, **52**, 655.

19. Newmeyer, D. D., Finlay, D. R., and Forbes, D. J. (1986) *In vitro* transport of a fluorescent nuclear protein and exclusion of non-nuclear proteins. *J. Cell Biol.*, **103**, 2091.

20. Finlay, D. R., Newmeyer, D. D., Price, T. M., and Forbes, D. J. (1987) Inhibition of *in vitro* nuclear import by a lectin that binds to nuclear pores. *J. Cell Biol.*, **104**, 189.

21. Dabauvalle, M.-C., Schultz, B., Scheer, U., and Peters, R. (1988) Inhibition of nuclear accumulation of karyophilic proteins in liver cells by microinjection of the lectin wheat germ agglutinin. *Exp. Cell Res.*, **174**, 291.

22. Baglia, F. A. and Maul, G. G. (1983) Nuclear ribonucleoprotein release and nucleoside triphosphatase activity are inhibited by antibodies directed against one nuclear matrix glycoprotein. *Proc. Natl Acad. Sci. USA*, **80**, 2285.

23. Dabauvalle, M.-C., Benavente, R., and Chaly, N. (1988) Monoclonal antibodies to a Mr 68000 pore complex glycoprotein interfere with nuclear protein uptake in Xenopus oocytes. *Chromosoma* (Berlin), **97**, 193.

24. Featherstone, C., Darby, M. K., and Gerace, L. (1988) A monoclonal antibody against the nuclear pore complex inhibits nucleocytoplasmic exchange of protein and RNA in vivo. *J. Cell Biol.*, **107**, 1289.

25. Greber, U. F. and Gerace, L. (1992) Nuclear protein import is inhibited by an antibody to a lumenal epitope of a pore complex glycoprotein. *J. Cell Biol.*, **116**, 15.

26. Akey, C. W. and Goldfarb, D. S. (1989) Protein import through the nuclear pore complex is a multistep process. *J. Cell Biol.*, **109**, 955.

27. Akey, C. W. (1990) Visualization of transport-related configurations of the nuclear pore transporter. *Biophys. J.*, **58**, 341.

28. Imamoto, N., Matsuoka, Y., Kurihara, T., Kohno, K., Miyagi, M., Sakiyama, F., Okada, Y., Tsunasawa, S., and Yoneda, Y. (1992) Antibodies to 70 kD heat shock cognate protein inhibit mediated nuclear import of karyophilic proteins. *J. Cell Biol.*, **119**, 1047.

29. Shi, Y. and Thomas, J. O. (1992) The transport of proteins into the nucleus requires the 70-kilodalton heat shock protein or its cytosolic cognate. *Mol. Cell. Biol.*, **12**, 2186.

30. Yang, J. and De Franco, D. B. (1994) Differential roles of heat shock protein 70 in the *in vitro* nuclear import of glucocorticoid receptor and simian virus 40 large tumor antigen. *Mol. Cell. Biol.*, **14**, 5088.

31. Feldherr, C. M., Kallebach, E., and Schultz, N. (1984) Movement of a karyophilic protein through the nuclear pore of oocytes. *J. Cell Biol.*, **99**, 2216.

32. Dworetzky, S. I., Lanford, R. E., and Feldherr, C. M. (1988) The effects of variations in the number and sequence of targeting signals on nuclear uptake. *J. Cell Biol.*, **107**, 1279.

33. Berrios, M. and Fisher, P. A. (1986) A myosin heavy chain-like polypeptide is associated with the nuclear envelope in higher eukaryotic cells. *J. Cell Biol.*, **103**, 711.

34. Berrios, M., Fisher, P. A., and Matz, E. C. (1991) Localization of a myosin heavy chain-like polypeptide to *Drosophila* nuclear pore complexes. *Proc. Natl Acad. Sci. USA*, **88**, 219.

35. Schroder, H. C., Rottmann, M., Bachmann, M., and Muller, W. E. G. (1986) Purification and characterization of the major nucleoside triphosphatase from rat liver nuclear envelopes. *J. Biol. Chem.*, **261**, 663.

36. Jarnik, M. and Aebi, U. (1991) Towards a more complete 3-D structure of the nuclear pore complex. *J. Struct. Biol.*, **107**, 237.

37. Meier, U. T. and Blobel, G. (1992) Nopp140 shuttles on tracks between nucleolus and cytoplasm. *Cell*, **70**, 127.

38. Sheetz, M. P., Steuer, E. R., and Schroer, T. A. (1989) The mechanism and regulation of fast axonal transport. *Trends Neurosci.*, **12**, 474.

39. Wozniak, R. W., Bartnik, E., and Blobel, G. (1989) Primary structure analysis of an integral membrane protein of the nuclear pore. *J. Cell Biol.*, **108**, 2083.

40. Hallberg, E., Wozniak, R. W., and Blobel, G. (1993) An integral membrane protein of the pore membrane domain of the nuclear envelope contains a nucleoporin-like region. *J. Cell. Biol.*, **122**, 513.

41. Wozniak, R. W. and Blobel, G. (1992) The single transmembrane segment of gp210 is sufficient for sorting to the pore membrane domain of the nuclear envelope. *J. Cell Biol.*, **119**, 1441.

42. Starr, C. M., D'Onofrio, M., Park, M. K., and Hanover, J. (1990) Primary sequence and heterologous expression of nuclear pore glycoprotein p62. *J. Cell Biol.*, **10**, 1861.

43. Carmo-Fonseca, M., Kern, H., and Hurt, E. C. (1991) Human nucleoporin p62 and the essential yeast nuclear pore protein NSP1 show sequence homology and a similar domain organization. *Eur. J. Biol.*, **55**, 17.

44. Nehrbass, U., Kern, H., Mutvei, A., Horstmann, B., Marshallsay, B., and Hurt, E. C. (1990) NSP1: a yeast nuclear envelope protein localized at the nuclear pores exerts its essential function by its carboxy-terminal domain. *Cell*, **61**, 979.

45. Davis, L. I. and Fink, G. R. (1990) The NUP1 gene encodes an essential component of the yeast nuclear pore complex. *Cell*, **61**, 965.

46. Loeb, J. D. J., Davis, L. I., and Fink, G. R. (1993) NUP2, a novel yeast nucleoporin, has functional overlap with other proteins of the nuclear pore complex. *Mol. Biol. Cell*, **4**, 209.

47. Sukegawa, J. and Blobel, G. (1993) A nuclear pore complex protein that contains zinc finger motifs, binds DNA, and faces the nucleoplasm. *Cell*, **72**, 29.

48. Fabre, E., Boelens, W. C., Wimmer, C., Mattaj, I. W., and Hurt, E. C. (1994) Nup145p is required for nuclear export of mRNA and binds homopolymeric RNA *in vitro* via a novel conserved motif. *Cell*, **78**, 275.

49. Wente, S. R., Rout, M. P., and Blobel, G. (1992) A new family of yeast nuclear pore complex proteins. *J. Cell Biol.*, **119**, 705.

50. Wimmer, C., Doyle, V., Grandi, P., Nehrbass, U., and Hurt, E. C. (1992) A new subclass of nucleoporins that functionally interacts with nuclear pore protein NSP1. *EMBO J.*, **11**, 5051.

51. Wente, S. R. and Blobel, G. (1993) A temperature-sensitive NUP116 null mutant forms a nuclear envelope seal over the yeast nuclear pore complex thereby blocking nucleo-cytoplasmic traffic. *J. Cell Biol.*, **123**, 275.

52. Finlay, D. R., Meier, E., Bradley, P., Horecka, J., and Forbes, D. J. (1991) A complex of nuclear pore proteins required for pore function. *J. Cell Biol.*, **114**, 169.

53. Grandi, P., Doye, V., and Hurt, E. C. (1993) Purification of NSP1 reveals complex formation with 'GLFG' nucleoporins and a novel nuclear pore protein NIC96. *EMBO J.*, **12**, 3061.

54. Finlay, D. R. and Forbes, D. J. (1990) Reconstitution of biochemically altered nuclear pores: transport can be eliminated and restored. *Cell*, **60**, 17.

55. Hurt, E. C. (1990) Targeting of a cytosolic protein to the nuclear periphery. *J. Cell Biol.*, **111**, 2829.

56. Bogerd, A. M., Hoffman, J. A., Amberg, D. C., Fink, G. R., and Davis, L. I. (1994) *nup1* mutants exhibit pleiotropic defects in nuclear pore complex function *J. Cell Biol.*, **127**, 319.

57. McKeon, F. D. (1991) Nuclear lamin proteins: domains required for nuclear targeting, assembly, and cell-cycle regulated dynamics. *Curr. Opin. Cell Biol.*, **3**, 82.

58. Kitten, G. T. and Nigg, E. A. (1991) The Caax motif is required for isoprenylation, carboxyl methylation and nuclear membrane association of lamin B2. *J. Cell Biol.*, **113**, 12.

59. Heald, R. and McKeon, F. (1990) Mutations of phosphorylation sites in lamin A that prevent nuclear lamina disassembly in motisis. *Cell*, **61**, 579.

60. Bachs, O., Agell, N., and Carafoli, E. (1992) Calcium and calmodulin function in the cell nucleus. *Biochim. Biophys. Acta*, **1113**, 259.

61. Divecha, N., Banfic, H., and Irvine, R. F. (1991) The polyphosphoinositide cycle exists in the nuclei of Swiss 3T3 cells under the control of a receptor (for IGF-I) in the plasma membrane, and stimulation of the cycle increases nuclear diacylglycerol and apparently induces translocation of protein kinase C to the nucleus. *EMBO J.*, **10**, 3207.

62. Hochstrasser, M., Ellison, J. J., Chau, V., and Varshavsky, A. (1991) The short-lived MAT alpha 2 transcriptional regulator is ubiquinated *in vivo*. *Proc. Natl Acad. Sci. USA*, **88**, 4606.

63. Ciechanover, A., Di Giuseppe, J. A., Bercovich, B., Orian, A., Richter, J. D., Schwartz, A. L., and Brodeur, G. M. (1991) Degradation of nuclear oncoproteins by the ubiquitin system *in vitro*. *Proc. Natl Acad. Sci. USA*, **88**, 139.

64. King, M. W., Roberts, J. M., and Eisenman, R. N. (1986) Expression of the c-myc proto-oncogene during development of *Xenopus laevis*. *Mol. Cell Biol.*, **6**, 4499.

65. Nurse, P. (1990) Universal control mechanism regulating onset of M-phase. *Nature*, **344**, 503.
66. Glotzer, M., Murray, A. W., and Kirschner, M. W. (1991) Cyclin is degraded by the ubiquitin pathway. *Nature*, **349**, 1332.
67. Cowie, A., De Villiers, J., and Kamen, R. (1986) Immortalization of rat embryo fibroblasts by mutant polyomavirus large T antigens deficient in DNA binding. *Mol. Cell. Biol.*, **6**, 4344.
68. Lyons, R. H., Ferguson, B. Q., and Rosenberg, M. (1987) Pentpeptide nuclear localization signal in adenovirus E1A. *Mol. Cell Biochem.*, **7**, 2451.
69. Kleinschmidt, J. A. and Seiter, A. (1988) Identification of domains involved in nuclear uptake and histone binding of nuclear protein N1/N2. *EMBO J.*, **7**, 1605.
70. Hall, M. N., Craik, C., and Hiraoka, Y. (1990) Homeodomain of yeast repressor alpha 2 contains a nuclear localization signal. *Proc. Natl Acad. Sci. USA*, **87**, 6954.
71. Rihs, H.-P. and Peters, R. (1989) Nuclear transport kinetics depend on phosphorylation-site-containing sequences flanking the karyophilic signal of the Simian virus 40 T-antigen. *EMBO J.*, **8**, 1479.
72. Schmidt-Zachmann, M. S., Dargemont, C., Kühn, L. C., and Nigg, E. A. (1993) Nuclear export of proteins: the role of nuclear retention. *Cell*, **74**, 493.
73. Jans, D. A. (1994) Nuclear signalling pathways for polypeptide ligands and their membrane-receptors? *FASEB J.*, **8**, 841.
74. Moreland, R. B., Langevin, G. L., Singer, R. H., Garcea, R. L., and Hereford, L. M. (1987) Amino acid sequences that determine the nuclear localization of yeast histone 2B. *Mol. Cell Biol.*, **7**, 4048.
75. Breeuwer, M. and Goldfarb, D. S. (1990) Facilitated nuclear transport of histone H1 and other small nucleophilic proteins. *Cell*, **60**, 999.
76. Roberts, B. (1989) Nuclear location signal-mediated protein transport. *Biochim. Biophys. Acta*, **1008**, 263.
77. Garcia-Bustos, J., Heitman, J., and Hall, M. N. (1991) Nuclear protein localization. *Biochim. Biophys. Acta*, **1071**, 83.
78. Dingwall, C. and Laskey, R. A. (1991) Nuclear targeting sequences—a consensus? *Trends Biochem. Sci.*, **16**, 478.
79. Chelsky, D., Ralph, R., and Jonak, G. (1989) Sequence requirements for synthetic peptide-mediated translocation to the nucleus. *Mol. Cell. Biol.*, **9**, 2487.
80. Lanford, R. E., Feldherr, C. M., White, R. G., Dunham, R. G., and Kanda, P. (1990) Comparison of diverse transport signals in synthetic peptide-induced nuclear transport. *Exp. Cell Res.*, **186**, 32.
81. Richardson, W. D., Roberts, B. L., and Smith, A. E. (1986) Nuclear location signals in polyoma virus large-T. *Cell*, **44**, 77.
82. Dingwall, C., Robbins, J., Dilworth, S. M., Roberts, B., and Richardson, W. D. (1988) The nucleoplasmin nuclear location sequence is larger and more complex than that of SV40 large T-antigen. *J. Cell Biol.*, **107**, 841.
83. Krohne, G., Wolin, S. L., McKeon, F. D., Franke, W. W., and Kirschner, M. W. (1987) Nuclear lamin L1 of *Xenopus laevis*: cDNA cloning, amino acid sequence and binding specificity of a member of the lamin B subfamily. *EMBO J.*, **6**, 3801.
84. Loewinger, L. and McKeon, F. (1988) Mutations in the nuclear lamin proteins resulting in their aberrant assembly in the cytoplasm. *EMBO J.*, **7**, 2301.
85. Abe, H., Nagaoka, R., and Obinata, T. (1993) Cytoplasmic localization and nuclear transport of cofilin in cultured myotubes. *Exp. Cell Res.*, **206**, 1.

86. Dang, C. V. and Lee, W. M. F. (1988) Identification of the human c-myc protein nuclear translocation signal. *Mol. Cell. Biol.*, **8**, 4048.

87. Wychowski, C., Benichou, D., and Girard, M. (1986) A domain of SV40 capsid polypeptide VP1 that specifies migration into the cell nucleus. *EMBO J.*, **5**, 2569.

88. Gharakhanian, E., Takahashi, J., and Kasamatsu, H. (1987) The carboxy 35 amino acids of SV40 VP3 are essential for its nuclear accumulation. *Virology*, **157**, 440.

89. Clever, H. and Kasamatsu, K. (1991) Simian virus 40 Vp2/3 small structural proteins harbor their own nuclear transport signal. *Virology*, **181**, 78.

90. Dang, C. V. and Lee, W. M. F. (1989) Nuclear and nucleolar targeting sequences of c-*erb*-A, c-*myb*, N-*myc*, p53, HSP70, and HIV *tat* proteins. *J. Biol. Chem.*, **264**, 18019.

91. Shaulsky, G., Goldfinger, N., Ben-Ze'ev, A., and Rotter, V. (1990) Nuclear localization of p53 is mediated by several nuclear localization signals and plays a role in tumorigenesis. *Mol. Cell Biol.*, **10**, 6565.

92. Blank, V., Kourilsky, P., and Israel, A. (1991) Cytoplasmic retention, DNA binding and processing of the NF-κB p50 precursor are controlled by a small region in its C-terminus. *EMBO J.*, **10**, 4159.

93. Henkel, T., Zabel, U., van Zee, K., Müller, J. M., Fanning, E., and Baeuerle, P. A. (1992) Intramolecular masking of the nuclear location signal and dimerization domain in the precursor for the p50 NF-κB subunit. *Cell*, **68**, 1121.

94. Zabel, U., Henkel, T., dos Santos Silva, M., and Baeuerle, P. A. (1993) Nuclear uptake control of NF-κB by MAD-3, an IκB protein present in the nucleus. *EMBO J.*, **12**, 201.

95. Gilmore, T. D. and Temin, H. M. (1988) v-*rel* oncoproteins in the nucleus and cytoplasm transform chicken spleen cells. *J. Virol.*, **62**, 733.

96. Greenspan, D., Palese, P., and Krystal, M. (1988) Two nuclear location signals in the influenza virus NS1 nonstructural protein. *J. Virol.*, **62**, 3020.

97. Chang, M.-F., Chang, S. C., Chang, C.-I., Wu, K., and Kang, H.-Y. (1992) Nuclear localization signals, but not putative leucine zipper motifs, are essential for nuclear transport of hepatitis delta antigen. *J. Virol.*, **66**, 6019.

98. Boulukos, K. E., Pognonec, P., Rabault, B., Begue, A., and Ghysael, J. (1989) Definition of an *Ets1* protein domain required for nuclear localization in cells and DNA-binding activity *in vitro*. *Mol. Cell. Biol.*, **9**, 5718.

99. Chida, K. and Vogt, P. K. (1992) Nuclear translocation of viral Jun but not of cellular Jun is cell cycle dependent. *Proc. Natl Acad. Sci. USA*, **89**, 4290.

100. Underwood, M. R. and Fried, H. M. (1990) Characterization of nuclear localizing sequences derived from yeast ribosomal protein L29. *EMBO J.*, **9**, 91.

101. Lanford, R. E., Kanda, P., and Kennedy, R. C. (1986) Induction of nuclear transport with a synthetic peptide homologous to the SV40 T antigen transport signal. *Cell*, **46**, 575.

102. Davey, J., Dimmock, N. J., and Colman, A. (1985) Identification of the sequence responsible for nuclear accumulation of influenza virus nucleoprotein in *Xenopus* oocytes. *Cell*, **40**, 667.

103. Ou, J.-H., Yeh, C.-T., and Yen, T. S. B. (1989) Transport of hepatitis B virus precore protein into the nucleus after cleavage of its signal peptide. *J. Virol.*, **63**, 5238.

104. Hannink, M. and Donoghue, D. J. (1986) Biosynthesis of the v-*sis* gene product: signal sequence cleavage, glycosylation, and proteolytic processing. *Mol. Cell. Biol.*, **6**, 1343.

105. Johnsson, A., Heldin, C.-H., Wasteson, A., Westermark, B., Deuel, T. F., Huang, J. S., Seeburg, P. H., Gray, A., Ullrich, A., Scrace, C., Stroobant, P., and Waterfield, W. D. (1984) *EMBO J.*, **3**, 921.

106. Maher, D. W., Lee, B. A., and Donoghue, D. J. (1989) The alternatively spliced exon of the platelet-derived growth factor A chain encodes a nuclear targeting signal. *Mol. Cell. Biol.*, **9**, 2251.

107. Zürcher, T., Pavlovic, J., and Staeheli, P. (1992) Nuclear localization of mouse Mx1 protein is necessary for inhibition of influenza virus. *J. Virol.*, **66**, 5059.

108. Manrow, R. E., Sburlati, A. R., Hanover, J. A., and Berger, S. L. (1991) Nuclear targeting of prothymosin alpha. *J. Biol. Chem.*, **266**, 3916.

109. Tinland, B., Koukolikova-Nicola, Z., Hall, M. N., and Hohn, B. (1992) The T-DNA-linked VirD2 protein contains two distinct functional nuclear localization signals. *Proc. Natl Acad. Sci. USA*, **89**, 7442.

110. Fischer-Fantuzzi, L. and Vesco, C. (1988) Cell-dependent efficiency of reiterated nuclear signals in a mutant simian virus 40 oncoprotein targeted to the nucleus. *Mol. Cell. Biol.*, **8**, 5495.

111. Roberts, B., Richardson, W. D., and Smith, A. E. (1987) The effect of protein context on nuclear location signal function. *Cell*, **50**, 465.

112. Zerrahn, J., Deppert, W., Weidemenn, D., Patschinsky, T., Richards, F., and Milner, J. (1992) Correlation between the conformation phenotype of p53 and its subcellular location. *Oncogene*, **7**, 1371.

113. Silver, P., Chiang, A., and Sadler, I. (1988) Mutations that alter both localization and production of a yeast nuclear protein. *Genes Dev.*, **2**, 707.

114. Nath, S. T. and Nayak, D. P. (1990) Function of two discrete regions is required for nuclear localization of polymerase basic protein 1 of A/WSN33 influenza virus (H1N1). *Mol. Cell Biol.*, **10**, 4139.

115. Mukaigawa, J. and Nayak, D. P. (1991) Two signals mediate nuclear localization of influenza virus (A/WSN/33) polymerase basic protein 2. *J. Virol.*, **65**, 245.

116. Morin, N., Delsert, C., and Klessig, D. F. (1989) Nuclear localization of the adenovirus DNA-binding protein: requirement for two signals and complementation during viral infection. *Mol. Cell. Biol.*, **9**, 4372.

117. Dingwall, C., Sharnick, S. V., and Laskey, R. A. (1982) A polypeptide domain that specifies migration of nucleoplasmin into the nucleus. *Cell*, **30**, 449.

118. Robbins, J., Dilworth, S. M., Laskey, R. A., and Dingwall, C. (1991) Two interdependent basic domains in nucleoplasmin nuclear targeting sequence: identification of a class of bipartite nuclear targeting sequence. *Cell*, **64**, 615.

119. Schreiber, J., Molinete, M., Boeuf, H., de Murcia, G., and Menissier-de Murcia, J. (1992) The human poly(ADP–ribose) polymerase nuclear localization signal is a bipartite element functionally separate from DNA binding and catalytic activity. *EMBO J.*, **11**, 3263.

120. Peculis, B. A. and Gall, J. G. (1992) Localization of the nucleolar protein NO38 in amphibian oocytes. *J. Cell Biol.*, **116**, 1.

121. Tratner, I. and Verma, I. M. (1991) Identification of a nuclear targeting sequence in the Fos protein. *Oncogene*, **6**, 2049.

122. Zacksenhaus, E., Bremner, R., Phillips, R. A., and Gallie, B. L. (1993) A bipartite nuclear localization signal in the retinoblastoma gene product and its importance for biological activity. *Mol. Cell. Biol.*, **13**, 4588.

123. Picard, D., Salser, S. J., and Yamamoto, K. R. (1988) A movable and regulable inactivation function within the steroid binding domain of the glucocorticoid receptor. *Cell*, **54**, 1073.

124. Cadepond, F., Gasc, J.-M., Delahaye, F., Jibaud, N., Schweizer-Groyer, G., Segard-Maurel, I., Evans, R., and Baulieu, E.-E. (1992) Hormonal regulation of the nuclear localization signals of the human glucocorticosteroid receptor. *Exp. Cell Res.*, **201**, 99.

125. Guiochon-Mantel, A., Lescop, P., Christin-Maitre, S., Loosfelt, H., Perrot-Applanat, M., and Milgrom, E. (1991) Nucleocytoplasmic shuttling of the progesterone receptor. *EMBO J.*, **10**, 3851.

126. Guiochon-Mantel, A., Delabre, K., Lescop, P., and Milgrom, E. (1994) Nuclear localization signals also mediate the outward movement of proteins from the nucleus. *Proc. Natl Acad. Sci. USA*, **81**, 7179.

127. Picard, D., Kumar, V., Chambon, P., and Yamamoto, K. R. (1990) Signal transduction by steroid hormones: nuclear localization is differentially regulated in estrogen and glucocorticoid receptors. *Mol. Biol. Cell*, **1**, 281.

128. Ylikomi, T., Bocquel, M. T., Berry, M., Gronemeyer, H., and Chambon, P. (1992) Cooperation of proto-signals for nuclear accumulation of estrogen and progesterone receptors. *EMBO J.*, **11**, 3681.

129. Van der Krol, A. R. and Chua, N.-H. (1991) The basic domain of plant B-ZIP proteins facilitates import of a reporter protein into plant nuclei. *Plant Cell*, **3**, 667.

130. Varagona, M. J., Schmidt, R. J., and Raikhel, N. V. (1992) Nuclear localization signal(s) required for nuclear targeting of the maize regulatory protein opaque-2. *Plant Cell*, **4**, 1213.

131. Howard, E. A., Zupan, J. R., Citovsky, V., and Zambryski, P. (1992) The VirD2 protein of *A. tumefaciens* contains a C-terminal bipartite nuclear localization signal: implications for nuclear uptake of DNA in plant cells. *Cell*, **68**, 109.

132. Citovsky, V., Zupan, J., Warnick, D., and Zambryski, P. (1992) Nuclear localization of *Agrobacterium* VirE2 protein in plant cells. *Science*, **256**, 1802.

133. Kambach, C. and Mattaj, I. W. (1992) Intracellular distribution of the U1A protein depends on active transport and nuclear binding to U1 snRNA. *J. Cell Biol.*, **118**, 11.

134. Nelson, M. and Silver, P. A. (1989) Context affects nuclear protein localization in *Saccharomyces cerevisiae*. *Mol. Cell. Biol.*, **9**, 384.

135. Yoneda, Y., Semba, T., Kaneda, Y., Noble, R. L., Matsuoka, Y., Kurihara, T., Okada, Y., and Imamoto, N. (1992) A long synthetic peptide containing a nuclear localization signal and its flanking sequences of SV40 T-antigen directs the transport of IgM into the nucleus efficiently. *Exp. Cell Res.*, **201**, 313.

136. Schmidt-Zachmann, M. S. and Nigg, E. A. (1993) Protein localization to the nucleolus: a search for targeting domains in nucleolin. *J. Cell Sci.*, **105**, 799.

137. Siomi, H., Shida, H., Nam, S. H., Nosaka, T., Maki, M., and Hatanaka, M. (1988) Sequence requirements for nucleolar localization of human T cell leukemia virus type I pX protein, which regulates viral RNA processing. *Cell*, **55**, 197.

138. Li, H. and Bingham, P. M. (1991) Arginine/serine-rich domains of the *su(w^a)* and *tra* RNA processing regulators target proteins to a subnuclear compartment implicated in splicing. *Cell*, **67**, 335.

139. Maeda, Y., Hisatake, K., Kondo, T., Hanada, K., Song, C.-Z., Nishimura, T., and Muramatsu, M. (1992) Mouse rRNA gene transcription factor mUBF requires both HMG-box1 and an acidic tail for nucleolar accumulation: molecular analysis of the nucleolar targeting mechanism. *EMBO J.*, **11**, 3695.

140. Voit, R., Schnapp, A., Kuhn, A., Rosenbauer, H., Hirschmann, P., Stunnenberg, H. G., and Grunt, I. (1992) The nucleolar transcription factor mUBF is phosphorylated by casein kinase II in the C-terminal hyperacidic tail which is essential for tansactivation. *EMBO J.*, **11**, 2211.

141. Lee, W.-C., Xue, Z., and Melese, T. (1991) The NSR1 gene encodes a protein that specifically binds nuclear localization sequences and has two RNA binding motifs. *J. Cell Biol.*, **113**, 1.

142. Lee, W.-C., Zabetakis, D., and Melese, T. (1992) NSR1 is required for pre-rRNA processing and for the proper maintenance of steady-state levels of ribosomal subunits. *Mol. Cell. Biol.*, **12,** 3865.

143. Carmo-Fonseca, M., Tollervey, D., Pepperkok, R., Barabino, S. M. L., Merdes, A., Brunner, C., Zamore, P., Green, M. R., Hurt, E., and Lamond, A. I. (1991) Mammalian nuclei contain foci which are highly enriched in components of the pre-mRNA splicing machinery. *EMBO J.*, **10,** 195.

144. Sawyers, C. L., McLaughlin, J., Goga, A., Havlik, M., and Witte, O. (1994) The nuclear tyrosine kinase c-Abl negatively regulates cell growth. *Cell*, **77,** 121.

145. Moreland, R. B., Lam, H. G., Hereford, L. M., and Fried, H. M. (1985) Identification of a nuclear localization signal of a yeast ribosomal protein. *Proc. Natl Acad. Sci. USA*, **82,** 6561.

146. Van Zhee, K., Appel, F., and Fanning, E. (1991) A hydrophobic protein sequence can override a nuclear localization signal independently of protein context. *Mol. Cell. Biol.*, **11,** 5137.

147. Gao, M. and Knipe, D. M. (1992) Distal protein sequences can affect the function of a nuclear localization signal. *Mol. Cell. Biol.*, **12,** 1330.

148. Sadler, I., Chiang, A., Kurihara, T., Rothblatt, J., Way, J., and Silver, P. A. (1989) A yeast gene important for assembly into the endoplasmic reticulum and the nucleus has homology to DnaJ, an *E. coli* heat shock protein. *J. Cell Biol.*, **109,** 2665.

149. Kiefer, P., Acland, P., Pappin, D., Peters, G., and Dickson, C. (1994) Competition between nuclear localization and secretory signals determines the subcellular fate of a single CUG-initiated form of FGF3. *EMBO J.*, **13,** 4126.

150. Li, J.-M., Hopper, A. K., and Martin, N. C. (1989) N2,N2-dimethylguanosine-specific tRNA methyltransferase contains both nuclear and mitochondrial targeting signals in *Saccharomyces cerevisiae*. *J. Cell Biol.*, **109,** 1411.

151. Tsuneoka, M., Imamoto, N. S., and Uchida, T. (1986) Monoclonal antibody against non-histone chromosomal protein high mobility group 1 co-migrates with high mobility group 1 into the nucleus. *J. Biol. Chem.*, **261,** 1829.

152. Zhao, L.-J. and Padmanabhan, R. (1988) Nuclear transport of adenovirus DNA polymerase is facilitated by interaction with preterminal protein. *Cell*, **55,** 1005.

153. Sommer, L., Hagenbüchle, O., Wellauer, P. K., and Strubin, M. (1991) Nuclear targeting of the transcription factor PTF1 is mediated by a protein subunit that does not bind to the PTF1 cognate sequence. *Cell*, **67,** 987.

154. Booher, R. N., Alfa, C. E., Hyams, J. S., and Beach, D. H. (1989) The fission yeast cdc2/cdc13/suc1 protein kinase: regulation of catalytic activity and nuclear localization. *Cell*, **58,** 485.

155. Ookata, K., Hisanaga, S., Okano, T., Tchibana, K., and Kishimoto, T. (1992) Relocation and distinct subcellular localization of p34^{cdc2}–cyclin B complex at meiosis reinitiation in starfish oocytes. *EMBO J.*, **11,** 1763.

156. Bandara, L. R., Adamczewski, J. P., Hunt, T., and La Thangue, N. B. (1991) Cyclin A and the retinoblastoma gene product complex with a common transcription factor. *Nature*, **352,** 249.

157. Colledge, W. H., Richardson, W. D., Edge, M. D., and Smith, A. E. (1986) Extensive mutagenesis of the nuclear localization signal of simian virus 40 large-T antigen. *Mol. Cell. Biol.*, **6,** 4136.

158. Martinez, J., Georgoff, I., Martinez, J., and Levine, A. J. (1991) Cellular localization and cell cycle regulation by a temperature-sensitive p53 protein. *Genes Dev.*, **5,** 151.

159. Metz, R. and Ziff, E. (1991) cAMP stimulates the C/EBP-related transcription factor rNFIL-6 to translocate to the nucleus and induce c-fos transcription. *Genes Dev.*, **5**, 1754.

160. Schindler, C., Shuai, K., Prezioso, V. R., and Darnell, Jr, J. E. (1992) Interferon-dependent tyrosine phosphorylation of a latent cytoplasmic transcription factor. *Science*, **257**, 809.

161. Nigg, E. A., Hilz, H., Eppenberger, H. M., and Dulty, F. (1985) Rapid and reversible translocation of the catalytic subunit of cAMP-dependent protein kinase type II from the Golgi complex to the nucleus. *EMBO J.*, **4**, 2801.

162. Pearson, D., Nigg, E. A., Nagamine, Y., Jans, D. A., and Hemmings, B. A. (1991) Mechanisms of cAMP-mediated gene induction; examination of renal epithelial cell mutants affected in the catalytic subunit of the cAMP-dependent protein kinase. *Exp. Cell Res.*, **192**, 315.

163. Hsu, S.-C., Qi, M., and De Franco, D. B. (1992) Cell cycle regulation of glucocorticoid receptor function. *EMBO J.*, **11**, 3457.

164. Pines, J. and Hunter, T. (1994) The differential localization of human cyclins A and B is due to a cytoplasmic retention signal in cyclin B. *EMBO J.*, **13**, 3772.

165. Norris, J. L. and Manley, J. L. (1992) Selective nuclear transport of the *Drosophila* morphogen *dorsal* can be established by a signaling pathway involving the transmembrane protein *Toll* and protein kinase A. *Genes Dev.*, **6**, 1654.

166. Whalen, A. M. and Steward, R. (1993) Dissociation of the dorsal–cactus complex and phosphorylation of the protein correlate with the nuclear localization of dorsal. *J. Cell Biol.*, **123**, 523.

167. Inoue, J., Kerr, L. D., Kakizuka, A., and Verma, I. M. (1992) IκBγ, a 70 kd protein identical to the C-terminal half of p110 NF-κB: a new member of the IκB family. *Cell*, **68**, 1109.

168. Wulczyn, F. G., Naumann, M., and Scheidereit, C. (1992) Candidate proto-oncogene bcl-3 encodes a subunit-specific inhibitor of transcription factor NF-κB. *Nature*, **358**, 597.

169. Jans, D. A., Ackermann, M., Bischoff, J. R., Beach, D. H., and Peters, R. (1991) p34^{cdc2}-mediated phosphorylation at T^{124} inhibits nuclear import of SV40 T-antigen proteins. *J. Cell Biol.*, **115**, 1203.

170. Beg, A. A., Ruben, S. M., Scheinman, R. I., Haskill, S., Rosen, C. A., and Baldwin, A. S. Jnr. (1992) IkB interacts with the nuclear localization sequences of the subunits of the NF-κB: a mechanism for cytoplasmic retention. *Genes Dev.*, **6**, 1899.

171. Riviere, Y., Blank, V., Kourilsky, P., and Israel, A. (1991) Processing of the precursor of NF-κB by the HIV-1 protease during acute infection. *Nature*, **350**, 625.

172. Beg, A. A., Finco, T. S., Nantermet, P. V., and Baldwin, A. S. Jr. (1993) Tumor necrosis factor and interleukin-1 lead to phosphorylation and loss of IκBα: a mechanism for NF-kB activation. *Mol. Cell. Biol.*, **13**, 3301.

173. Traeckner, E. B.-M., Wilk, S., and Baeuerle, P. A. (1994) A proteasome inhibitor prevents activation of NF-κB and stabilizes a newly phosphorylated form of IκB-α that is still bound to NF-κB. *EMBO J.*, **13**, 5433.

174. Jans, D. A., Moll, T., Nasmyth, K., and Jans, P. (1995) Cyclin-dependent kinase-site regulated signal-dependent nuclear localization of the SWI5 yeast transcription factor in mammalian cells. *J. Biol. Chem.*, **270**, 17064.

175. Hamel, P. A., Cohen, B. L., Sorce, L. M., Gallie, B. L., and Phillips, R. A. (1990) Hyperphosphorylation of the retinoblastoma gene product is determined by domains outside the simian virus 40 large-T-binding regions. *Mol. Cell. Biol.*, **10**, 6586.

176. Templeton, D. J., Park, S. H., Lanier, L., and Weinberg, R. A. (1991) Nonfunctional mutants of the retinoblastoma protein are characterized by defects in phosphorylation, viral oncoprotein association, and nuclear tethering. *Proc. Natl Acad. Sci. USA*, **88**, 3833.

177. Templeton, D. J. (1992) Nuclear binding of purified retinoblastoma gene product is determined by cell cycle-regulated phosphorylation. *Mol. Cell. Biol.*, **12,** 435.

178. Hennekes, H., Peter, M., Weber, K., and Nigg, E. A. (1993) Phosphorylation of protein kinase C sites inhibits nuclear import of Lamin B_2. *J. Cell Biol.*, **120,** 1293.

179. Ohta, Y., Nishida, E., Sakai, H., and Miyamoto, E. (1989) Dephosphorylation of cofilin accompanies heat shock-induced nuclear accumulation of cofilin. *J. Biol. Chem.*, **264,** 16143.

180. Rihs, H.-P., Jans, D. A., Fan, H., and Peters, R. (1991) The rate of nuclear cytoplasmic protein transport is determined by a casein kinase II site flanking the nuclear localization sequence of the SV40 T-antigen. *EMBO J.*, **10,** 633.

181. Jans, D. A. and Jans, P. (1994) Negative charge at the casein kinase II site flanking the nuclear localization signal of the SV40 large T-antigen is mechanistically important for enhanced nuclear import. *Oncogene*, **9,** 2961.

182. Mosialos, G., Hamer, P., Capobianco, A. J., Laursen, R. A., and Gilmore, T. D. (1991) A protein-kinase-A recognition sequence is structurally linked to transformation by p59$^{v\text{-}rel}$ and cytoplasmic retention of p68$^{c\text{-}rel}$. *Mol. Cell Biol.*, **11,** 5867.

183. Kidd, S. (1992) Characterization of the Drosophila *cactus* locus and analysis of interactions between *cactus* and dorsal proteins. *Cell*, **71,** 623.

184. Standiford, D. M. and Richter, J. D. (1992) Analysis of a developmentally regulated nuclear localization signal in *Xenopus*. *J. Cell Biol.*, **118,** 991.

185. Yoneda, Y., Imamoto-Sonobe, N., Matsuoka, Y., Iwamoto, R., Kiho, Y., and Uchida, T. (1988) Antibodies to asp-asp-asp-glu-asp can inhibit transport of nuclear proteins into the nucleus. *Science*, **242,** 275.

186. Adam, S. A., Lobl, Th. J., Mitchell, M. A., and Gerace, L. (1989) Identification of specific binding proteins for a nuclear location sequence. *Nature*, **227,** 276.

187. Benditt, J. O., Meyer, C., Fasold, H., Barnard, F. C., and Riedel, N. (1989) Interaction of a nuclear location signal with isolated nuclear envelopes and identification of signal-binding proteins by photoaffinity labelling. *Proc. Natl Acad. Sci. USA*, **86,** 9327.

188. Li, R. and Thomas, J. O. (1989) Identification of a human protein that interacts with nuclear localization signals. *J. Cell Biol.*, **109,** 2623.

189. Lee, W.-C. and Melese, T. (1989) Identification and characterization of a nuclear localization sequence-binding protein in yeast. *Proc. Natl Acad. Sci. USA*, **86,** 8808.

190. Silver, P., Sadler, I., and Osborne, M. A. (1989) Yeast proteins that recognize nuclear localization signals. *J. Cell Biol.*, **109,** 983.

191. Meier, U. T. and Blobel, G. (1990) A nuclear localization signal binding protein in the nucleolus. *J. Cell Biol.*, **111,** 2235.

192. Imamoto-Sonobe, N., Matsuoka, Y., Semba, T., Okada, Y., Uchida, T., and Yoneda, Y. (1990) A protein recognized by antibodies to asp-asp-asp-glu-asp shows specific binding activities to heterogeneous nuclear transport signals. *J. Biol. Chem.*, **265,** 16504.

193. Li, R., Shi, Y., and Thomas, J. O. (1992) Intracellular distribution of a nuclear localization signal binding protein. *Exp. Cell Res.*, **202,** 355.

194. Stochaj, U., Osborne, M., Kurihara, T., and Silver, P. A. (1991) A yeast protein that binds nuclear localization signals: purification, localization, and antibody inhibition of binding activity. *J. Cell Biol.*, **113,** 1243.

195. Stochaj, U. and Silver, P. A. (1992) A conserved phosphoprotein that specifically binds nuclear localization sequences is involved in nuclear import. *J. Cell Biol.*, **117,** 473.

196. Adam, S. A. and Gerace, L. (1991) Cytosolic factors that specifically bind nuclear location signals are receptors for nuclear transport. *Cell*, **66,** 837.

197. Adam, S. A., Sterne Marr, R., and Gerace, L. (1990) Nuclear import in permeabilized mammalian cells require soluble cytoplasmic factors. *J. Cell Biol.*, **111**, 807.
198. Dingwall, C. and Laskey, R. A. (1992) The nuclear membrane. *Science*, **258**, 942.
199. Adam, E. J. H. and Adam, S. A. (1994) Identification of cytosolic factors required for nuclear location sequence-mediated binding to the nuclear envelope. *J. Cell Biol.*, **125**, 547.
200. Moore, M. S. and Blobel, G. (1993) The GTP-binding protein Ran/TC4 is required for protein import into the nucleus. *Nature*, **365**, 661.
201. Melchior, F., Paschal, B., Evans, J., and Gerace, L. (1993) Inhibition of nuclear protein import by nonhydrolyzable analogues of GTP and identification of the small GTPase Ran/TC4 as an essential transport factor. *J. Cell Biol.*, **123**, 1649.
202. Newmeyer, D. D. and Forbes, D. J. (1990) An *N*-ethylmaleimide-sensitive cytosolic factor necessary for nuclear protein import: requirement in signal-mediated binding to the nuclear pore. *J. Cell Biol.*, **110**, 547.
203. Bataille, N., Helser, T., and Fried, H. M. (1990) Cytoplasmic transport of ribosomal subunits microinjected into the *Xenopus laevis* oocyte nucleus: a generalized, facilitated process. *J. Cell Biol.*, **111**, 1571.
204. Borer, R. A., Lehner, C. F., Eppenberger, H. M., and Nigg, E. A. (1989) Major nucleolar proteins shuttle between nucleus and cytoplasm. *Cell*, **56**, 379.
205. Mandell, R. B. and Feldherr, C. M. (1991) Identification of two HSP70-related *Xenopus* oocyte proteins that are capable of recycling across the nuclear envelope. *J. Cell Biol.*, **111**, 1775.
206. Pinol-Roma, S. and Dreyfuss, G. (1992) Shuttling of pre-mRNA binding proteins between nucleus and cytoplasm. *Nature*, **355**, 730.
207. Martin, K. and Helenius, A. (1991) Nuclear transport of Influenza virus ribonucleo-proteins: the viral matrix protein (M1) promotes export and inhibits import. *Cell*, **67**, 117.
208. Flach, J., Bossie, M., Vogel, J., Corbett, A., Jinks, T., Willins, D. A., and Silver, P. A. (1994) A yeast RNA-binding protein shuttle between the nucleus and the cytoplasm. *Mol. Cell. Biol.*, **14**, 8399.
209. Jans, D. A. (1992) The mobile receptor hypothesis revisited: a mechanistic role for hormone receptor lateral mobility in signal transduction. *Biochim. Biophys. Acta*, **1113**, 271.
210. Scheidtmann, K.-H., Echle, B., and Walter, G. (1982) Simian virus 40 large T-antigen is phosphorylated at multiple sites clustered in two separate regions. *Virology*, **44**, 116.
211. Shiozaki, K. and Yanagida, M. (1992) Functional dissection of the phosphorylated termini of fission yeast DNA topoisomerase II. *J. Cell Biol.*, **119**, 1023.
212. Raikhel, N. V. (1992) Nuclear targeting in plants. *Plant Physiol.*, **100**, 1627.

3 | Targeting and import of peroxisomal proteins

CHRISTOPHER J. DANPURE

1. Introduction

Progress in our understanding of the mechanisms by which proteins are targeted to and imported into peroxisomes has been slow compared with the rate at which the processes involved in mitochondrial and endoplasmic reticulum (ER) protein translocation have been elucidated (see Chapters 1 and 4, respectively). This has been due partly to the greater technical difficulty of studying peroxisomes (especially *in vitro*) and partly to a general ignorance of the importance of peroxisomes to cellular and organismal viability. However, in recent years there has been a considerable increase in interest in, and a parallel increase in understanding of, the way peroxisomes and related organelles (i.e. glyoxysomes and glycosomes) are formed. This has been spurred on largely by the desire to explain the molecular basis of the human genetic disorders of peroxisome biogenesis, such as Zellweger syndrome (ZS), neonatal adrenoleukodystrophy (NALD), infantile Refsum disease (IRD), and rhizomelic chondrodysplasia punctata (RCDP) (Table 1) (1), and the mechanism by which certain hypolipidaemic drugs and xenobiotics cause phenomenal expansion of the peroxisomal compartmental volume (2–5). This increased interest is well reflected in the numerous reviews written in recent years on various aspects of peroxisomal, glyoxysomal, and glycosomal biogenesis in general and protein targeting in particular (6–27).

2. Overview of peroxisome, glyoxysome, and glycosome biogenesis

Early studies suggested that peroxisomes and glyoxysomes (specialized peroxisomes found in the endosperm and cotyledons of some higher plants) were formed by budding off from the ER (for review of evidence see refs 28, 29). This hypothesis arose from the findings that there appeared to be membranous continuities between the two compartments (30), and that newly synthesized peroxisomal and glyoxysomal enzymes, such as catalase (31) and malate synthase (32), were present in rat liver and castor bean microsomes, respectively. In addition, it was shown that

Table 1 Peroxisomal targeting defects in human peroxisomal diseases

Disease	Enzyme deficiency	Peroxisomal targeting abnormality	Mutations	References
Type 1				
Zellweger syndrome	Multiple	Failure to import both PTS1 and PTS2 proteins	*PAF1* (*PMP35*), *PMP70*, and *PXR1* genes	103, 184, 193
Neonatal adrenoleukodystrophy	Multiple	Failure to import PTS1 proteins (some patients) or both PTS1 and PTS2 proteins (some patients)	*PXR1* gene	184
Infantile Refsum disease	Multiple	Failure to import both PTS1 and PTS2 proteins		
Hyperpipicolic acidaemia	Multiple			
Type 2				
Rhizomelic chondrodysplasia punctata	Multiple	Failure to import PTS2 proteins (some patients) or both PTS1 and PTS2 proteins (some patients)		
Type 3				
X-linked adrenoleukodystrophy	VLCFA–CoA synthase	Failure to import VLCFA–CoA synthase?	*ALD* gene	195, 196
Pseudo-Zellweger syndrome	3-Oxoacyl-CoA thiolase			
Pseudo-neonatal adrenoleukodystrophy	Acyl-CoA oxidase			
Bi(tri)functional enzyme deficiency	Bi(tri)functional enzyme			
Primary hyperoxaluria type 1	Alanine:glyoxylate aminotransferase	AGT mistargeted from peroxisomes to mitochondria	*AGT* (*AGXT*) gene	170, 171
Acatalasaemia	Catalase			

Diseases are classified into three types: type 1, no recognizable peroxisomes, multiple enzyme deficiencies; type 2, peroxisomes recognizable but small and sparse, multiple enzyme deficiencies; type 3, normal peroxisomes recognizable, single enzyme deficiencies. Data for the peroxisome targeting abnormalities were taken from (219) for the type 1 and type 2 diseases and from (169, 220) for primary hyperoxaluria type 1. The targeting abnormality in X-linked adrenoleukodystrophy has been suggested as a possibility (195).

glyoxysomal membrane proteins in castor bean endosperm were similar to ER proteins (33). Gradually evidence has accumulated which has cast doubt on this hypothesis, culminating in the landmark review by Lazarow and Fujiki (6). As pointed out by these authors, modern evidence is incompatible with the 'ER-budding' hypothesis, the characteristics of the translocation of proteins to peroxisomes differing from the equivalent processes in the ER in almost all major respects.

For example, unlike proteins destined for the ER (see Chapter 4), peroxisomal proteins (whether soluble matrix enzymes (34, 35), matrical core enzymes (34), or membrane proteins (36)) are synthesized on free cytoplasmic ribosomes and imported post-translationally (37), as are glyoxysomal (38) and glycosomal (39, 40) matrix proteins (for a review see ref. 6). With only a few exceptions (see Section 4), neither matrical (41) nor membranous (36) peroxisomal proteins are subjected to any post-translational processing. Peroxisomal proteins are not glycosylated (see ref. 6). In addition, using more discriminating cell-fractionation procedures, peroxi-somal membrane proteins have been shown to be very different from ER membrane proteins (42).

It is now generally thought that new peroxisomes form by division or budding of pre-existing peroxisomal structures, proteins being imported into both the peroxi-somal matrix and membrane directly from the cytosol and not routed through the ER. An inevitable consequence of this hypothesis is that peroxisomes must already exist before new peroxisomes can be formed. Direct evidence in support of this has been difficult to obtain, although recent studies on the genetic complementation of peroxisome-deficient cell lines is beginning to shed some light on the topic. Although peroxisomes are plentiful in oleate-fed *Saccharomyces cerevisiae*, they are difficult to find in glucose-repressed cells. However, using immunofluorescence microscopy, Thieringer *et al.* (43) were always able to detect at least one thiolase immunoreactive organelle (presumably a peroxisome) in glucose-fed cells. This was interpreted by the authors to mean that all cells must have at least one peroxisome if additional organelles are to be induced in that cell by oleate. This is compatible with the observation that some cytoplasmic component (possibly a peroxisome) is necessary for the restoration of peroxisome biogenesis in so-called 'nuclear' hybrids between normal and peroxisome-deficient Chinese hamster ovary (CHO) cells (44). Recently, however, studies on temperature-sensitive *Hansenula polymorpha* mutants have indicated that in peroxisome-deficient cells (grown at 43°C) new peroxisomes can form on lowering the temperature to 37°C (45). The idea that peroxisomes can appear where none existed before is a novel concept, a full explanation for which will have to wait for a full molecular analysis of the mutant. However, the minimal requirement for peroxisomal formation might be a structure (a pre-organelle) that is not recognizable as a peroxisome. Some basic membranous components, not necessarily formed into a vesicle, might be all that is required.

3. Protein targeting

3.1 C-terminal targeting sequences

3.1.1 C-terminal sequences recognized by mammalian cells

Firefly luciferase (FFL) is a peroxisomal protein present in the cells of the lantern organ of *Photinus pyralis*. It is also localized in the peroxisomes when expressed heterologously in transfected monkey kidney (CV-1) cells (46). This important find-ing has allowed the development of an easily manipulatable heterologous system which has been invaluable in the identification of sequences that are necessary and sufficient for the targeting and import of peroxisomal proteins (47). Expression of various FFL deletion and insertion mutants in CV-1 cells (48) indicated that the C-terminus and some N-terminal/internal sequences of FFL were necessary for peroxisomal import. Expression of various bacterial chloramphenicol acetyltransferase (CAT)–FFL fusion proteins showed that the C-terminal 12 amino acids of FFL were sufficient to direct the otherwise cytosolic protein CAT to the peroxisomes (48). On the other hand, the other rather ill-defined N-terminal/internal regions shown to be necessary for FFL peroxisomal import were not sufficient to direct CAT to peroxisomes.

The C-terminus was also shown to be important for the peroxisomal targeting of a number of mammalian peroxisomal proteins (49). For example the C-terminal 28 amino acids of human catalase were sufficient to target CAT fusion proteins to peroxisomes, as were the C-terminal 15 residues of rat bifunctional protein, the C-terminal 14 amino acids of pig D-amino acid oxidase, and the C-terminal 15 amino acids of rat acyl-CoA oxidase. Sequence comparisons of the C-termini of all these mammalian peroxisomal proteins and FFL demonstrated the presence of the tripeptides SKL or SHL. In all except catalase, this tripeptide comprised the last three amino acids. The SHL of catalase was separated from the C-terminus by a further eight amino acids. Mutation of the last three amino acids of rat bifunctional protein from SKL to SNL abolished peroxisomal targeting (49).

Further studies (50) showed that only the last three amino acids of FFL (i.e SKL) were needed to target the protein to peroxisomes. In the same study, it was also shown that S(–3) could be replaced by A or C, K(–2) could be replaced by R or H, but the terminal L could not be substituted by any of the amino acids tested. These studies resulted in the formulation of the mammalian consensus C-terminal peroxisomal targeting sequence (PTS) of S(A/C)K(R/H)L. In order to function as a PTS, this tripeptide motif needed to be located at the extreme C-terminus, as the addition of one or two extra amino acids abolished its ability to target FFL to peroxisomes (50). The requirement for the SKL-like PTS motif to be located at the extreme C-terminus would appear to call into question the role of the SHL tripeptide in catalase. Follow-up studies, using CAT fusions, defined the requirements for peroxisomal targeting even more precisely (51). C-terminal SKL, SRL, SHL, AKL, CKL, and CRL were efficient PTSs; SKM, SRM, AHL, AKM, and ARM were much less efficient, and SHM, AHM, CHL, CKM, CRM, and CHM did not work at all (Fig. 1, Table 2).

Parallel studies to those described above, using an *in vitro* rat liver peroxisomal import system (52), showed that the PTS of rat acyl-CoA oxidase resided in the C-terminal five amino acids, although as with FFL some internal sequences were also shown to be necessary, but not sufficient, for peroxisomal import (52). *In vitro* import of dihydrofolate reductase (DHFR) and CAT fusion proteins showed that only the C-terminal tripeptide of acyl-CoA oxidase (i.e. SKL) was necessary to direct these proteins to the peroxisomes (53). Similar to the studies on FFL, it was shown that S(–3) could be replaced by A, K(–2) could be replaced by R or H, but the substitution of the terminal L by E abolished peroxisomal targeting. The α-carboxyl group of the C-terminal L must be free, as amidation abolished targeting. Also in this *in vitro* import system, it was shown that the peroxisomal targeting/import machinery was saturable, in so far as peroxisomal import could be inhibited by the presence of a synthetic decapeptide consisting of the last 10 amino acids of acyl-CoA oxidase (53).

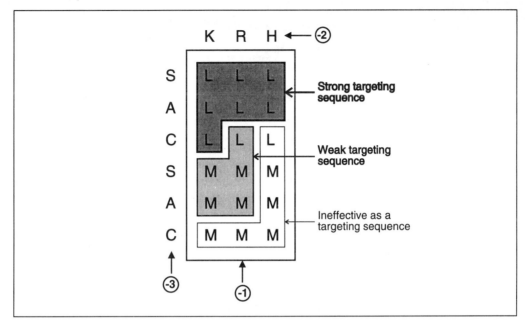

Fig. 1 Relative efficiency of various C-terminal tripeptide combinations in targeting the reporter protein chloramphenicol acetyltransferase to peroxisomes of monkey kidney cells. Amino acids are those located –3, –2, –1 from the C-terminus. Data taken from ref. 51.

3.1.2 C-terminal sequences recognized by yeasts

Heterologous expression of *Candida tropicalis* trifunctional enzyme showed that it was targeted and imported into the peroxisomes of other yeasts, *Candida albicans* and *S. cerevisiae* (54). Deletion of the C-terminal tripeptide AKI abolished targeting in both yeasts. However, when A(–3) was substituted by G, or when K(–2) was substituted by Q, targeting was abolished only in *S. cerevisiae*, whereas targeting in *C. albicans* remained unimpaired (54).

A number of *H. polymorpha* peroxisomal enzymes possess C-terminal SKL-like sequences, for example catalase (SKI), dihydroxyacetone synthase (NKL), and methanol oxidase (ARF) (55, 56). However, none conform to the minimal C-terminal PTS identified in mammalian systems (see Section 3.1.1). Substitution of the C-terminal SKI in catalase by SKS abolished peroxisomal targeting (55). Expression of mouse DHFR fusions and bacterial β-lactamase fusions showed that NKL and ARF, as well as SKL, could function as PTSs in *H. polymorpha* (56).

The C-terminus of peroxisomal citrate synthase (CIT2) in *S. cerevisiae* is SKL, whereas the C-terminus of mitochondrial citrate synthase (CIT1) is SKN (57). The SKL is essential for the peroxisomal targeting of CIT2, its deletion resulting in mistargeting to the mitochondria, presumably due to the presence of an, as yet, unidentified cryptic mitochondrial targeting sequence (MTS). Addition of SKL to the C-terminus of the mature CIT1 protein (i.e. without its N-terminal MTS) caused it to

Table 2 Consensus peroxisomal targeting sequences (PTSs)

Amino acid residue

C-Terminal PTS1	−3	−2	−1
Mammals[a]	S	K	L
	A	R	M
	C	H	
Trypanosomes[b]	S	K	L
	A	R	M
	C	H	I
	G	M	Y
	H	N	
	N	Q	
	P	S	
	T		

N-Terminal PTS2	0[d]	+1	+2	+3	+4	+5	+6	+7	+8	+9
Mammals, yeasts, trypanosomes, and plants[c]	$(X)_n$	R	L	X	X	X	X	X	H Q	L

[a]Data taken from ref. 51; [b], data taken from ref. 72; [c], data taken from ref. 24; [d], for the N-terminal PTS2 the amino acid residues are numbered relative to the nth residue.

be targeted to the peroxisomes (57). *S. cerevisiae* catalase A appears to contain two PTSs (58). One resides in the C-terminal six amino acids, ending in SKF, which has been shown not to work in mammalian cells (50). Although these amino acids can direct the peroxisomal import of DHFR-fusion proteins, they can be deleted from catalase A without interfering with import. Therefore, although the C-terminus is sufficient, it is not necessary. The N-terminal third of the protein (see Section 3.3) appeared to contain sufficient information to target the protein to peroxisomes.

3.1.3 C-terminal sequences recognized by plant cells

C-terminal PTSs seem to be important also for the peroxisomal targeting of a number of plant proteins. For example, the C-terminal hexapeptide of spinach glycolate oxidase (RAVARL) is sufficient to target a reporter protein (β-glucuronidase) to peroxisomes in transgenic tobacco plants (59). Whether only the ARL tripeptide is necessary is not known. The C-terminal tripeptide SKL of cottonseed glyoxysomal malate synthase is necessary for its peroxisomal import in tobacco cells in culture (60). The C-termini of the oilseed rape glyoxysomal enzymes isocitrate lyase and malate synthase are SRM and SRL respectively. When expressed in transgenic *Arabidopsis thaliana*, both proteins are targeted to peroxisomes in the cells of the leaves and roots (61). Expression of various deletion mutants and CAT fusions showed that these tripeptides were the PTSs of these proteins, showing that at least one aspect of glyoxysomal and peroxisomal protein targeting is the same, and reinforcing the suggestion that these organelles differ only as a result of developmental alterations in gene expression or protein stability (61). These results differ from studies on castor bean glyoxysomal isocitrate lyase using an *in vitro* import system (see Section 3.3) (62), which showed that C-terminal deletions had no effect on glyoxysomal import.

3.1.4 C-terminal sequences recognized by kinetoplastids

Glycosomes are unusual peroxisome-like organelles found in certain unicellular parasites, such as the kinetoplastids *Trypanosoma* and *Leishmania*. Antibodies raised against a synthetic dodecapeptide possessing an SKL C-terminus were able to detect the presence of cross-reacting proteins within the glycosomes of *Trypanosoma brucei* (63). This was taken to indicate the presence of SKL-like (targeting?) motifs in glycosomal proteins as well as peroxisomal and glyoxysomal proteins. Indeed, SKL-like sequences have been found in a number, but certainly not all, glycosomal enzymes (for a review see ref. 24). For example, AKL is found at the C-terminus of *T. brucei* glyceraldehyde-3-phosphate dehydrogenase (GAPDH) (64), ARL in *Trypanosoma cruzi* GAPDH (65), SKM in *Leishmania mexicana* GAPDH (Michels *et al.*, personal communication quoted in (24)) and SHL in *T. brucei* glucose-phosphate isomerase (66).

The glycosomal 3-phosphoglycerate kinases (PGKs) in *T. brucei* and *Crithidia fasciculata* have 20 and 38 amino acid C-terminal extensions, respectively, when compared with their cytosolic homologues (67, 68). In *C. fasciculata* PGK the glycosomal and cytosolic isozymes were found to be virtually identical with the exception of the C-terminal extension in the glycosomal enzyme. This observation led Swinkels *et al.* (68) to suggest that the glycosomal targeting sequence resides within this extension and not with any internal 'hot spots' as previously suggested (see Section 3.3).

The C-terminus of *T. brucei* PGK terminates in SSL (67), which does not conform to the minimal sequence requirements for SKL-mediated peroxisomal targeting in mammalian cells (see Section 3.1.1) due to the loss of the basic residue at position -2 (50). Expression of CAT–NRWSSL (69), FFL–SSL, and β-glucuronidase–SSL (70) fusion proteins in *T. brucei* has shown that SSL is sufficient for glycosomal import. CAT–SKL is also targeted to glycosomes when expressed in *T. brucei*, but not as efficiently as the CAT fusion protein containing the C-terminal 21 amino acids of *T. brucei* glycosomal PGK (71). The apparent degeneracy of the C-terminal glycosomal targeting sequence is amazing. FFL expressed in *T. brucei* can still be targeted to the glycosomes when the S(−3) is replaced, by A, C, G, H, N, P, or T; or when K(−2) is replaced by H, M, N, Q, R, or S; or when the terminal L is replaced by I, M, or Y (72) (Table 2). It is very difficult to understand how such a variable sequence can offer any specificity at all.

3.2 N-terminal targeting sequences

Many peroxisomal proteins do not possess a mammalian SKL-like C-terminal motif or even the less restrictive yeast-type C-terminal tripeptide motifs (see ref. 24). Therefore, clearly other PTSs are involved in these proteins. The best characterized of these 'other' PTSs is that possessed by 3-ketoacyl-CoA thiolase. Rat liver thiolase

Table 3 N-terminal mutations causing mistargeting of peroxisomal proteins to the mitochondria

Enzyme	N-terminal sequence	Intracellular location		
		P	M	C
Rat thiolase B[a]	MH<u>RL</u>QVVLG<u>HL</u>AGRPESSSALQAAPC↓	+		
	Q	+		+
	R		+	+
	K		+	+
	L		+	+
	V		+	+
Human AGT[b]	MASHKLLVTPPKALLKPLSIPNQLLL	+		
	L	+/−		+

[a]Data taken from ref. 76; [b] data taken from ref. 170. The underlined thiolase residues comprise the putative PTS2 consensus and the arrow indicates the cleavage site. Although the P11→L substitution is all that is required to target AGT to mitochondria in isolated organelle import systems (221), an additional G170→R substitution is required in whole cell systems (99, 169, 170). P, peroxisomes; M, mitochondria; C, cytosol.

exists as two forms—A and B—with 36 amino acid and 26 amino acid N-terminal extensions, respectively, which are cleaved upon import into peroxisomes (73). Deletion of these leader sequences abolished peroxisomal targeting when thiolase was expressed in mammalian CV-H cells and, when appended to the N-terminus of CAT, the prepieces were able to direct the fusion proteins to peroxisomes (74). Deletion analysis showed that only the first 11 amino acids of the 26 amino acid presequence were necessary (74). Parallel studies also showed that the PTS was contained within the leader sequences of thiolase by expressing N-terminal DHFR fusions in CHO cells (75). However, in this system not only did deletion of amino acids −25 to −23, or amino acids −22 to −13, from the 26 amino acid presequence abolish targeting, but so did deletion of amino acids −12 to −2 which left the first 14 amino acids intact. Processing of peroxisomal proteins is uncommon (see Section 4) and is more typical of the mitochondrial (see Chapter 1) or ER (see Chapter 4) protein import machinery. In fact there are distinct similarities between the N-terminal PTS of thiolase and N-terminal MTSs. Substitution of H(−17) by S, N, D, or E abolished the peroxisomal targeting of thiolase, but its substitution by R, K, L, or V all caused the protein to be mistargeted to the mitochondria (76) (Table 3). This remarkable finding has led to speculation about the relationship between the peroxisomal and mitochondrial targeting processes (76). The thiolase of cucumber cotyledon glyoxysomes also appears to have an N-terminal leader sequence which is cleaved following import (77). However, its role in targeting has yet to be established. *S. cerevisiae* thiolase also possesses an N-terminal PTS2 but, unlike the homologous enzymes in mammals and plants, it is not cleaved (78) (see Section 4).

To what extent N-terminal PTSs are found amongst other peroxisomal proteins is not clear. Watermelon glyoxysomal malate dehydrogenase has an N-terminal presequence with characteristics of a PTS2 (79–82) as does *H. polymorpha* amine oxidase (83). *T. brucei* glycosomal fructose biphosphate aldolase might also possess an N-terminal presequence (84), but evidence that it contains the necessary topogenic information is currently lacking.

Attempts have been made to identify common motifs in the N-terminal pre-sequences of peroxisomal proteins (24), which have led to the formulation of the consensus targeting sequence of $(X)_n$ RLXXXXXH/QL$(X)_n$ (Table 2). Because no obvious overall sequence similarity exists, it is impracticable to screen for the presence of N-terminal PTSs using antibodies, as has been done fairly successfully for C-terminal PTSs (see above). Relatively few heterologous import systems have been used to study N-terminal peroxisomal protein targeting and import. However, one such system has shown that watermelon malate dehydrogenase expressed in *H. polymorpha* targets to the peroxisomes (82, 85).

3.3 Other ill-defined, ambiguous, and multiple targeting sequences

The C-terminal SKL-type PTS is often referred to as a type 1 PTS or PTS1 (as it was the first to be identified), while the N-terminal thiolase-type PTS is called a type 2 PTS or PTS2 (for obvious reasons). Many peroxisomal proteins, however, do not obviously fit into either category.

C. *tropicalis* peroxisomal acyl-CoA oxidase (product of the *POX4* gene) has no C-terminal SKL-like motif and no N-terminal leader sequence. Using an *in vitro* isolated peroxisomal import assay, it was shown that the C-terminal half of the protein contained sufficient information for import (86). Subsequent work showed that two internal sequences (between amino acids 1 and 118, and between amino acids 309 and 427) were important, and could target DHFR fusions to peroxisomes (87). Neither of these sequences have been characterized further.

A number of peroxisomal enzymes possess an SKL-like tripeptide near to but not at the C-terminus (e.g. mammalian catalases (49)). One such protein is the amine oxidase of *H. polymorpha* which has SRL + nine other amino acids at the C-terminus, the last three amino acids being CGK (88) which does not conform to any recognized consensus motif. *H. polymorpha* amine oxidase is not imported into the peroxisomes of *S. cerevisiae in vivo*, but when SKL or SRL was added to its C-terminus, peroxisomal targeting was found, albeit not very efficiently (89). Whether the authentic SRL or any other part of the C-terminus of *H. polymorpha* amine oxidase is actually involved in peroxisomal targeting is unclear, but in any case replacement of the last eight amino acids by five others appeared to have no effect on targeting (90). Recent studies (83) have suggested that the amine oxidase PTS actually resides at the N-terminus (see Section 3.2).

Although *S. cerevisiae* catalase A possesses a C-terminal SSNSKF sequence that is sufficient to target fusion proteins to peroxisomes, it cannot be considered to be the PTS for catalase A, in the normal sense of the phrase, because its deletion had no effect on the targeting of the enzyme (58). The N-terminal third of the protein also contains information sufficient to target fusion proteins to peroxisomes. Therefore *S. cerevisiae* catalase A appears to contain two sequences which are sufficient, but possibly not necessary, to target the protein to peroxisomes. The converse of this finding is that in some contexts an SKL-like C-terminal tripeptide appears to be

Table 4 The C-terminal tripeptides (PTS1s?) of AGT

Natural C-terminal tripeptides[a]	Animal	Natural subcellular distribution[c]			Subcellular disribution in vitro[d]	
		P	M	C	P	C
KKL	Human	+			+	
	Marmoset	+	+			
SQL	Rabbit	+			+	
NKL	Rat	+	+		+	
	Cat	+/–	+			
HRL	Guinea pig	+		+		
Artificial C-terminal tripeptides[b]						
SKL	None				+	
SSL	None				+	+
SEL	None				+	
DEL	None				+	
LLL	None				+	

[a] The natural C-terminal tripeptides are those found in the species indicated under the heading 'Animal'; [b] the artificial C-terminal tripeptides have not been found naturally in any species so far; [c], the natural subcellular distribution of AGT in the marmoset and rat results from the use of multiple transcription and translation initiation sites so that the cleavable N-terminal mitochondrial targeting sequence is either included or excluded from the open reading frame (96, 222, 223); [d], the subcellular distribution in vitro refers to the distribution in micro-injected human fibroblasts or transfected COS-1 cells of human AGT in which the C-terminal KKL has been modified as indicated (99). P, peroxisomes; M, mitochondria; C, cytosol.

necessary, but not sufficient, for correct targeting. The C-terminal SKL of FFL is necessary for peroxisomal targeting in *S. cerevisiae*, but it is not sufficient to target DHFR fusions (91). Presumably there are other sequences, as yet unidentified, in FFL that are necessary for SKL-mediated peroxisomal targeting in *S. cerevisiae* which are lacking in DHFR. Alternatively, DHFR could contain inhibitory sequences/structures. In this respect it is interesting to note that DHFR has been reported to possess cryptic MTSs (92). The *H. polymorpha* peroxisomal matrix protein PER1p also appears to possess more than one PTS (in this case a C-terminal PTS1 and an N-terminal PTS2) (93).

Mammalian alanine:glyoxylate aminotransferase (AGT) is another example of a matrix protein with an ill-defined PTS (Table 4). In all five species so far studied, the C-terminus of AGT has only a two out of three match with the mammalian consensus PTS1 motif (see Section 3.1.1). For example, in humans where AGT is normally targeted only to the peroxisomes the C-terminus is KKL (94, 95), as it is in the marmoset which targets half to peroxisomes and half to mitochondria (96). The C-terminus is SQL in the rabbit which, like the human, targets AGT only to the peroxisomes (96), whereas in the rat, which targets AGT to both organelles, and the cat, which targets approximately 10% to the peroxisomes and about 90% to the mitochondria, the C-terminus is NKL (97, 98). The C-terminus of guinea-pig

AGT, which is partly peroxisomal and partly cytosolic, is HRL (unpublished observations). None of these C-terminal tripeptides would be expected to act as PTSs as they do not conform to the minimal consensus sequence for mammals. However, recent studies have shown that AGT, in the human at least, is imported by the PTS1 import machinery and, therefore, presumably contains a PTS1 (99). Although the C-terminal KKL is necessary for the peroxisomal import of human AGT, it is not sufficient to target CAT or FFL to peroxisomes in human fibroblasts, monkey kidney cells, or *T. brucei* (50, 72, 99). Other, as yet unidentified, sequences in AGT, both contiguous and non-contiguous with the C-terminus, appear to be necessary for its peroxisomal import. Despite the apparent context specificity of the AGT PTS1, it is remarkably degenerate. Not only is the prototypical PTS1 (i.e. SKL) able to target human AGT to peroxisomes, so is SSL which has only been shown to work so far with trypanosome glycosomes (Table 4) (99). However, introduction of acidic amino acids into the C-terminal tripeptide of human AGT does abolish targeting.

With respect to the presence of other sequences/structures which might mediate peroxisomal import, it should be noted that internal sequences in both FFL (48) and rat liver acyl-CoA oxidase (52) (both of which possess a C-terminal PTS1) were found to be necessary, if not sufficient, for peroxisomal targeting and import. The role of these rather ill-defined internal regions has been attributed to conformational alterations, following mutation/deletion, which lead directly or indirectly to the decreased availability of the C-terminal PTS, or possibly to other structural modifications incompatible with peroxisomal import.

It is of potential concern that the identification of PTSs might be dependent on the experimental system chosen, that is whether *in vivo* or *in vitro*. For example, castor bean glyoxysomal isocitrate lyase ends in ARM. Extensive C-terminal deletions, leaving only the N-terminal 168 amino acids, had no effect on its import into sunflower cotyledon glyoxysomes *in vitro* (62). By a process of elimination this would place the PTS in the N-terminal third of the molecule. However, experiments *in vivo* using *A. thaliana* expressing oilseed rape glyoxysomal isocitrate lyase (with the similar C-terminus SRM) showed that this C-terminal motif was the PTS (61).

The finding that a number of glycosomal enzymes possessed two internal positively charged regions that were absent from the cytosolic homologues led to the suggestion that these basic 'hotspots' could be involved in glycosomal protein targeting (67, 100). However, this idea lost favour when it was found that such sequences were not present in other glycosomal proteins and molecular modelling showed that the basic 'hotspots' of *T. brucei* and *T. cruzei* glyceraldehyde-3-phosphatase would be unlikely to be available to fulfil a targeting role (65).

Many glycosomal enzymes possess C-termini that do not conform even to the degenerate kinetoplastid sequence requirements (see Section 3.1.4) (for a list see ref. 24). Recently, a second glycosomal PGK-like protein has been characterized in *T. brucei* (101). This protein has a C-terminus identical to the cytosolic form of the enzyme (i.e. DKE) but does contain a large insertion in the N-terminal domain. The location of the targeting sequences in these glycosomal proteins currently remains a mystery.

So far, no targeting sequences have been conclusively identified in any peroxisomal membrane proteins (PMPs) (see Section 7.3). No PMPs possess thiolase-like PTS2s and, apart from *Candida boidinii* PMP20 which has a C-terminal AKL (102), none possess C-terminal PTS1 motifs. For example, the C-termini of human PMP35 (103), rat PMP70 (104), *S. cerevisiae* PMP48 (105), and *C. boidinii* PMP47 (106) are NAL, FGS, FKP, and AKE, respectively. The most extensive investigations have been carried out on *C. boidinii* PMP47 (107). PMP47 is a 423 amino acid protein with six predicted membrane-spanning domains. Expression of DHFR–PMP47 fusion proteins in *S. cerevisiae* has indicated that the necessary targeting information is contained within the 69 amino acid region (residues 199–267) containing the fourth and fifth membrane domains (107).

Clearly there is still a lot to be learnt about PTSs, not only because many peroxisomal, glyoxysomal, and glycosomal proteins do not contain any of the so-far recognized targeting motifs (i.e. PTS1s and PTS2s, see ref. 24), but also because the presence of such motifs does not necessarily mean that they are involved in peroxisomal targeting.

3.4 Evolutionary conservation of peroxisomal, glyoxysomal, and glycosomal protein targeting

Although, at a structural level, a simple C-terminal PTS1 motif would appear to be rather unlikely to provide the degree of specificity required of a targeting sequence, its general importance in peroxisomal targeting and import now seems to be unarguable, although its exact role is far from clear. Although in a recent review (24) only six out of 13 different mammalian peroxisomal proteins contained a C-terminal SKL-like consensus sequence as defined by Gould *et al.* (50), antibodies raised against a synthetic dodecapeptide containing a C-terminal SKL cross-reacted with 15–20 rat liver peroxisomal proteins but few proteins from other subcellular compartments (108). In addition, similar antibodies could detect peroxisomal and glyoxysomal proteins in *Pichia pastoris*, *Neurospora crassa*, and germinating castor bean seeds, as well as glycosomes of *T. brucei* (63). However, proteins in the hydrogenosomes of *Trichomonas vaginalis* did not cross-react, indicating to the authors that these organelles are not related to peroxisomes. Counter to this suggestion is a recent finding that anti-SKL antibodies were able to detect hydrogenosomal proteins in the anaerobic fungus *Neocallimastix* sp. *L2* (109). Extrapolations about the evolutionary relatedness of intracellular organelles based solely on antibody cross-reactivity of their constituents clearly has to be treated with caution, especially as anti-SKL antibodies failed to pick up any peroxisomal proteins in *C. tropicalis* (63). There are other indications that the requirements for peroxisomal import in *Candida* might be very different from those in *S. cerevisiae*. For example, anti-AKI antibodies raised against the C-terminal 12 amino acids of *C. tropicalis* trifunctional protein recognized multiple peroxisomal proteins in *C. tropicalis*, *C. albicans*, and *Yarrowia lipolytica*, but reacted only weakly with peroxisomal proteins in *S. cerevisiae*, and not at all with a rat liver peroxisomal proteins (110).

It would appear very likely that not only do the specific sequence requirements of the C-terminal PTSs vary between different organisms, but so do the context specificities in which they operate. The minimal consensus PTS1 appears to be most constrained in mammalian cells, much less constrained in yeasts, and almost completely unrestrained in *Trypanosoma*. For example, a number of yeast C-terminal PTSs have I or F at the extreme C-terminus rather than L (54–56). These amino acids appear not to allow efficient peroxisomal targeting in mammalian cells (50). Rat liver epoxide hydrolase possesses the C-terminal tripeptide SKI (111). That this is a poor PTS in mammalian cells is demonstrated by the fact that only a small proportion is targeted to the peroxisomes, while most remains in the cytosol in rat liver. Whether epoxide hydrolase would target efficiently to peroxisomes in yeast cells remains to be seen. Most of the amino acid substitutions to the C-terminus of FFL which do not affect its targeting to glycosomes in *T. brucei* (72) would probably abolish peroxisomal targeting in mammalian cells.

Only three out of 22 yeast peroxisomal proteins listed in a recent review (24) possessed the mammalian C-terminal SKL consensus sequence. This is probably partly a reflection of the greater degeneracy of this motif in yeasts and possibly the greater importance of peroxisomal targeting mechanisms that utilize PTSs other than those based on the PTS1 motif. In the same review (24), four out of seven plant peroxisomal/glyoxysomal proteins possessed a C-terminal SKL-like motif, although two out of five plant catalases had a similar sequence near, but not at, the C-terminus (see also ref. 112).

The presence or absence of C-terminal SKL-like motifs does not necessarily indicate the presence or absence of similar targeting or import machinery, although clearly common mechanisms do exist. The fact that the insect protein FFL is targeted to peroxisomes in mammalian cells (46), yeast, and plants (113) and to glycosomes in trypanosomes (72) suggests that at least one mechanism for the targeting of peroxisomal proteins has been conserved throughout eukaryotic evolution. Similarly, the C-terminus of *Candida boidinii* PMP20 is sufficient to target CAT fusions to peroxisomes in CV-1 cells (113). However, the context sensitivity of the PTS1 motif appears to differ between different cell types. For example, the C-terminal 104 amino acids of FFL can direct DHFR fusions to the peroxisomes when expressed in *S. cerevisiae* and mutations in the C-terminal SKL abolish targeting (91). Therefore SKL would appear to be essential. However, SKL on its own was not sufficient to target DHFR fusions to peroxisomes.

Subtle differences with respect to the functioning of the C-terminal tripeptide motifs also occur within different yeasts. For example, although *C. tropicalis* trifunctional protein is targeted to peroxisomes when expressed in both *C. albicans* and *S. cerevisiae*, and deletion of the C-terminal AKI abolished targeting in both yeasts, when A(−3) or K(−2) were substituted by G or Q, respectively, targeting was still abolished in *S. cerevisiae* but remained unaffected in *C. albicans* (54). These differences between the *Candida* genera and *S. cerevisiae* might apply to other peroxisomal proteins and the yeast peroxisomal import machinery in general, as shown by the poor cross-reactivity of an anti-*C. tropicalis* peroxisomal protein antibody with peroxisomal proteins in *S. cerevisiae* (110) (see above).

It has been suggested that the ability of a particular sequence, such as the C-terminal SKL-like motif, to target a protein to peroxisomes in different cell types is indicative of the evolutionary conservation of the peroxisomal import machinery. However, just because a protein is targeted to the peroxisomes in a heterologous system does not necessarily imply that other components of the translocation pathway for the same protein are conserved. For example, although expression of *S. cerevisiae* catalase A in catalase-deficient *H. polymorpha* mutants leads to it being targeted to peroxisomes where it folds, oligomerizes, and becomes functionally active (114), when *S. cerevisiae* are transformed with *H. polymorpha* alcohol oxidase, even though the enzyme is also targeted to the peroxisomes it fails to oligomerize and remains functionally inactive (115). At the present time it is not possible to predict whether a heterologously expressed peroxisomal enzyme will be functionally active or not. Even though humans and *S. cerevisiae* would be expected to be evolutionarily very distant, human catalase expressed in this yeast not only targets correctly but also oligomerizes and acquires catalytic activity (116). Watermelon malate dehydrogenase targets to peroxisomes in *H. polymorpha* where it appears to fold sufficiently accurately to become enzymically active (85). However, unlike the situation in watermelon, it is not cleaved (82) (see Section 4).

4. Proteolytic processing

In contrast to mitochondrial protein translocation (see Chapter 1) where proteolytic removal of the N-terminal MTS is the norm, import of peroxisomal proteins is rarely accompanied by any significant processing. However, there are a few peroxisomal proteins which are cleaved following import.

The N-terminal presequences of mammalian peroxisomal thiolases are cleaved following import into peroxisomes (73, 117, 118) (see Section 3.2). Although these leader sequences (and implicitly the PTS2s) are removed during peroxisomal import, cleavage is not an obligatory step in the import process. This is best demonstrated by the finding that in fibroblasts from some patients with disorders of peroxisome biogenesis (e.g. ZS and RCDP) thiolase, although still imported into residual peroxisome-like structures (ghosts), is present only in the immature precursor form (119, 120). Presumably, the putative peroxisomal peptidase responsible for thiolase presequence cleavage in normal cells (see below) in one of the large group of peroxisomal enzymes that is not imported in these diseases. Similarly, thiolase expressed in peroxisome-deficient CHO cells is also only present in the unprocessed immature form (121). The PTS2 of *S. cerevisiae* thiolase is not cleaved even in cells with normal peroxisomal function (78), again demonstrating the absence of any relationship between cleavage and import.

Cleavage of the watermelon glyoxysomal malate dehydrogenase PTS2 leads to a decrease in molecular mass from 38 to 33 kDa (79–81). The exact role of this cleavage is currently unknown, but appears unrelated to the ability to acquire a functional conformation as demonstrated by the studies on its heterologous expression in *H. polymorpha* (82, 85) (see Section 3.4).

Rat liver acyl-CoA oxidase is also processed following peroxisomal import (35, 52) to produce a change of molecular mass from 75 kDa to 53 kDa and 22 kDa. Again, processing does not occur in peroxisome-deficient CHO cells (121). The functional reason for this processing is not known, as the PTS of acyl-CoA oxidase resides in the C-terminal SKL (52, 53) which is not cleaved. Glyoxysomal and peroxisomal catalases in pumpkin (122, 123) and sunflower cotyledons (124) appear to be processed following import in a rather idiosyncratic fashion. Inside the organelles, the catalase oligomers appear to exist in various combinations of processed (55 kDa) and unprocessed (59 kDa) subunits. The functional significance of this, if any, is unknown.

The situation with sterol-carrier protein 2 (SCP2) appears to be very confusing. In rat (125), mouse (126), and human (127) livers, as well as in castor bean endosperm (128), SCP2 is synthesized as a larger precursor which is processed to the mature size following import into the peroxisomes or glyoxysomes. In mammalian livers, the 20 amino acid N-terminal presequence probably has no role in peroxisomal targeting as the C-terminus possesses an SKL-like tripeptide motif (AKL) which is likely to be the PTS. This is even more complicated by the presence of multiple cross-reacting proteins, multiple transcripts, possibly multiple genes. The N-terminal presequence might even be involved with mitochondrial import rather than peroxisomal import (126).

Until recently, the evidence for the presence of an intraperoxisomal processing peptidase, similar to those found in mitochondria (129), had only been indirect. A leucine aminopeptidase has been identified in the peroxisomal matrix of pea leaves, but its role in signal-sequence cleavage is unknown (130). A protease called insulin-degrading enzyme has been localized to peroxisomes in transfected CHO cells (131). Despite the apparently inappropriate name of this protease, it seems to play a role in the cleavage of the thiolase leader sequence.

5. Import energetics

Unlike the situation with mitochondrial protein import, the energetics of peroxisomal protein import are only poorly understood. Import of the peroxisomal matrix enzyme acyl-CoA oxidase into rat liver peroxisomes *in vitro* was shown to require ATP hydrolysis (132). GTP could not substitute for ATP. In addition, the absence of any effect of a variety of ionophores indicated that an electrochemical potential across the peroxisomal membrane was not necessary (132). The findings have been confirmed using other systems. For example, when CHO cells are depleted of cellular ATP, the peroxisomal import of micro-injected FFL is inhibited (133). Dissipation of the peroxisomal membrane potential in streptolysin-O-permeabilized CHO cells had no effect on the peroxisomal import of FFL or artificial albumin–SKL peptide conjugates (134). Also in the latter study, GTP was shown not to be necessary for import. Possibly counter to the suggestion that a potential gradient is unnecessary is the finding that proton ionophores inhibit the peroxisomal import and oligomerization of alcohol oxidase in *C. boidinii* (135). However, the mechanism of action in the latter case is unknown. Recent studies have shown that neither ATP nor GTP is necessary for the import of PMP22 into isolated rat liver peroxisomes (136). Whether this difference in ATP requirement reflects generalized differences between peroxisomal matrix and membrane proteins remains to be established.

With respect to the requirement for ATP hydrolysis, at least for some matrix proteins, it is interesting to note that peroxisomal membranes in rat liver (137) and various yeasts, including *H. polymorpha* (138, 139), have been shown to possess ATPases. What role these play in peroxisomal protein import is unknown.

6. Folding, oligomerization, and possible interaction with molecular chaperones

It is well established that proteins are imported post-translationally into mitochondria as unfolded monomeric polypeptides (see Chapter 1). Cytosolic molecular chaperones of the hsp70 type are largely responsible for maintaining such polypeptides in import competent conformations. In addition, intramitochondrial molecular chaperones (hsp70 + hsp60) are required for completion of the import process, and the controlled folding, oligomerization, and acquisition of functional activity of proteins in the mitochondrial matrix. Polypeptide unfolding appears to be a general requirement for the translocation of proteins across membranes (140, 141). Although the conformational requirements for peroxisomal protein import are largely unknown, recent evidence suggests that they are significantly different to the requirements for the transport of proteins to other intracellular compartments.

Enzymically active mammalian catalase is a homotetramer. However, in rat liver it is imported into peroxisomes as a monomer. It oligomerizes, acquires its prosthetic group (haem), and becomes catalytically active within the peroxisomal matrix (6, 142). Similarly, *C. boidinii* and *H. polymorpha* alcohol oxidases, which exist as homo-octamers, are imported as monomers, and oligomerize and acquire their prosthetic group (FAD) within the peroxisomal matrix (143, 144). In a heterologous expression system, *H. polymorpha* alcohol oxidase is imported into *S. cerevisiae* peroxisomes in monomeric form but does not then oligomerize in the foreign environment (115). Catalase in the leaves of sunflower cotyledons also appears to tetramerize after peroxisomal import (124). However, there appears to be no general requirement for peroxisomal proteins to be imported before oligomerization can take place. Many peroxisomal proteins can not only fold and oligomerize in the cytosol, but they can also acquire completely normal catalytic activity. For example, many, but not all, peroxisomal matrix enzymes are enzymically active (and have presumably folded and oligomerized properly) in the cytosol of patients suffering from a variety of disorders of peroxisomal biogenesis (145–148). Numerous cell fusions between complementary peroxisome-deficient cell lines have shown that these normally peroxisomal cytosolic enzymes can be subsequently imported into the newly established peroxisomal population (149–155). Although it has often been assumed that peroxisomal proteins would have to unfold prior to import, very little direct evidence is available. In this respect, it is interesting to note that import of cytosolic catalase into the newly formed peroxisomes of fused complementary cell lines from pan-peroxisome disease patients can be inhibited by 3-amino-1,2,4-triazole (156). This has been attributed to a drug-induced irreversible cross-linking of the catalase. Counter to this evidence that peroxisomal proteins need to be unfolded for import to occur is the remarkable finding that multi-dodecapeptide-SKL conjugates with serum albumin micro-injected into various mammalian cells find their way into the peroxisomal matrix (157). Such an artificial multichain hybrid would not be expected to be able to 'unfold' in the normal sense of the word. This is confirmed by the recent observation (158) that chemically cross-linked albumin–SKL or IgG–SKL conjugates, in which the native disulfide bonding is maintained, are still imported into peroxisomes efficiently. Perhaps some peroxisomal proteins need to unfold more than others before import can take place. When temperature-sensitive peroxisome-deficient mutants of *H. polymorpha* are grown at 43°C various peroxisomal enzymes, such as catalase, alcohol oxidase, and dihydroxyacetone synthase, accumulate in the cytosol where they are assembled into catalytically active enzymes (45). Upon lowering the temperature to 37°C, new peroxisomes form, but neither peroxisomal matrix nor membrane proteins already existing in the cytosol are imported. One of many possible explanations is that these proteins can not easily be unfolded and therefore remain import incompetent.

Recent experiments in which epitope-tagged oligomeric PTS⁻ subunits have been co-expressed with the normal PTS⁺ untagged subunits has shown that, at least in some systems, protein oligomers can be imported into peroxisomes. Using such an approach in *S. cerevisiae*, thiolase appears to be imported as a dimer (159) and CAT–AKL as a trimer (160).

Up until recently it was not known whether newly synthesized peroxisomal proteins interacted with any macromolecular components of the cytosol. However, recent studies using micro-injection and semipermeabilized cell systems have shown that cytosolic hsp70s interact with albumin–SKL dodecapeptide conjugates in mammalian cells (161). Although the role of these heat-shock proteins (hsps) is uncertain, it has been suggested that they are concerned with the presentation of the PTS to the putative import receptor, rather than the maintenance of unfolded conformations *per se* (161). Other cytosolic factors might also be necessary for the import of peroxisomal proteins. Studies using streptolysin-O-permeabilized CHO cells have shown that peroxisomal import of FFL and artificial albumin–SKL peptide conjugates is dependent upon a SKL-specific interaction with an NEM-insensitive cytosolic factor (134). Whatever this factor is, and whatever its role, it appears to function normally in cells from patients with disorders of peroxisome biogenesis (162).

The recent observations that under some conditions some putative PTS receptors are cytosolic (see Section 7) raises the possibility of other macromolecular interactions in the cytosol involving peroxisomal proteins. However, as yet, such interactions have not been demonstrated.

How proteins fold properly within the peroxisomal matrix (if they are not already folded prior to import) is a mystery equal to that of how they manage to fold properly in the cytosol and whether they need to unfold prior to import. Currently, there is no evidence for there being intraperoxisomal molecular chaperones. Although it had been suggested previously that the peroxisomal matrix of rat liver contained a homologue of the α-subunit of mitochondrial F_1-ATPase (163), a protein that, based on certain sequence similarities to hsp60 (164), was proposed to be involved in organelle biogenesis, subsequent studies by the same authors cast doubt on its presence in peroxisomes (165). Interestingly, GroEL, which is a bacterial homologue of mitochondrial hsp60, binds to *H. polymorpha* alcohol oxidase *in vitro* and prevents it from aggregating (166). Release of functional enzyme is stimulated by ATP and GroES. Whether any parallels to this occur *in vivo* remains to be seen.

The requirements for the unfolding/folding/oligomerization of mitochondrial proteins at various points in the translocation pathway have been compared with the putative analogous requirements for peroxisomal proteins (165). In this context, proteins that target to, are imported into, and acquire functional activity within both compartments (such as SCP2 and AGT) are of especial interest. In the case of AGT, the protein is not only targeted to more than one intracellular compartment, but also the specific intracellular localization depends on the species concerned. Some animals target AGT to the peroxisomes, some mainly to the mitochondria, and some more evenly to both organelles (167, 168). In humans, AGT is normally exclusively peroxisomal. But in patients with the autosomal recessive disease primary hyperoxaluria type 1 (PH1) (Table 1), the combination of two naturally occurring point mutations (P11→L and G170→R) leads AGT to be mistargeted to the mitochondria (Table 3) (169, 170). The observation that AGT is catalytically active in both organelles implies that the protein has folded very similarly in each organelle. If mitochondrial molecular chaperones are required for the correct folding of AGT inside the mitochondrial matrix (assumed), what fulfils that function (if anything) within the peroxisomal matrix? Similarly, if AGT can fold normally (presumed) and become catalytically active in the cytosol, how is it kept unfolded (if necessary) for import into mitochondria? At present, the answers to both these rhetorical questions are unknown.

With respect to the intraperoxisomal folding of AGT, it is interesting to note that in a small group of PH1 patients, the AGT appears to aggregate within the peroxisomal matrix to produce core-like structures (171). In all such patients, the AGT gene contains a mutation which causes a G41→R amino acid substitution. Whether this mutation results in an impaired interaction of AGT with any putative peroxisomal matrix factors (molecular chaperones?) that mediate its correct folding remains to be seen.

7. Import receptors, membrane proteins, and other proteins involved in peroxisome assembly

By analogy with the processes involved in the translocation of proteins into other intracellular compartments, such as mitochondria, ER, and nuclei, the targeting sequence of peroxisomal proteins (i.e. PTS1s and PTS2s) are presumed to interact first with receptors, which determine the specificity of protein import, and second with the membrane translocation machinery, which allows passage of the proteins through the peroxisomal membrane. Although elucidation of these more distal parts of the putative peroxisomal import pathway has lagged behind the characterization of PTSs, rapid advances have been made over the past few years.

7.1   Proteins involved in yeast peroxisome assembly

Many proteins have been shown to be involved in peroxisomal biogenesis by genetic complementation of the various peroxisome assembly defects generated in yeasts, such as *S. cerevisiae* (11–13, 172), *P. pastoris* (173, 174), *H. polymorpha* (15, 175–177) and *Y. lipolytica* (178) (Table 5). Although the functions of most of these proteins have not yet been determined, some (especially those associated with the peroxisomal membrane) are likely to be components of the membrane translocation machinery, while others appear to be good candidates for PTS1 and PTS2 receptors.

The phenotype in *S. cerevisiae pas10* (179), *P. pastoris pas8* (180), and *H. polymorpha per3* (181) mutants is characterized by a specific failure to import PTS1-containing proteins. The import of the PTS2-containing protein thiolase is normal. PAS10p, PAS8p, and PER3p are homologous proteins belonging to the tetratricopeptide repeat (TPR) protein superfamily, which includes the mitochondrial import receptors *S. cerevisiae* MAS70 and *N. crassa* MOM72. They are good candidates for PTS1 import receptors and, as expected for such a role, *S. cerevisiae* PAS10p (182) and *P. pastoris* PAS8p (180, 183) have both been shown to bind to SKL-containing peptides via their TPR domains. Surprisingly, the subcellular distribution of these putative PTS1 receptors varies between different yeasts. *P. pastoris* PAS8p is bound to the cytosolic face of the peroxisomal membrane (183), whereas *S. cerevisiae* PAS10p appears to be mainly cytosolic (H. Tabak, personal communication quoted in (184)). *H. polymorpha* PER3p is found in both the cytosol and the peroxisomal matrix (181).

In *S. cerevisiae pas7/peb1* mutants (185, 186), the phenotype is the opposite of that found in *pas10* mutants, PTS1-containing proteins are imported normally, whereas the PTS2-containing protein thiolase is not imported. Although evidence for a direct interaction between PAS7p/PEB1p and thiolase is currently lacking, PAS7p/PEB1p must be considered as a good candidate for a PTS2 receptor. PAS7p/PEB1p is a member of the WD-40 protein superfamily, intriguingly members of which are often found functionally associated with members of the TPR protein superfamily.

As with the putative PTS1 receptor, it is not at all clear how this putative PTS2 receptor might operate. Depending on the specific expression system used, one laboratory has shown that N-terminal c-*myc* epitope-tagged PAS7p is mainly cytosolic with only a small amount associated with the peroxisomes (185), whereas another laboratory has shown that C-terminal haemagglutinin epitope-tagged PEB1p is mainly located in the peroxisomal matrix (186). It has been suggested that the receptor either shuttles between the cytosol and the peroxisomal membrane

Table 5 Yeast and mammalian genes involved in peroxisome assembly

Gene	Size of gene Product (kDa)	Structural motif and protein family	Subcellular location	Homology group	Putative function	References
Yeasts						
S. cerevisiae						
PAS1	1043 aa (117)	AAA		A		13, 224, 225
PAS2	183 aa (21)	UBC	P			13, 226
PAS3	441 aa (48)		P(memb-c)			105
PAS4		C_3HC_4 Zn finger				13, 172
PAS5		C_3HC_4 Zn finger				13, 172
PAS7 (PEB1)	375 aa (42)	WD-40	C + p or P(matrix)		PTS2 receptor	185, 186
PAS8	1030 aa (116)	AAA		A		227
PAS10	612 aa (69)	TPR	C*	B	PTS1 receptor	179, 182
PAS12		CaaX				13
P. pastoris						
PAS1	1157 aa (127)	AAA		A		228
PAS4	204 aa (24)	UBC	P(memb-c)			229
PAS5	1165 aa (127)	AAA		A		230
PAS8	580 aa (68)	TPR	P(memb-c)	B	PTS1 receptor	180, 183
PER3	713 aa (81)		P(memb)			231
H. polymorpha						
PER1	650 aa (71)		P (matrix)			93
PER3	569 aa (64)	TPR	C + P (matrix)	B	PTS1 receptor?	181
PER8	295 aa (34)	C_3HC_4 Zn finger	P(memb)			232
Y. lipolytica						
PAY2	404 aa (42)		P(memb)			233
PAY4	1025 aa (112)	AAA		A		234
Mammals						
Human						
PAF1	305 aa (35)	C_3HC_4 Zn finger	P(memb)			103
PMP70	659 aa (70)	Mdr-ABC	P(memb)			191, 193
PXR1 (PTS1R)	602/642aa (67/70)	TPR	C + p(memb) or P(memb-c)	B	PTS1 receptor	184, 200, 201
Rat						
PAF1	305 aa (35)	C_3HC_4 Zn finger	P(memb)			187, 189
PMP70	659 aa (70)	Mdr-ABC	P(memb)			104
Hamster						
PAF1	304 aa (35)	C_3HC_4 Zn finger	P(memb)			189–190

AAA, putative ATPase associated with diverse cellular activities; CaaX, containing CaaX box isoprenylation motif; Mdr-ABC, multidrug resistance -ATP binding cassette transporter family; TPR, tetratricopeptide repeat superfamily; UBC, ubiquitin-conjugating enzyme family; WD-40, WD-40 β-transducin superfamily.

The heading 'Homology group' indicates the sequence similarity with other polypeptides involved in peroxisome biogenesis: A, AAA family of putative ATPases with similarity to Cdc48 and Sec 18-NSF (13, 224); B, TPR family with similarity to *S. cerevisiae* MAS70 and *N. crassa* MOM72, putative PTS1 receptors. P, peroxisomes (upper or lower case represents the relative amounts); P(matrix), peroxisomal matrix; P(membrane), peroxisomal membrane; P(memb-c), cytosolic face of peroxisomal membrane; C, cytosol.
*H. Tabak unpublished observation quoted in ref. 184.

transporting thiolase from the former to the latter (185), or that it acts by pulling thiolase already bound to the peroxisomal membrane into the peroxisome interior (186) (Fig. 2).

7.2 Proteins involved in mammalian peroxisome biogenesis

Far fewer mammalian proteins involved in peroxisomal biogenesis have been identified than yeast proteins (Table 5). However, two peroxisomal membrane proteins (PMP35 and PMP70) have been shown to play major roles. The importance of mammalian PMP35, which is encoded by the PAF1 gene, in the assembly of peroxisomes is demonstrated by the fact that the expression of human liver PAF1 in some ZS fibroblasts (103) and rat liver PAF1 in some peroxisome-deficient CHO cell lines (187–189) restores normal peroxisomal biogenesis. In addition, mutations have been found in the *PAF1* gene in a ZS patient (103) (Table 1) and in peroxisome-deficient CHO cell lines (188, 190). The *PAF1* gene product (PMP35) contains a sequence that is similar to the zinc finger motifs found in some DNA-binding proteins (188). However, the exact role of this protein in peroxisomal biogenesis is unknown.

Mammalian PMP70 appears to have sequence similarities to members of the multidrug-resistance-related ATP-binding protein superfamily (104, 191). Its exact role is unclear, but it might be involved in peroxisomal membrane transport of polypeptides (192). The importance of PMP70 in peroxisomal biogenesis is confirmed by the finding of two different mutations in the *PMP70* gene in two different ZS patients in the same complementation group (193) (Table 1). There would appear to be some functional redundancy between PMP35 and PMP70. Expression of either PMP35 or PMP70 in a peroxisome-deficient CHO cell line appeared to restore normal peroxisome biogenesis (194). The ability of PMP70 to restore apparent normality is especially remarkable as the cell line used had previously been shown to have a mutation in the *PAF1* (*PMP35*) gene. The exact role of PMP70 is also complicated by the finding that the deficient gene in X-linked adrenoleuko-dystrophy, which is not a generalized disorder of peroxisome biogenesis but rather appears to be associated with very-long-chain-fatty-acid(VLCFA)–CoA synthetase deficiency (Table 1), encodes a protein (ALDP) with a high level of sequence similarity to PMP70 (195, 196). ALDP is also located in the peroxisomal membrane (197, 198) and it has been speculated that it might form homodimers or heterodimers with PMP70 and be involved with the peroxisomal import of VLCFA–CoA synthase, homodimers of PMP70 being involved in import of other peroxisomal matrix proteins (199).

Recently, a human homologue (*PXR1/PTS1R*) of the yeast PTS1 receptor gene has been identified and partly characterized (184, 200, 201). A missense mutation has been found in *PXR1* in an NALD patient which led to a specific failure of PTS1 protein import, but surprisingly a nonsense mutation in the same gene in a ZS patient prevented the import of both PTS1 proteins and thiolase (184) (Table 1). The latter observation has led to the suggestion that in humans, but not yeast, the PTS1

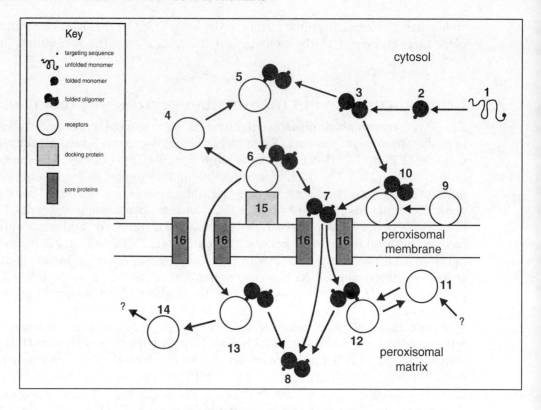

receptor (a TPR protein) and the yet to be identified PTS2 receptor (probably a WD-40 protein) might interact. As expected for a PTS1 receptor, PXR1p/PTS1Rp binds to SKL-containing peptides (184, 201).

Yet again, the manner in which PXR1p/PTS1Rp functions as a PTS1 import receptor is unclear due to differences of opinion regarding its subcellular distribution. Some workers (184, 201) have suggested that the receptor is mainly cytosolic, with only a small proportion being bound to the peroxisomal membrane. This distribution is similar to that suggested for the *S. cerevisiae* homologue PAS10p. On the other hand, other workers (200) have provided evidence that the receptor is entirely confined to the peroxisomal membrane, with the SKL-binding moiety exposed on the cytoplasmic face. This distribution is more similar to that found with the *P. pastoris* homologue PAS8p. The possible dual subcellular distribution of the PTS1 receptor has led some workers (184, 201) to suggest that it might shuttle between the cytosol and the peroxisomal membrane in a manner similar to that suggested for the *S. cerevisiae* PTS2 receptor (PAS7p) (185) (Fig. 2). So far, a candidate for a PTS2 receptor in mammals has not materialized.

The confusion with regard to the mode of operation of the putative PTS1 and PTS2 receptors arises mainly from disagreements about their intracellular compartmentalization. However, as has been suggested previously (184), a variable distribution is consistent with the 'cytosol → peroxisome membrane → cytosol'

Fig. 2 Speculative models of peroxisomal protein import. *Folding and oligomerization*: Newly synthesized peroxisomal matrix proteins existing as unfolded monomers [1] fold into their native subunit conformations [2] in the cytosol possibly with the help of cytosolic factors (e.g. molecular chaperones). The monomers [2] oligomerize, where appropriate, in the cytosol to form functional catalytically active enzymes [3] (see text Section 6). *Import scheme 1*: Either the oligomers [3] or the folded [2] or unfolded [1] monomers interact with cytosolic receptors [4] via their PTSs to form ligand–receptor complexes [5]. For simplicity, only the interaction of the oligomer with the receptor by way of one PTS is shown. The ligand–receptor complex [5] is transported to the peroxisomal membrane where it interacts with a docking protein [15] to form a ligand–receptor-docking protein complex [6]. The ligand and the receptor detach from the docking complex, the released receptor [4] being available for reuse. The released ligand [7], possibly following interaction with other membrane proteins, is then translocated through the putative peroxisomal pore [16] and enters the peroxisomal interior already in its fully folded functional conformation [8] (on the assumption that it is the folded oligomer that binds to the cytosolic receptor). If the translocated peroxisomal protein [8] possesses a PTS2 then the latter will be cleaved (in mammals and plants but not yeasts) in the peroxisomal matrix. A 'receptor shuttling' model such as this has been suggested for both the *S. cerevisiae* putative PTS2 receptor (PAS7p) (185) and the mammalian PTS1 receptor (PXR1p) (184, 201) to explain their mainly cytosolic localization with some association with the peroxisomal membrane. *Import scheme 2*: The ligand [3] binds to a peroxisomal membrane receptor [9] to form a ligand–receptor complex [10]. The ligand is released [7] and, possibly following interaction with other membrane proteins, passes through the putative membrane pore [16] into the peroxisomal interior [8]. A 'classical membrane receptor' model such as this is suggested by the observations that the *P. pastoris* PTS1 receptor (PAS8p) (180, 183) and the human PTS1 receptor (according to one laboratory at least (200)) have been shown to be located mainly on the cytoplasmic face of the peroxisomal membrane. *Import scheme 3*: After the ligand [7] has entered the putative membrane pore [16] translocation is facilitated by interaction with an intra-peroxisomal receptor [11] which somehow pulls the ligand inside. The resulting complex [12] then dissociates in the peroxisomal matrix to give free receptor [11] and free ligand [8]. Presumably this intraperoxisomal receptor [11] must have been imported by some separate mechanism. A 'pulling in' model such as this has been proposed for the *S. cerevisiae* putative PTS2 receptor (PEB1p) which has been found (by one laboratory at least) to be mainly located in the peroxisomal matrix (186). *Import scheme 4*: The peroxisomal matrix localization of some putative receptors, such as *S. cerevisiae* PEB1p (186) or *H. polymorpha* PER3p (181) could be explained by the import of the whole ligand–receptor complex [13] through the putative membrane pore [16]. Following separation of the ligand [8] and receptor [14] in the peroxisomal matrix, the receptor [14] could either be degraded or returned to the cytosol to continue the shuttling cycle. *Membrane docking and pore proteins*: There are numerous candidates for the hypothetical docking [15] and pore [16] proteins, whose existence is predicted by analogy with the better-studied systems of mitochondrial and ER protein translocation. Some possibilities are PMP35 (PAF1) and PMP70 in mammals and the numerous yeast PMPs encoded by some of the *PAS/PER/PAY* genes (see Table 5).

---

shuttling hypothesis (Fig. 2). The subcellular distribution of the receptors would depend on the particular stage of the shuttling cycle and would be likely to be strongly influenced by dietary/metabolic stimuli in a cell-specific manner. The presence of receptors in the peroxisomal matrix is more difficult to fit into the shuttling model of protein translocation unless it is envisaged that the receptor accompanies the ligand (i.e. the peroxisomal protein to be imported) all the way into the peroxisomal interior (see Fig. 2).

7.3 Miscellaneous peroxisomal membrane proteins (PMPs) of uncertain function

In addition to a number of the proteins encoded by the yeast *PAS/PER/PEB/PAY* genes (see Table 5), many other proteins have been identified in the peroxisomal membranes of diverse cell types (20, 33, 202–209), but the functions of most are unknown. Some PMPs appear to be ATPases (138, 139), while others might be involved in pore formation and the control of peroxisomal permeability (210–212). Three GTP-binding proteins of molecular masses 25, 27 and 29 kDa have been found in the peroxisomal membranes of rat liver (213). Although their function is unknown, it has been suggested that they might be involved in peroxisomal pro-liferation (213). Several other enzymes, such as dihydroxyacetone-phosphate acyltransferase, alkyl-dihydroxyacetone phosphate synthase, and very-long-chain acyl-CoA synthase, have been located to the peroxisomal membrane in mammals, but as yet the PMPs responsible for these activities have yet to be identified (214). A number of genes encoding peroxisomal membrane proteins have been cloned and sequenced, including those for *C. boidinii* PMP20 (102) and PMP47 (106), *S. cerevisiae* PMP27 (215) and PMP48 (105), and mammalian PMP22 (216), PMP35 (187), PMP70 (104, 191), and ALDP (195, 197). Although the functions of many of these are unclear, several have been shown to be involved in peroxisomal biogenesis (see Sections 7.1 and 7.2). *C. boidinii* PMP47 has sequence homology to mitochondrial solute carrier proteins and therefore might be involved in the peroxisomal mem-brane transport of metabolic intermediates (217).

Glyoxysomal membrane proteins have, in general, been only poorly studied. However, it has recently been shown that a peptide containing the C-terminal 12 amino acids of rat acyl-CoA oxidase binds specifically to castor bean endosperm glyoxysome membranes *in vitro* (218). Two types of binding sites were identified (high and low affinity). This binding was saturable, reversible, and showed many of the characteristics expected of the interaction with import receptors (218). However, the specific glyoxysomal PMPs to which the peptide bound were not identified.

8. Conclusions

Over the past decade the importance of the functions of peroxisomes (and related organelles) has become increasingly recognized at both the cellular and organismal levels. Although peroxisomes do not appear to be essential for viability in some cell types grown under certain conditions (e.g. cultured human fibroblasts and CHO cells, or yeasts grown on glucose), they are essential in other circumstances (e.g. *S. cerevisiae* and *P. pastoris* grown on oleic acid or *P. pastoris* and *H. polymorpha* grown on methanol). At the level of the whole organism, the importance of peroxisomes is amply demonstrated by the extreme severity of the disorders of peroxisomal bio-genesis, as typified by ZS.

The desire to provide molecular explanations for this group of disorders has been largely responsible for the current intense interest in peroxisome biogenesis. The increased attention that this field of study has received in recent years has led to a marked improvement in our understanding of the various processes involved in the targeting and import of peroxisomal, glyoxysomal and, to a lesser extent, glyco-somal proteins. There is no doubt that even greater advances will be made over the next few years as : (i) more peroxisome assembly genes are cloned and have functions attributed to their translation products, especially in mammalian systems; (ii) more mutant genes are identified in patients with disorders of peroxisome biogenesis; (iii) the enigmas of context specificity, degeneracy, and redundancy of the various PTSs are solved; and (iv) new improved *in vitro* isolated organelle or semi-intact cell systems are developed which overcome the problems associated with the current techniques.

References

1. Lazarow, P. B. and Moser, H. W. (1995) Disorders of peroxisome biogenesis. In *The metabolic and molecular bases of inherited disease*. Scriver, C. R., Beaudet, A. L., Sly, W. S., and Valle, D. (ed.). McGraw-Hill, New York, pp. 2287–324.

2. Green, S. (1992) Receptor-mediated mechanisms of peroxisome proliferators. *Biochem. Pharmacol.*, **43,** 393.

3. Reddy, J. K. and Rao, M. S. (1992) Peroxisome proliferation and hepatocarcinogenesis. *IARC Sci. Publ.*, 225.

4. Gibson, G. G. (1993) Peroxisome proliferators: paradigms and prospects. *Toxicol. Lett.*, **68,** 193.

5. Reddy, J. K., Rao, M. S., Lalwani, N. D., Reddy, M. K., Nemali, M. R., and Alvares, K. (1987) Induction of hepatic peroxisome proliferation by xenobiotics. In *Peroxisomes in biology and medicine*. Fahimi, H. D. and Sies, H. (ed.). Springer-Verlag, Berlin, pp. 255–62.

6. Lazarow, P. B. and Fujiki, Y. (1985) Biogenesis of peroxisomes. *Annu. Rev. Cell Biol.*, **1,** 489.

7. Lazarow, P. B. (1989) Peroxisome biogenesis. *Curr. Opin. Cell Biol.*, **1,** 630.

8. Borst, P. (1986) How proteins get into microbodies (peroxisomes, glyoxysomes, glycosomes). *Biochim. Biophys. Acta*, **866,** 179.

9. Borst, P. (1989) Peroxisome biogenesis revisited. *Biochim. Biophys. Acta*, **1008,** 1.

10. Osumi, T. and Fujiki, Y. (1990) Topogenesis of peroxisomal proteins. *BioEssays*, **12,** 217.

11. Kunau, W. H. (1992) Peroxisomal biogenesis in *Saccharomyces cerevisiae*. *Prog. Clin. Biol. Res.*, **375,** 9.

12. Kunau, W. H. and Hartig, A. (1992) Peroxisome biogenesis in *Saccharomyces cerevisiae*. *Antonie Van Leeuwenhoek*, **62,** 63.

13. Kunau, W. H., Beyer, A., Franken, T., Gotte, K., Marzioch, M., Saidowsky, J., Skaletz Rorowski, A., and Wiebel, F. F. (1993) Two complementary approaches to study peroxisome biogenesis in *Saccharomyces cerevisiae*: forward and reversed genetics. *Biochimie*, **75,** 209.

14. Veenhuis, M. (1992) Peroxisome biogenesis and function in *Hansenula polymorpha*. *Cell Biochem. Funct.*, **10,** 175.

15. Veenhuis, M., van der Klei, I. J., Titorenko, V., and Harder, W. (1992) *Hansenula polymorpha*: an attractive model organism for molecular studies of peroxisome biogenesis and function. *FEMS Microbiol. Lett.*, **79,** 393.

16. Aitchison, J. D., Nuttley, W. M., Szilard, R. K., Brade, A. M., Glover, J. R., and Rachubinski, R. A. (1992) Peroxisome biogenesis in yeast. *Mol. Microbiol.*, **6,** 3455.

17. Kindl, H. (1992) Plant peroxisomes: recent studies on function and biosynthesis. *Cell Biochem. Funct.*, **10,** 153.

18. Kindl, H. (1993) Fatty acid degradation in plant peroxisomes: function and biosynthesis of the enzymes involved. *Biochimie*, **75,** 225.

19. Baker, A. (1996) Biogenesis of plant peroxisomes. In *Membranes: specialised functions in*

plant cells. Bowles, D., Knox, J. P., and Smallwood, M. (ed.). Bios Scientific Publishers Ltd., Oxford, UK. Chapter 24 (in press).

20. Just, W. W. and Soto, U. (1992) Biogenesis of peroxisomes in mammals. *Cell Biochem. Funct.*, **10**, 159.

21. Fahimi, H. D., Baumgart, E., and Volkl, A. (1993) Ultrastructural aspects of the biogenesis of peroxisomes in rat liver. *Biochimie*, **75**, 201.

22. Opperdoes, F. R. (1988) Glycosomes may provide clues to the import of peroxisomal proteins. *Trends Biochem. Sci.*, **13**, 255.

23. Opperdoes, F. R. and Michels, P. A. (1993) The glycosomes of the Kinetoplastida. *Biochimie*, **75**, 231.

24. de Hoop, M. J. and Ab, G. (1992) Import of proteins into peroxisomes and other microbodies. *Biochem. J.*, **286**, 657.

25. Subramani, S. (1992) Targeting of proteins into the peroxisomal matrix. *J. Membr. Biol.*, **125**, 99.

26. Subramani, S. (1993) Protein import into peroxisomes and biogenesis of the organelle. *Annu. Rev. Cell Biol.*, **9**, 445.

27. Roggenkamp, R. (1992) Targeting signals for protein import into peroxisomes. *Cell Biochem. Funct.*, **10**, 193.

28. De Duve, C. and Baudhuin, P. (1966) Peroxisomes (microbodies and related particles). *Physiol. Rev.*, **46**, 323.

29. Beevers, H. (1982) Glyoxysomes in higher plants. *Ann. NY Acad. Sci.*, **386**, 243.

30. Novikoff, A. B. and Shin, W. (1964) The endoplasmic reticulum in the Golgi zone and its relations to microbodies, Golgi apparatus and autophagic vacuoles in rat liver cells. *J. Microsc.*, **3**, 187.

31. Higashi, T. and Peters, T. J. (1963) Studies on rat liver catalase. II. Incorporation of C14-leucine into catalase of liver cell fractions *in vivo*. *J. Biol. Chem.*, **238**, 3952.

32. Gonzalez, E. and Beevers, H. (1976) Role of the endoplasmic reticulum in glyoxysome formation in castor bean endosperm. *Plant Physiol.*, **57**, 406.

33. Bowden, L. and Lord, J. M. (1976) Similarities in the polypeptide composition of glyoxysomal and endoplasmic-reticulum membranes from castor-bean endosperm. *Biochem. J.*, **154**, 491.

34. Goldman, B. M. and Blobel, G. (1978) Biogenesis of peroxisomes: intracellular site of synthesis of catalase and uricase. *Proc. Natl Acad. Sci. USA*, **75**, 5066.

35. Miura, S., Mori, M., Takiguchi, M., Tatibana, M., Furuta, S., Miyazawa, S., and Hashimoto, T. (1984) Biosynthesis and intracellular transport of enzymes of peroxisomal beta-oxidation. *J. Biol. Chem.*, **259**, 6397.

36. Fujiki, Y., Rachubinski, R. A., and Lazarow, P. B. (1984) Synthesis of a major integral membrane polypeptide of rat liver peroxisomes on free polysomes. *Proc. Natl Acad. Sci. USA*, **81**, 7127.

37. Fujiki, Y. and Lazarow, P. B. (1985) Post-translational import of fatty acyl-CoA oxidase and catalase into peroxisomes of rat liver *in vitro*. *J. Biol. Chem.*, **260**, 5603.

38. Zimmermann, R. and Neupert, W. (1980) Biogenesis of glyoxysomes. Synthesis and intracellular transfer of isocitrate lyase. *Eur. J. Biochem.*, **112**, 225.

39. Clayton, C. E. (1987) Import of fructose bisphosphate aldolase into the glycosomes of *Trypanosoma brucei*. *J. Cell Biol.*, **105**, 2649.

40. Hart, D. T., Baudhuin, P., Opperdoes, F. R., and De Duve, C. (1987) Biogenesis of the glycosome in *Trypanosoma brucei*: the synthesis, translocation and turnover of glycosomal polypeptides, *EMBO J.*, **6**, 1403.

41. Robbi, M. and Lazarow, P. B. (1982) Peptide mapping of peroxisomal catalase and its precursor. Comparison to the primary wheat germ translation product. *J. Biol. Chem.*, **257**, 964.

42. Fujiki, Y., Fowler, S., Shio, H., Hubbard, A. L., and Lazarow, P. B. (1982) Polypeptide and phospholipid composition of the membrane of rat liver peroxisomes: comparison with endoplasmic reticulum and mitochondrial membranes. *J. Cell Biol.*, **93**, 103.

43. Thieringer, R., Shio, H., Han, Y. S., Cohen, G., and Lazarow, P. B. (1991) Peroxisomes in *Saccharomyces cerevisiae*: immunofluorescence analysis and import of catalase A into isolated peroxisomes. *Mol. Cell Biol.*, **11**, 510.

44. Allen, L. A., Morand, O. H., and Raetz, C. R. (1989) Cytoplasmic requirement for peroxisome biogenesis in Chinese hamster ovary cells. *Proc. Natl Acad. Sci. USA*, **86**, 7012.

45. Waterham, H. R., Titorenko, V. I., Swaving, G. J., Harder, W., and Veenhuis, M. (1993) Peroxisomes in the methylotrophic yeast *Hansenula polymorpha* do not necessarily derive from pre-existing organelles. *EMBO J.*, **12**, 4785.

46. Keller, G. A., Gould, S., Deluca, M., and Subramani, S. (1987) Firefly luciferase is targeted to peroxisomes in mammalian cells. *Proc. Natl Acad. Sci. USA*, **84**, 3264.

47. Gould, S. J. and Subramani, S. (1988) Firefly luciferase as a tool in molecular and cell biology. *Anal. Biochem.*, **175**, 5.

48. Gould, S. G., Keller, G. A., and Subramani, S. (1987) Identification of a peroxisomal targeting signal at the carboxy terminus of firefly luciferase. *J. Cell Biol.*, **105**, 2923.

49. Gould, S. J., Keller, G. A., and Subramani, S. (1988) Identification of peroxisomal targeting signals located at the carboxy terminus of four peroxisomal proteins. *J. Cell Biol.*, **107**, 897.

50. Gould, S. J., Keller, G. A., Hosken, N., Wilkinson, J., and Subramani, S. (1989) A conserved tripeptide sorts proteins to peroxisomes. *J. Cell Biol.*, **108**, 1657.

51. Swinkels, B. W., Gould, S. J., and Subramani, S. (1992) Targeting efficiencies of various permutations of the consensus C-terminal tripeptide peroxisomal targeting signal. *FEBS Lett.*, **305**, 133.

52. Miyazawa, S., Osumi, T., Hashimoto, T., Ohno, K., Miura, S., and Fujiki, Y. (1989) Peroxisome targeting signal of rat liver acyl-coenzyme A oxidase resides at the carboxy terminus. *Mol. Cell Biol.*, **9**, 83.

53. Miura, S., Kasuya Arai, I., Mori, H., Miyazawa, S., Osumi, T., Hashimoto, T., and Fujiki, Y. (1992) Carboxyl-terminal consensus Ser–Lys–Leu-related tripeptide of peroxisomal proteins functions *in vitro* as a minimal peroxisome-targeting signal. *J. Biol. Chem.*, **267**, 14405.

54. Aitchison, J. D., Murray, W. W., and Rachubinski, R. A. (1991) The carboxyl-terminal tripeptide Ala–Lys–Ile is essential for targeting *Candida tropicalis* trifunctional enzyme to yeast peroxisomes. *J. Biol. Chem.*, **266**, 23197.

55. Didion, T. and Roggenkamp, R. (1992) Targeting signal of the peroxisomal catalase in the methylotrophic yeast *Hansenula polymorpha. FEBS Lett.*, **303**, 113.

56. Hansen, H., Didion, T., Thiemann, A., Veenhuis, M., and Roggenkamp, R. (1992) Targeting sequences of the two major peroxisomal proteins in the methylotrophic yeast *Hansenula polymorpha. Mol. Gen. Genet.*, **235**, 269.

57. Singh, K. K., Small, G. M., and Lewin, A. S. (1992) Alternative topogenic signals in peroxisomal citrate synthase of *Saccharomyces cerevisiae. Mol. Cell. Biol.*, **12**, 5593.

58. Kragler, F., Langeder, A., Raupachova, J., Binder, M., and Hartig, A. (1993) Two independent peroxisomal targeting signals in catalase A of *Saccharomyces cerevisiae. J. Cell Biol.*, **120**, 665.

59. Volokita, M. (1991) The carboxy-terminal end of glycolate oxidase directs a foreign protein into tobacco leaf peroxisomes. *Plant J.*, **1**, 361.

60. Banjoko, A. and Trelease, R. N. (1993) The C-terminal tripeptide S–K–L of cottonseed malate synthase is necessary for import *in vivo* in tobacco suspension-cultured cells. *Plant Physiol*, **102** (Suppl. 1), 481.

61. Olsen, L. J., Ettinger, W. F., Damsz, B., Matsudaira, K., Webb, M. A., and Harada, J. J. (1993) Targeting of glyoxysomal proteins to peroxisomes in leaves and roots of a higher plant. *Plant Cell*, **5**, 941.

62. Behari, R. and Baker, A. (1993) The carboxyl terminus of isocitrate lyase is not essential for import into glyoxysomes in an *in vitro* system. *J. Biol. Chem.*, **268**, 7315.

63. Keller, G. A., Krisans, S., Gould, S. J., Sommer, J. M., Wang, C. C., Schliebs, W., Kunau, W., Brody, S., and Subramani, S. (1991) Evolutionary conservation of a microbody targeting signal that targets proteins to peroxisomes, glyoxysomes, and glycosomes. *J. Cell Biol.*, **114**, 893.

64. Michels, P. A., Poliszczak, A., Osinga, K. A., Misset, O., Van Beeumen, J., Wierenga, R. K., Borst, P., and Opperdoes, F. R. (1986) Two tandemly linked identical genes code for the glycosomal glyceraldehyde-phosphate dehydrogenase in *Trypanosoma brucei*. *EMBO J.*, **5**, 1049.

65. Kendall, G., Wilderspin, A. F., Ashall, F., Miles, M. A., and Kelly, J. M. (1990) *Trypanosoma cruzi* glycosomal glyceraldehyde-3-phosphate dehydrogenase does not conform to the 'hotspot' topogenic signal model. *EMBO J.*, **9**, 2751.

66. Marchand, M., Kooystra, U., Wierenga, R. K., Lambier, A., Van Beeumen, J., Opperdoes, F. R., and Michels, P. A. (1989) Glucosephosphate isomerase from *Trypanosoma brucei*. Cloning and characterization of the gene and analysis of the enzyme. *Eur. J. Biochem.*, **184**, 455.

67. Osinga, K. A., Swinkels, B. W., Gibson, W. C., Borst, P., Veeneman, G. H., Van Boom, J. H., Michels, P. A., and Opperdoes, F. R. (1985) Topogenesis of microbody enzymes: a sequence comparison of the genes for the glycosomal (microbody) and cytosolic phosphoglycerate kinases of *Trypanosoma brucei*. *EMBO J.*, **4**, 3811.

68. Swinkels, B., Evers, R., and Borst, P. (1988) The topogenic signal of the glycosomal (microbody) phosphoglycerate kinase of *Crithidia fasciculata* resides in a carboxy-terminal extension. *EMBO J.*, **7**, 1159.

69. Blattner, J., Swinkels, B., Dorsam, H., Prospero, T., Subramani, S., and Clayton, C. (1992) Glycosome assembly in trypanosomes: variations in the acceptable degeneracy of a COOH-terminal microbody targeting signal. *J. Cell Biol.*, **119**, 1129.

70. Sommer, J. M., Peterson, G., Keller, G. A., Parsons, M., and Wang, C. C. (1993) The C-terminal tripeptide of glycosomal phosphoglycerate kinase is both necessary and sufficient for import into the glycosomes of *Trypanosoma brucei*. *FEBS Lett.*, **316**, 53.

71. Fung, K. and Clayton, C. (1991) Recognition of a peroxisomal tripeptide entry signal by the glycosomes of *Trypanosoma brucei*. *Mol. Biochem. Parasitol.*, **45**, 261.

72. Sommer, J. M., Cheng, Q. L., Keller, G. A., and Wang, C. C. (1992) *In vivo* import of firefly luciferase into the glycosomes of *Trypanosoma brucei* and mutational analysis of the C-terminal targeting signal. *Mol. Biol. Cell*, **3**, 749.

73. Hijikata, M., Wen, J. K., Osumi, T., and Hashimoto, T. (1990) Rat peroxisomal 3-ketoacyl-CoA thiolase gene. Occurrence of two closely related but differentially regulated genes. *J. Biol. Chem.*, **265**, 4600.

74. Swinkels, B. W., Gould, S. J., Bodnar, A. G., Rachubinski, R. A., and Subramani, S. (1991) A novel, cleavable peroxisomal targeting signal at the amino-terminus of the rat 3-ketoacyl-CoA thiolase. *EMBO J.*, **10**, 3255.

75. Osumi, T., Tsukamoto, T., Hata, S., Yokota, S., Miura, S., Fujiki, Y., Hijikata, M., Miyazawa, S., and Hashimoto, T. (1991) Amino-terminal presequence of the precursor of peroxisomal 3-ketoacyl-CoA thiolase is a cleavable signal peptide for peroxisomal targeting. *Biochem. Biophys. Res. Commun.*, **181**, 947.

76. Osumi, T., Tsukamoto, T., and Hata, S. (1992) Signal peptide for peroxisomal targeting: replacement of an essential histidine residue by certain amino acids converts the amino-terminal presequence of peroxisomal 3-ketoacyl-CoA thiolase to a mitochondrial signal peptide. *Biochem. Biophys. Res. Commun.*, **186**, 811.

77. Preisig Muller, R. and Kindl, H. (1993) Thiolase mRNA translated *in vitro* yields a peptide with a putative N-terminal presequence. *Plant Mol. Biol.*, **22**, 59.

78. Glover, J. R., Andrews, D. W., Subramani, S., and Rachubinski, R. A. (1994) Mutagenesis of the amino targeting signal of *Saccharomyces cerevisiae* 3-ketoacyl-CoA thiolase reveals conserved amino acids required for import into peroxisomes *in vivo*. *J. Biol. Chem.*, **269**, 7558.

79. Gietl, C. (1990) Glyoxysomal malate dehydrogenase from watermelon is synthesized with an amino-terminal transit peptide. *Proc. Natl Acad. Sci. USA*, **87**, 5773.

80. Gietl, C. (1991) Organelle targeting of malate dehydrogenase isoenzymes in watermelon. In *Molecular approaches to compartmentation and metabolic regulation*. Huang, A. H. and Taiz, L. (ed.). Am. Soc. Plant Physiologists, Rockville, pp. 138–50.

81. Yamaguchi, J., Mori, H., and Nishimura, M. (1987) Biosynthesis and intracellular transport of glyoxysomal malate dehydrogenase in germinating pumpkin cotyledons. *FEBS Lett.*, **213**, 329.

82. Gietl, C., Faber, K. N., van der Klei, I. J., and Veenhuis, M. (1994) Mutational analysis of the N-terminal topogenic signal of watermelon glyoxysomal malate dehydrogenase using the heterologous host *Hansenula polymorpha*. *Proc. Natl Acad. Sci. USA*, **91**, 3151.

83. Faber, K. N., Keizer Gunnink, I., Pluim, D., Harder, W., Ab, G., and Veenhuis, M. (1995) The N-terminus of amine oxidase of *Hansenula polymorpha* contains a peroxisomal targeting signal. *FEBS Lett.*, **357**, 115.

84. Clayton, C. E. (1985) Structure and regulated expression of genes encoding fructose biphosphate aldolase in *Trypanosoma brucei*. *EMBO J.*, **4**, 2997.

85. van der Klei, I. J., Faber, K. N., Keizer Gunnink, I., Gietl, C., Harder, W., and Veenhuis, M. (1993) Watermelon glyoxysomal malate dehydrogenase is sorted to peroxisomes of the methylotrophic yeast, *Hansenula polymorpha*. *FEBS Lett.*, **334**, 128.

86. Small, G. M. and Lazarow, P. B. (1987) Import of the carboxy-terminal portion of acyl-CoA oxidase into peroxisomes of *Candida tropicalis*. *J. Cell Biol.*, **105**, 247.

87. Small, G. M., Szabo, L. J., and Lazarow, P. B. (1988) Acyl-CoA oxidase contains two targeting sequences each of which can mediate protein import into peroxisomes. *EMBO J.*, **7**, 1167.

88. Bruinenberg, P. G., Evers, M., Waterham, H. R., Kuipers, J., Arnberg, A. C., and Ab, G. (1989) Cloning and sequencing of the peroxisomal amine oxidase gene from *Hansenula polymorpha*. *Biochim. Biophys. Acta*, **1008**, 157.

89. de Hoop, M. J., Valkema, R., Kienhuis, C. B., Hoyer, M. A., and Ab, G. (1992) The peroxisomal import signal of amine oxidase from the yeast *Hansenula polymorpha* is not universal. *Yeast*, **8**, 243.

90. Faber, K. N., Haima, P., de Hoop, M. J., Harder, W., Veenhuis, M., and Ab, G. (1993) Peroxisomal amine oxidase of *Hansenula polymorpha* does not require its SRL-containing C-terminal sequence for targeting. *Yeast*, **9**, 331.

91. Distel, B., Gould, S. J., Voorn Brouwer, T., van der Berg, M., Tabak, H. F., and Subramani,

S. (1992) The carboxyl-terminal tripeptide serine–lysine–leucine of firefly luciferase is necessary but not sufficient for peroxisomal import in yeast. *New Biol.*, **4**, 157.

92. Hurt, E. C. and Schatz, G. (1987) A cytosolic protein contains a cryptic mitochondrial targeting signal. *Nature*, **325**, 499.

93. Waterham, H. R., Titorenko, V. I., Haima, P., Cregg, J. M., Harder, W., and Veenhuis, M. (1994) The *Hansenula polymorpha PER1* gene is essential for peroxisome biogenesis and encodes a peroxisomal matrix protein with both carboxy- and amino-terminal targeting signals. *J. Cell Biol.*, **127**, 737.

94. Takada, Y., Kaneko, N., Esumi, H., Purdue, P. E., and Danpure, C. J. (1990) Human peroxisomal L-alanine: glyoxylate aminotransferase. Evolutionary loss of a mitochondrial targeting signal by point mutation of the initiation codon. *Biochem. J.*, **268**, 517.

95. Nishiyama, K., Berstein, G., Oda, T., and Ichiyama, A. (1990) Cloning and nucleotide sequence of cDNA encoding human liver serine-pyruvate aminotransferase. *Eur. J. Biochem.*, **194**, 9.

96. Purdue, P. E., Lumb, M. J., and Danpure, C. J. (1992) Molecular evolution of alanine:glyoxylate aminotransferase 1 intracellular targeting. Analysis of the marmoset and rabbit genes. *Eur. J. Biochem.*, **207**, 757.

97. Oda, T., Miyajima, H., Suzuki, Y., and Ichiyama, A. (1987) Nucleotide sequence of the cDNA encoding the precursor for mitochondrial serine:pyruvate aminotransferase of rat liver. *Eur. J. Biochem.*, **168**, 537.

98. Lumb, M. J., Purdue, P. E., and Danpure, C. J. (1994) Molecular evolution of alanine/glyoxylate aminotransferase 1 intracellular targeting: analysis of the feline gene. *Eur. J. Biochem.*, **221**, 53.

99. Motley, A., Lumb, M. J., Oatey, P. B., Jennings, P. R., De Zoysa, P. A., Wanders, R. J., Tabak, H. F., and Danpure, C. J. (1995) Mammalian alanine:glyoxylate aminotransferase 1 is imported into peroxisomes via the PTS1 translocation pathway. Increased degeneracy and context specificity of the mammalian PTS1 motif and implications for the peroxisome-to-mitochondrion mistargeting of AGT in primary hyperoxaluria type 1. *J. Cell Biol.*, **131**, 95.

100. Wierenga, R. K., Swinkels, B., Michels, P. A., Osinga, K., Misset, O., Van Beeumen, J., Gibson, W. C., Postma, J. P., Borst, P., Opperdoes, F. R., and Hol, W. G. (1987) Common elements on the surface of glycolytic enzymes from *Trypanosoma brucei* may serve as topogenic signals for import into glycosomes. *EMBO J.*, **6**, 215.

101. Alexander, K. and Parsons, M. (1991) A phosphoglycerate kinase-like molecule localized to glycosomal microbodies: evidence that the topogenic signal is not at the C-terminus. *Mol. Biochem. Parasitol.*, **46**, 1.

102. Garrard, L. J. and Goodman, J. M. (1989) Two genes encode the major membrane-associated protein of methanol-induced peroxisomes from *Candida boidinii*. *J. Biol. Chem.*, **264**, 13929.

103. Shimozawa, N., Tsukamoto, T., Suzuki, Y., Orii, T., Shirayoshi, Y., Mori, T., and Fujiki, Y. (1992) A human gene responsible for Zellweger syndrome that affects peroxisome assembly. *Science*, **255**, 1132.

104. Kamijo, K., Taketani, S., Yokota, S., Osumi, T., and Hashimoto, T. (1990) The 70-kDa peroxisomal membrane protein is a member of the Mdr (P-glycoprotein)-related ATP-binding protein superfamily. *J. Biol. Chem.*, **265**, 4534.

105. Hohfeld, J., Veenhuis, M., and Kunau, W. H. (1991) *PAS3*, a *Saccharomyces cerevisiae* gene encoding a peroxisomal integral membrane protein essential for peroxisome biogenesis. *J. Cell Biol.*, **114**, 1167.

106. McCammon, M. T., Dowds, C. A., Orth, K., Moomaw, C. R., Slaughter, C. A., and Goodman, J. M. (1990) Sorting of peroxisomal membrane protein PMP47 from *Candida boidinii* into peroxisomal membranes of *Saccharomyces cerevisiae*. *J. Biol. Chem.*, **265**, 20098.

107. McCammon, M. T., McNew, J. A., Willy, P. J., and Goodman, J. M. (1994) An internal region of the peroxisomal membrane protein PMP47 is essential for sorting to peroxisomes. *J. Cell Biol.*, **124**, 915.

108. Gould, S. J., Krisans, S., Keller, G. A., and Subramani, S. (1990) Antibodies directed against the peroxisomal targeting signal of firefly luciferase recognize multiple mammalian peroxisomal proteins. *J. Cell Biol.*, **110**, 27.

109. Marvin Sikkema, F. D., Kraak, M. N., Veenhuis, M., Gottschal, J. C., and Prins, R. A. (1993) The hydrogenosomal enzyme hydrogenase from the anaerobic fungus *Neocallimastix* sp. L2 is recognized by antibodies, directed against the C-terminal microbody protein targeting signal SKL. *Eur. J. Cell Biol.*, **61**, 86.

110. Aitchison, J. D., Szilard, R. K., Nuttley, W. M., and Rachubinski, R. A. (1992) Antibodies directed against a yeast carboxyl-terminal peroxisomal targeting signal specifically recognize peroxisomal proteins from various yeasts. *Yeast*, **8**, 721.

111. Arand, M., Knehr, M., Thomas, H., Zeller, H. D., and Oesch, F. (1991) An impaired peroxisomal targeting sequence leading to an unusual bicompartmental distribution of cytosolic epoxide hydrolase. *FEBS Lett.*, **294**, 19.

112. Gonzalez, E. (1991) The C-terminal domain of plant catalases. Implications for a glyoxysomal targeting sequence. *Eur. J. Biochem.*, **199**, 211.

113. Gould, S. J., Keller, G. A., Schneider, M., Howell, S. H., Garrard, L. J., Goodman, J. M., Distel, B., Tabak, H., and Subramani, S. (1990) Peroxisomal protein import is conserved between yeast, plants, insects and mammals. *EMBO J.*, **9**, 85.

114. Hansen, H. and Roggenkamp, R. (1989) Functional complementation of catalase-defective peroxisomes in a methylotrophic yeast by import of the catalase A from *Saccharomyces cerevisiae*. *Eur. J. Biochem.*, **184**, 173.

115. Distel, B., Veenhuis, M., and Tabak, H. F. (1987) Import of alcohol oxidase into peroxisomes of *Saccharomyces cerevisiae*. *EMBO J.*, **6**, 3111.

116. de Hoop, M. J., Holtman, W. L., and Ab, G. (1993) Human catalase is imported and assembled in peroxisomes of *Saccharomyces cerevisiae*. *Yeast*, **9**, 59.

117. Bodnar, A. G. and Rachubinski, R. A. (1990) Cloning and sequence determination of cDNA encoding a second rat liver peroxisomal 3-ketoacyl-CoA thiolase. *Gene*, **91**, 193.

118. Hijikata, M., Ishii, N., Kagamiyama, H., Osumi, T. and Hashimoto, T. (1987) Structural analysis of cDNA for rat peroxisomal 3-ketoacyl-CoA thiolase. *J. Biol. Chem.*, **262**, 8151.

119. Balfe, A., Hoefler, G., Chen, W. W., and Watkins, P. A. (1990) Aberrant subcellular localization of peroxisomal 3-ketoacyl-CoA thiolase in the Zellweger syndrome and rhizomelic chondrodysplasia punctata. *Pediatr. Res.*, **27**, 304.

120. van Roermund, C. W., Brul, S., Tager, J. M., Schutgens R. B., and Wanders, R. J. (1991) Acyl-CoA oxidase, peroxisomal thiolase and dihydroxyacetone phosphate acyltransferase: aberrant subcellular localization in Zellweger syndrome. *J. Inherit. Metab. Dis.*, **14**, 152.

121. Tsukamoto, T., Yokota, S., and Fujiki, Y. (1990) Isolation and characterization of Chinese hamster ovary cell mutants defective in assembly of peroxisomes. *J. Cell Biol.*, **110**, 651.

122. Yamaguchi, J., Nishimura, M., and Akazawa, T. (1984) Maturation of catalase precursor proceeds to a different extent in glyoxysomes and leaf peroxisomes of pumpkin cotyledons. *Proc. Natl Acad. Sci. USA*, **81**, 4809.

123. Yamaguchi, J. and Nishimura, M. (1984) Purification of glyoxysomal catalase and immunochemical comparison of glyoxysomal and leaf peroxisomal catalase in germinating pumpkin cotyledons. *Plant Physiol.*, **74**, 261.

124. Eising, R., Trelease, R. N., and Ni, W. T. (1990) Biogenesis of catalase in glyoxysomes and leaf-type peroxisomes of sunflower cotyledons. *Arch. Biochem. Biophys.*, **278**, 258.

125. Fujiki, Y., Tsuneoka, M., and Tashiro, Y. (1989) Biosynthesis of nonspecific lipid transfer protein (sterol carrier protein 2) on free polyribosomes as a larger precursor in rat liver. *J. Biochem. Tokyo*, **106**, 1126.

126. Moncecchi, D., Pastuszyn, A., and Scallen, T. J. (1991) cDNA sequence and bacterial expression of mouse liver sterol carrier protein-2. *J. Biol. Chem.*, **266**, 9885.

127. Yamamoto, R., Kallen, C. B., Babalola, G. O., Rennert, H., Billheimer, J. T., and Strauss, J. F. (1991) Cloning and expression of a cDNA encoding human sterol carrier protein 2. *Proc. Natl Acad. Sci. USA*, **88**, 463.

128. Tsuboi, S., Osafune, T., Tsugeki, R., Nishimura, M., and Yamada, M. (1992) Nonspecific lipid transfer protein in castor bean cotyledon cells: subcellular localization and a possible role in lipid metabolism. *J. Biochem.*, **111**, 500.

129. Kalousek, F., Hendrick, J. P., and Rosenberg, L. E. (1988) Two mitochondrial matrix proteases act sequentially in the processing of mammalian matrix enzymes. *Proc. Natl Acad. Sci. USA*, **85**, 7536.

130. Corpas, F. J., Palma, J. M., and del Rio, L. A. (1993) Evidence for the presence of proteolytic activity in peroxisomes. *Eur. J. Cell Biol.*, **61**, 81.

131. Authier, F., Bergeron, J. J., Ou, W. J., Rachubinski, R. A., Posner, B. I., and Walton, P. A. (1995) Degradation of the cleaved leader peptide of thiolase by a peroxisomal proteinase. *Proc. Natl Acad. Sci. USA*, **92**, 3859.

132. Imanaka, T., Small, G. M., and Lazarow, P. B. (1987) Translocation of acyl-CoA oxidase into peroxisomes requires ATP hydrolysis but not a membrane potential. *J. Cell Biol.*, **105**, 2915.

133. Soto, U., Pepperkok, R., Ansorge, W., and Just, W. W. (1993) Import of firefly luciferase into mammalian peroxisomes *in vivo* requires nucleoside triphosphates. *Exp. Cell Res.*, **205**, 66.

134. Wendland, M. and Subramani, S. (1993) Cytosol-dependent peroxisomal protein import in a permeabilized cell system. *J. Cell Biol.*, **120**, 675.

135. Bellion, E. and Goodman, J. M. (1987) Proton ionophores prevent assembly of a peroxisomal protein. *Cell*, **48**, 165.

136. Diestelkotter, P. and Just, W. W. (1993) *In vitro* insertion of the 22-kD peroxisomal membrane protein into isolated rat liver peroxisomes. *J. Cell Biol.*, **123**, 1717.

137. Shimizu, S., Imanaka, T., Takano, T., and Ohkuma, S. (1992) Major ATPases on clofibrate-induced rat liver peroxisomes are not associated with 70 kDa peroxisomal membrane protein (PMP70). *J. Biochem. Tokyo*, **112**, 733.

138. Douma, A. C., Veenhuis, M., Waterham, H. R., and Harder, W. (1990) Immunocytochemical demonstration of the peroxisomal ATPase of yeasts. *Yeast*, **6**, 45.

139. Douma, A. C., Veenhuis, M., Sulter, G. J., and Harder, W. (1987) A proton-translocating adenosine triphosphatase is associated with the peroxisomal membrane of yeasts. *Arch. Microbiol.*, **147**, 42.

140. Eilers, M. and Schatz, G. (1988) Protein unfolding and the energetics of protein translocation across biological membranes. *Cell*, **52**, 481.

141. Hannavy, K., Rospert, S., and Schatz, G. (1993) Protein import into mitochondria: a paradigm for the translocation of polypeptides across membranes. *Curr. Opin. Cell Biol.*, **5**, 694.

142. Lazarow, P. B. and De Duve, C. (1973) The synthesis and turnover of rat liver peroxisomes. V. Intracellular pathway of catalase synthesis. *J. Cell Biol.*, **59**, 507.

143. Goodman, J. M., Scott, C. W., Donahue, P. N., and Atherton, J. P. (1984) Alcohol oxidase assembles post-translationally into the peroxisome of *Candida boidinii*. *J. Biol. Chem.*, **259**, 8485.

144. van der Klei, I. J., Harder, W., and Veenhuis, M. (1991) Biosynthesis and assembly of alcohol oxidase, a peroxisomal matrix protein in methylotrophic yeasts: a review. *Yeast*, **7**, 195.

145. Wanders, R. J., Strijland, A., van Roermund, C. W., van den Bosch, H., Schutgens, R. B., Tager, J. M., and Schram, A. W. (1987) Catalase in cultured skin fibroblasts from patients with the cerebro–hepato–renal (Zellweger) syndrome: normal maturation in peroxixome-deficient cells. *Biochim. Biophys. Acta*, **923**, 478.

146. Wanders, R. J., Kos, M., Roest, B., Meijer, A. J., Schrakamp, G., Heymans, H. S., Tegelaers, W. H., van den Bosch, H., Schutgens, R. B., and Tager, J. M. (1984) Activity of peroxisomal enzymes and intracellular distribution of catalase in Zellweger syndrome. *Biochem. Biophys. Res. Commun.*, **123**, 1054.

147. Wanders, R. J., Schutgens, R. B., and Tager, J. M. (1985) Peroxisomal matrix enzymes in Zellweger syndrome: activity and subcellular localization in liver. *J. Inherit. Metab. Dis.*, **8** (Suppl. 2), 151.

148. Danpure, C. J., Fryer, P., Griffiths, S., Guttridge, K. M., Jennings, P. R., Allsop, J., Moser, A. B., Naidu, S., Moser, H. W., MacCollin M., and DeVivo, D. C. (1994) Cytosolic compartmentalization of hepatic alanine:glyoxylate aminotransferase in patients with aberrant peroxisomal biogenesis and its effect on oxalate metabolism. *J. Inherit. Metab. Dis.*, **17**, 27.

149. Brul, S., Westerveld, A., Strijland, A., Wanders, R. J., Schram, A. W., Heymans, H. S., Schutgens, R. B., van den Bosch, H., and Tager, J. M. (1988) Genetic heterogeneity in the cerebrohepatorenal (Zellweger) syndrome and other inherited disorders with a generalized impairment of peroxisomal functions. A study using complementation analysis. *J. Clin. Invest.*, **81**, 1710.

150. Brul, S., Wiemer, E. A., Westerveld, A., Strijland, A., Wanders, R. J., Schram, A. W., Heymans, H. S., Schutgens, R. B., van den Bosch, H., and Tager, J. M. (1988) Kinetics of the assembly of peroxisomes after fusion of complementary cell lines from patients with the cerebro–hepato–renal (Zellweger) syndrome and related disorders. *Biochem. Biophys. Res. Commun.*, **152**, 1083.

151. McGuinness, M. C., Moser, A. B., Moser, H. W., and Watkins, P. A. (1990) Peroxisomal disorders: complementation analysis using beta-oxidation of very long chain fatty acids. *Biochem. Biophys. Res. Commun.*, **172**, 364.

152. McGuinness, M. C., Moser, A. B., Poll The, B. T., and Watkins, P. A. (1993) Complementation analysis of patients with intact peroxisomes and impaired peroxisomal beta-oxidation. *Biochem. Med. Metab. Biol.*, **49**, 228.

153. Poll The, B. T., Skjeldal, O. H., Stokke, O., Demaugre, F., and Saudubray, J. M. (1990) Complementation analysis of peroxisomal disorders and classical Refsum. *Prog. Clin. Biol. Res.*, **321**, 537.

154. Roscher, A. A., Hoefler, S., Hoefler, G., Paschke, E., Paltauf, F., Moser, A., and Moser, H. (1989) Genetic and phenotypic heterogeneity in disorders of peroxisome biogenesis—a complementation study involving cell lines from 19 patients. *Pediatr. Res.*, **26**, 67.

155. Yajima, S., Suzuki, Y., Shimozawa, N., Yamaguchi, S., Orii, T., Fujiki, Y., Osumi, T., Hashimoto, T., and Moser, H. W. (1992) Complementation study of peroxisome-

deficient disorders by immunofluorescence staining and characterization of fused cells. *Hum. Genet.*, **88,** 491.

156. Middelkoop, E., Strijland, A., and Tager, J. M. (1991) Does aminotriazole inhibit import of catalase into peroxisomes by retarding unfolding? *FEBS Lett.*, **279,** 79.
157. Walton, P. A., Gould, S. J., Feramisco, J. R., and Subramani, S. (1992) Transport of microinjected proteins into peroxisomes of mammalian cells: inability of Zellweger cell lines to import proteins with the SKL tripeptide peroxisomal targeting signal. *Mol. Cell Biol.*, **12,** 531.
158. Walton, P. A., Hill, P. E., and Subramani, S. (1995) Import of stably folded proteins into peroxisomes. *Mol. Biol. Cell*, **6,** 675.
159. Glover, J. R., Andrews, D. W., and Rachubinski, R. A. (1994) *Saccharomyces cerevisiae* peroxisomal thiolase is imported as a dimer. *Proc. Natl Acad. Sci. USA*, **91,** 10541.
160. McNew, J. A. and Goodman, J. M. (1994) An oligomeric protein is imported into peroxisomes *in vivo*. *J. Cell Biol.*, **127,** 1245.
161. Walton, P. A., Wendland, M., Subramani, S., Rachubinski, R. A., and Welch, W. J. (1994) Involvement of 70-kD heat-shock proteins in peroxisomal import. *J. Cell Biol.*, **125,** 1037.
162. Wendland, M. and Subramani, S. (1993) Presence of cytoplasmic factors functional in peroxisomal protein import implicates organelle-associated defects in several human peroxisomal disorders. *J. Clin. Invest.*, **92,** 2462.
163. Cuezva, J. M., Santaren, J. P., Gonzalez, P., Valcarce, C., Luis, A. M., and Izquierdo, J. M. (1990) Immunological detection of the mitochondrial F1–ATPase alpha subunit in the matrix of rat liver peroxisomes. A protein involved in organelle biogenesis? *FEBS Lett.*, **270,** 71.
164. Luis, A. M., Alconada, A., and Cuezva, J. M. (1990) The alpha regulatory subunit of the mitochondrial F1–ATPase complex is a heat-shock protein. Identification of two highly conserved amino acid sequences among the alpha-subunits and molecular chaperones. *J. Biol. Chem.*, **265,** 7713.
165. Cuezva, J. M., Flores, A. I., Liras, A., Santaren, J. F., and Alconada, A. (1993) Molecular chaperones and the biogenesis of mitochondria and peroxisomes. *Biol. Cell*, **77,** 47.
166. Evers, M. E., Langer, T., Harder, W., Hartl, F. U., and Veenhuis, M. (1992) Formation and quantification of protein complexes between peroxisomal alcohol oxidase and GroEL. *FEBS Lett.*, **305,** 51.
167. Danpure, C. J., Guttridge, K. M., Fryer, P., Jennings, P. R., Allsop, J., and Purdue, P. E. (1990) Subcellular distribution of hepatic alanine:glyoxylate aminotransferase in various mammalian species. *J. Cell. Sci.*, **97,** 669.
168. Danpure, C. J., Fryer, P., Jennings, P. R., Allsop, J., Griffiths, S., and Cunningham, A. (1994) Evolution of alanine:glyoxylate aminotransferase 1 peroxisomal and mitochondrial targeting. A survey of it subcellular distribution in the livers of various representatives of the classes Mammalia, Aves and Amphibia. *Eur. J. Cell Biol.*, **64,** 295.
169. Danpure, C. J., Cooper, P. J., Wise, P. J., and Jennings, P. R. (1989) An enzyme trafficking defect in two patients with primary hyperoxaluria type 1: peroxisomal alanine/glyoxylate aminotransferase rerouted to mitochondria. *J. Cell Biol.*, **108,** 1345.
170. Purdue, P. E., Takada, Y., and Danpure, C. J. (1990) Identification of mutations associated with peroxisome-to-mitochondrion mistargeting of alanine/glyoxylate aminotransferase in primary hyperoxaluria type 1. *J. Cell Biol.*, **111,** 2341.
171. Danpure, C. J., Purdue, P. E., Fryer, P., Griffiths, S., Allsop, J., Lumb, M. J., Guttridge, K. M., Jennings, P. R., Scheinman, J. I., Mauer, S. M., and Davidson, N. O. (1993)

Enzymological and mutational analysis of a complex primary hyperoxaluria type 1 phenotype involving alanine:glyoxylate aminotransferase peroxisome-to-mitochondrion mistargeting and intraperoxisomal aggregation. *Am. J. Hum. Genet.*, **53,** 417.

172. Erdmann, R. and Kunau, W. H. (1992) A genetic approach to the biogenesis of peroxisomes in the yeast *Saccharomyces cerevisiae*. *Cell Biochem. Funct.*, **10,** 167.

173. Liu, H., Tan, X., Veenhuis, M., McCollum, D., and Cregg, J. M. (1992) An efficient screen for peroxisome-deficient mutants of *Pichia pastoris*. *J. Bacteriol.*, **174,** 4943.

174. Gould, S. J., McCollum, D., Spong, A. P., Heyman, J. A., and Subramani, S. (1992) Development of the yeast *Pichia pastoris* as a model organism for a genetic and molecular analysis of peroxisome assembly. *Yeast*, **8,** 613.

175. Titorenko, V. I., Waterham, H. R., Haima, P., Harder, W., and Veenhuis, M. (1992) Peroxisome biogenesis in *Hansenula polymorpha*: different mutations in genes, essential for peroxisome biogenesis, cause different peroxisomal mutant phenotypes. *FEMS Microbiol. Lett.*, **74,** 143.

176. Titorenko, V. I., Waterham, H. R., Cregg, J. M., Harder, W., and Veenhuis, M. (1993) Peroxisome biogenesis in the yeast *Hansenula polymorpha* is controlled by a complex set of interacting gene products. *Proc. Natl Acad. Sci. USA*, **90,** 7470.

177. Cregg, J. M., Van Klei, I. J., Sulter, G. J., Veenhuis, M., and Harder, W. (1990) Peroxisome-deficient mutants of *Hansenula polymorpha*. *Yeast*, **6,** 87.

178. Nuttley, W. M., Brade, A. M., Gaillardin, C., Eitzen, G. A., Glover, J. R., Aitchison, J. D., and Rachubinski, R. A. (1993) Rapid identification and characterization of peroxisomal assembly mutants in *Yarrowia lipolytica*. *Yeast*, **9,** 507.

179. Van Der Leij, I., Franse, M. M., Elgersma, Y., Distel, B., and Tabak, H. F. (1993) PAS10 is a tetratricopeptide repeat protein that is essential for the import of most matrix proteins into peroxisomes of *Saccharomyces cerevisiae*. *Proc. Natl Acad. Sci. USA*, **90,** 11782.

180. McCollum, D., Monosov, E., and Subramani, S. (1993) The pas8 mutant of *Pichia pastoris* exhibits the peroxisomal protein import deficiencies of Zellweger syndrome cells—the PAS8 protein binds to the COOH-terminal tripeptide peroxisomal targeting signal, and is a member of the TPR protein family. *J. Cell Biol.*, **121,** 761.

181. van der Klei, I. J., Hilbrands, R. E., Swaving, G. J., Waterham, H. R., Vrieling, E. G., Titorenko, V. I., Cregg, J. M., Harder, W., and Veenhuis, M. (1995) The *Hansenula polymorpha PER3* gene is essential for the import of PTS1 proteins into the peroxisomal ʹmatrix. *J. Biol. Chem.*, **270,** 17229.

182. Brocard, C., Kragler, F., Simon, M. M., Schuster, T., and Hartig, A. (1994) The tetratricopeptide repeat-domain of the PAS10 protein of *Saccharomyces cerevisiae* is essential for binding the peroxisomal targeting signal-SKL. *Biochem. Biophys. Res. Commun.*, **204,** 1016.

183. Terlecky, S. R., Nuttley, W. M., McCollum, D., Sock, E., and Subramani, S. (1995) The *Pichia pastoris* peroxisomal protein PAS8p is the receptor for the C-terminal tripeptide peroxisomal targeting signal. *EMBO J.*, **14,** 3627.

184. Dodt, G., Braverman, N., Wong, C., Moser, A., Moser, H. W., Watkins, P., Valle, D., and Gould, S. J. (1995) Mutations in the PTS1 receptor gene, *PXR1*, define complementation group 2 of the peroxisome biogenesis disorders. *Nature Genet.*, **9,** 115.

185. Marzioch, M., Erdmann, R., Veenhuis, M., and Kunau, W. H. (1994) PAS7 encodes a novel yeast member of the WD-40 protein family essential for import of 3-oxoacyl-CoA thiolase, a PTS2-containing protein, into peroxisomes. *EMBO J.*, **13,** 4908.

186. Zhang, J. W. and Lazarow, P. B. (1995) PEB1 (PAS7) in *Saccharomyces cerevisiae* encodes a hydrophilic, intra-peroxisomal protein that is a member of the WD repeat family and is essential for the import of thiolase into peroxisomes. *J. Cell Biol.*, **129,** 65.

187. Tsukamoto, T., Miura, S., and Fujiki, Y. (1991) Restoration by a 35K membrane protein of peroxisome assembly in a peroxisome-deficient mammalian cell mutant. *Nature*, **350**, 77.

188. Thieringer, R. and Raetz, C. R. (1993) Peroxisome-deficient Chinese hamster ovary cells with point mutations in peroxisome assembly factor-1. *J. Biol. Chem.*, **268**, 12631.

189. Tsukamoto, T., Shimozawa, N., and Fujiki, Y. (1994) Peroxisome assembly factor 1: nonsense mutation in a peroxisome-deficient Chinese hamster ovary cell mutant and deletion analysis. *Mol. Cell Biol.*, **14**, 5458.

190. Allen, L. A., Hope, L., Raetz, C. R., and Thieringer, R. (1994) Genetic evidence supporting the role of peroxisome assembly factor (PAF)-1 in peroxisome biogenesis. Polymerase chain reaction detection of a missense mutation in PAF-1 of Chinese hamster ovary cells. *J. Biol. Chem.*, **269**, 11734.

191. Kamijo, K., Kamijo, T., Ueno, I., Osumi, T., and Hashimoto, T. (1992) Nucleotide sequence of the human 70 kDa peroxisomal membrane protein: a member of ATP-binding cassette transporters. *Biochim. Biophys. Acta*, **1129**, 323.

192. Gartner, J. and Valle, D. (1993) The 70 kDa peroxisomal membrane protein: an ATP-binding cassette transporter protein involved in peroxisome biogenesis. *Semin. Cell Biol.*, **4**, 45.

193. Gartner, J., Moser, H., and Valle, D. (1992) Mutations in the 70K peroxisomal membrane protein gene in Zellweger syndrome. *Nat. Genet.*, **1**, 16.

194. Gartner, J., Obie, C., Watkins, P., and Valle, D. (1994) Restoration of peroxisome biogenesis in a peroxisome-deficient mammalian cell line by expression of either the 35 kDa or the 70 kDa peroxisomal membrane proteins. *J. Inherit. Metab. Dis.*, **17**, 327.

195. Aubourg, P., Mosser, J., Douar, A. M., Sarde, C. O., Lopez, J., and Mandel, J. L. (1993) Adrenoleukodystrophy gene: unexpected homology to a protein involved in peroxisome biogenesis. *Biochimie*, **75**, 293.

196. Mosser, J., Douas, A. M., Sarde, C. O., Kioschis, P., Feil, R., Moser, H., Poustka, A. M., Mandel, J. L., and Aubourg, P. (1993) Putative X-linked adrenoleukodystrophy gene shares unexpected homology with ABC transporters. *Nature*, **361**, 726.

197. Contreras, M., Mosser, J., Mandel, J. L., Aubourg, P., and Singh, I. (1994) The protein coded by the X-adrenoleukodystrophy gene is a peroxisomal integral membrane protein. *FEBS Lett.*, **344**, 211.

198. Mosser, J., Lutz, Y., Stoeckel, M. E., Sarde, C. O., Kretz, C., Douar, A. M., Lopez, J., Aubourg, P., and Mandel, J. L. (1994) The gene responsible for adrenoleukodystrophy encodes a peroxisomal membrane protein. *Hum. Mol. Genet.*, **3**, 265.

199. Valle, D. and Gartner, J. (1993) Human genetics. Penetrating the peroxisome (news; comment). *Nature*, **361**, 682.

200. Fransen, M., Brees, C., Baumgart, E., Vanhooren, J. C., Baes, M., Mannaerts, G. P., and Van Veldhoven, P. P. (1995) Identification and characterization of the putative human peroxisomal C-terminal targeting signal import receptor. *J. Biol. Chem.*, **270**, 7731.

201. Wiemer, E. A., Nuttley, W. M., Bertolaet, B. L., Li, X., Francke, U., Wheelock, M. J., Anne, U. K., Johnson, K. R., and Subramani, S. (1995) Human peroxisomal targeting signal-1 receptor restores peroxisomal protein import in cells from patients with fatal peroxisomal disorders. *J. Cell Biol.*, **130**, 51.

202. Goodman, J. M., Maher, J., Silver, P. A., Pacifico, A., and Sanders, D. (1986) The membrane proteins of the methanol-induced peroxisome of *Candida boidinii*. Initial characterization and generation of monoclonal antibodies. *J. Biol. Chem.*, **261**, 3464.

203. Sulter, G. J., Looyenga, L., Veenhuis, M., and Harder, W. (1990) Occurrence of peroxiso-

mal membrane proteins in methylotrophic yeasts grown under different conditions. *Yeast*, **6,** 35.

204. Sulter, G. J., Harder, W., and Veenhuis, M. (1993) Structural and functional aspects of peroxisomal membranes in yeasts. *FEMS Microbiol. Rev.*, **11,** 285.

205. Wilson, G. N. and King, T. E. (1991) Structure and variability of mammalian peroxisomal membrane proteins. *Biochem. Med. Metab. Biol.*, **46,** 235.

206. Causeret, C., Bentejac, M., and Bugaut, M. (1993) Proteins and enzymes of the peroxisomal membrane in mammals. *Biol. Cell*, **77,** 89.

207. Santos, M. J., Kawada, M. E., Espeel, M., Figueroa, C., Alvarez, A., Hidalgo, U., and Metz, C. (1994) Characterization of human peroxisomal membrane proteins. *J. Biol. Chem.*, **269,** 24890.

208. Erdmann, R. and Blobel, G. (1995) Giant peroxisomes in oleic acid-induced *Saccharomyces cerevisiae* lacking the peroxisomal membrane protein Pmp27p. *J. Cell Biol.*, **128,** 509.

209. Corpas, F. J., Bunkelmann, J., and Trelease, R. N. (1994) Identification and immunochemical characterization of a family of peroxisome membrane proteins (PMPs) in oilseed glyoxysomes. *Eur. J. Cell Biol.*, **65,** 280.

210. Van Veldhoven, P. P., Just, W. W., and Mannaerts, G. P. (1987) Permeability of the peroxisomal membrane to cofactors of beta-oxidation. Evidence for the presence of a pore-forming protein. *J. Biol. Chem.*, **262,** 4310.

211. Douma, A. C., Veenhuis, M., Sulter, G. J., Waterham, H. R., Verheyden, K., Mannaerts, G. P., and Harder, W. (1990) Permeability properties of peroxisomal membranes from yeasts. *Arch. Microbiol.*, **153,** 490.

212. Sulter, G. J., Verheyden, K., Mannaerts, G., Harder, W., and Veenhuis, M. (1993) The *in vitro* permeability of yeast peroxisomal membranes is caused by a 31 kDa integral membrane protein. *Yeast*, **9,** 733.

213. Verheyden, K., Fransen, M., Van Veldhoven, P. P., and Mannaerts, G. P. (1992) Presence of small GTP-binding proteins in the peroxisomal membrane. *Biochim. Biophys. Acta*, **1109,** 48.

214. van den Bosch, H., Schutgens, R. B., Wanders, R. J., and Tager, J. M. (1992) Biochemistry of peroxisomes. *Annu. Rev. Biochem.*, **61,** 157.

215. Marshall, P. A., Krimkevich, Y. I., Lark, R. H., Dyer, J. M., Veenhuis, M., and Goodman, J. M. (1995) Pmp27 promotes peroxisomal proliferation. *J. Cell Biol.*, **129,** 345.

216. Kaldi, K., Diestelkotter, P., Stenbeck, G., Auerbach, S., Jakle, U., Magert, H. J., Wieland, F. T., and Just, W. W. (1993) Membrane topology of the 22 kDa integral peroxisomal membrane protein. *FEBS Lett.*, **315,** 217.

217. Jank, B., Habermann, B., Schweyen, R. J., and Link, T. A. (1993) PMP47, a peroxisomal homologue of mitochondrial solute carrier proteins. *Trends Biochem. Sci.*, **18,** 427.

218. Wolins, N. E. and Donaldson, R. P. (1994) Specific binding of the peroxisomal protein targeting sequence to glyoxysomal membranes. *J. Biol. Chem.*, **269,** 1149.

219. Slawecki, M. L., Dodt, G., Steinberg, S., Moser, A. B., Moser, H. W., and Gould, S. J. (1995) Identification of three distinct peroxisomal protein import defects in patients with peroxisome biogenesis disorders. *J. Cell. Sci.*, **108,** 1817.

220. Danpure, C. J., Jennings, P. R., Fryer, P., Purdue, P. E., and Allsop, J. (1994) Primary hyperoxaluria type 1: genotypic and phenotypic heterogeneity. *J. Inherit. Metab. Dis.*, **17,** 487.

221. Purdue, P. E., Allsop, J., Isaya, G., Rosenberg, L. E., and Danpure, C. J. (1991) Mistargeting of peroxisomal L-alanine:glyoxylate aminotransferase to mitochondria in primary

hyperoxaluria patients depends upon activation of a cryptic mitochondrial targeting sequence by a point mutation. *Proc. Natl Acad. Sci. USA*, **88**, 10900.

222. Oda, T., Funai, T., and Ichiyama, A. (1990) Generation from a single gene of two mRNAs that encode the mitochondrial and peroxisomal serine:pyruvate aminotransferase of rat liver. *J. Biol. Chem.*, **265**, 7513.

223. Danpure, C. J. (1995) How can the products of a single gene be localized to more than one intracellular compartment? *Trends Cell Biol.*, **5**, 230.

224. Erdmann, R., Wiebel, F. F., Flessau, A., Rytka, J., Beyer, A., Frohlich, K. U., and Kunau, W. H. (1991) *PAS1*, a yeast gene required for peroxisome biogenesis, encodes a member of a novel family of putative ATPases. *Cell*, **64**, 499.

225. Krause, T., Kunau, W. H., and Erdmann, R. (1994) Effect of site-directed mutagenesis of conserved lysine residues upon Pas1 protein function in peroxisome biogenesis. *Yeast*, **10**, 1613.

226. Wiebel, F. F. and Kunau, W. H. (1992) The Pas2 protein essential for peroxisome biogenesis is related to ubiquitin conjugating enzymes. *Nature*, **359**, 73.

227. Voorn Brouwer, T., Van Der Leij, I., Hemrika, W., Distel, B., and Tabak, H. F. (1993) Sequence of the *PAS8* gene, the product of which is essential for biogenesis of peroxisomes in *Saccharomyces cerevisiae. Biochim. Biophys. Acta*, **1216**, 325.

228. Heyman, J. A., Monosov, E., and Subramani, S. (1994) Role of the *PAS1* gene of *Pichia pastoris* in peroxisome biogenesis. *J. Cell Biol.*, **127**, 1259.

229. Crane, D. I., Kalish, J. E., and Gould, S. J. (1994) The *Pichia pastoris PAS4* gene encodes a ubiquitin-conjugating enzyme required for peroxisome assembly. *J. Biol. Chem.*, **269**, 21835.

230. Spong, A. P. and Subramani, S. (1993) Cloning and characterization of *PAS5*: a gene required for peroxisome biogenesis in the methylotrophic yeast *Pichia pastoris. J. Cell Biol.*, **123**, 535.

231. Liu, H., Tan, X., Russell, K. A., Veenhuis, M., and Cregg, J. M. (1995) *PER3*, a gene required for peroxisome biogenesis in *Pichia pastoris*, encodes a peroxisomal membrane protein involved in protein import. *J. Biol. Chem.*, **270**, 10940.

232. Tan, X., Waterham, H. R., Veenhuis, M., and Cregg, J. M. (1995) The *Hansenula polymorpha PER8* gene encodes a novel peroxisomal integral membrane protein involved in proliferation. *J. Cell Biol.*, **128**, 307.

233. Eitzen, G. A., Aitchison, J. D., Szilard, R. K., Veenhuis, M., Nuttley, W. M., and Rachubinski, R. A. (1995) The *Yarrowia lipolytica* gene *PAY2* encodes a 42-kDa peroxisomal integral membrane protein essential for matrix protein import and peroxisome enlargement but not for peroxisome membrane proliferation. *J. Biol. Chem.*, **270**, 1429.

234. Nuttley, W. M., Brade, A. M., Eitzen, G. A., Veenhuis, M., Aitchison, J. D., Szilard, R. K., Glover, J. R., and Rachubinski, R. A. (1994) *PAY4*, a gene required for peroxisome assembly in the yeast *Yarrowia lipolytica*, encodes a novel member of a family of putative ATPases. *J. Biol. Chem.*, **269**, 556.

4 | Protein translocation into the endoplasmic reticulum

KARIN RÖMISCH and ANN CORSI

1. Introduction

In eukaryotic cells all non-organellar protein synthesis starts on free ribosomes in the cytoplasm (1). Nascent chains destined for the secretory pathway are marked as such by an N-terminal signal peptide (2, 3). The signal peptide is recognized and bound by a ribonucleoprotein complex, the signal recognition particle (SRP), as soon as it emerges from the ribosome (4). Binding of SRP to the signal peptide retards or stops translation (5) and allows for targeting of the SRP–nascent chain–ribosome complex to the endoplasmic reticulum (ER) membrane (6). Interaction of SRP with its receptor in the ER membrane subsequently leads to its dissociation from the nascent chain (5, 7). Translation resumes and the secretory protein is translocated into the ER where its signal peptide is cleaved, and glycosylation, folding, and oligomerization take place (8–11). In the past two decades a large body of work has focused on the elucidation of the first step of the secretory pathway, translocation into the ER. The results allow us to draw a detailed picture of the components involved and their respective functions.

2. The signal hypothesis

Morphological studies by Palade and co-workers in the 1950s and 1960s showed that the entrance for the secretory pathway in eukaryotic cells is at the level of the ER (for review see 1), that is, mRNAs for eukaryotic secretory proteins are translated on ER membrane-bound ribosomes (12). Since no differences had been found between free and membrane-bound ribosomes, Blobel and Sabatini (13) contended that it was the mRNA for the secretory protein that encoded an N-terminal signal responsible for the location of the nascent chain–ribosome complex at the ER membrane. This signal, after its translation, would be recognized and bound by a soluble factor mediating the targeting of the ribosome–nascent chain complex to the ER membrane. A more detailed version of this 'signal hypothesis' was published by Blobel and Dobberstein in 1975 (3) and postulated that at the ER membrane the signal peptide emerging from the ribosome triggers the assembly of several membrane proteins into a 'tunnel' which allows translocation of the nascent chain into the ER lumen. The authors further hypothesized that the protein components of the tunnel should bind to the large ribosomal subunit and thus ensure continuity of the protein-translocating channel in the ER membrane with the nascent chain-containing channel in the large ribosomal subunit. After termination of translocation, detachment of the ribosome from the membrane would cause the tunnel to dissociate into its subunits thus maintaining the principal function of the ER membrane as a diffusion barrier.

3. The signal peptide

In 1972, Milstein and co-workers discovered that secretory proteins are indeed synthesized as cytosolic precursor proteins with a cleavable N-terminal extension (2). Their results, and work by Blobel and Dobberstein (3), confirmed the central postulate of the signal hypothesis that secretory proteins differ from cytosolic proteins by initially containing an N-terminal signal peptide which mediates their association with the ER membrane (13). Meanwhile, a large number of natural and synthetic signal peptides has been characterized (for review see refs 14, 15). They vary in length (from 15–70 amino acids) and primary sequence, but have a common overall structure (Fig. 1): the N-terminal part (n region) has a variable length and generally carries a net positive charge. It is followed by a central hydrophobic core (h region) of 6–15 amino acids which is essential for signal peptide function. The transition between the hydrophobic and the C-terminal region (c region) of signal peptides is often marked by a helix-breaking residue such as proline, glycine, or serine. The 4–6 amino acids constituting the c region define the cleavage site for signal peptidase: small amino acids like alanine and glycine are generally found at amino acid positions –3 and –1 relative to the cleavage site (16). Both the length of the hydrophobic core and the nature of the helix-breaking residue adjacent to it can influence signal peptide processing (17, 18).

The signal peptide of most soluble secretory proteins is cleaved after its translocation into the lumen of the ER (2, 3). Notable exceptions to this rule are proteins such as ovalbumin (19) or plasminogen activator inhibitor II (20) which retain their signal peptide.

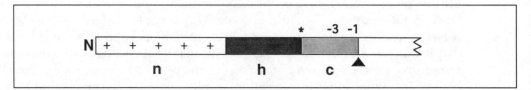

Fig. 1 The signal peptide. The characteristics of most signal peptides are shown. + signs in the N-terminal region (n) depict positive charges. The central dark grey region is the hydrophobic core (h). The asterisk depicts the position of a helix-breaking residue such as proline at the h–c boundary. The light grey C-terminal region (c) defines the signal peptidase cleavage site which is marked by the arrowhead. Small residues in positions –3 and –1 are required for signal peptide cleavage. The portion following the c region represents the N-terminus of the mature protein.

The general structure of signal peptides is evolutionarily conserved and some signal peptides can be functionally exchanged between species; for example mammalian proteins can be secreted from *Escherichia coli* (21) and yeast proteins can be translocated into mammalian microsomes (22). However, this is not universally true, and species-specific variations in signal peptides have been analysed both statistically (23) and experimentally (24–26). Eukaryotic signal peptides tend to be shorter than their bacterial counterparts and carry less positive net charge in their n region. Eukaryotic h regions are more hydrophobic than those of bacteria (23) and translocation into mammalian ER requires a higher degree of hydrophobicity than translocation into the yeast ER (25). Variations in the c regions probably reflect variations in signal peptidase specificity since secretory proteins are often processed incorrectly or not at all if expressed in heterologous systems (16, 23).

Due to the absence of primary sequence homology, it has been suggested that a common secondary structure is important for signal peptide function (27, 28). Synthetic signal peptides can adopt multiple conformations, depending on their environment (29, 30). To date most of the available data is derived from studies on *E. coli* signal peptides (leader peptides) and their mutant counterparts. The general notion that emerges is that functional leader peptides adopt α-helical or β-sheet structures in polar solutions or in contact with phospholipids, whereas they are less ordered in aqueous solvents (29). However, data concerning the secondary structure of mammalian signal peptides are contradictory (30) and the physicochemical structure of yeast signal peptides has not yet been analysed. Therefore, general conclusions about the influence of secondary structure on the functionality of signal peptides await further experimentation.

4. Membrane proteins and topology

Membrane proteins destined for organelles of the secretory pathway or the plasma membrane are initially inserted into the ER membrane (31). They contain stretches of amino acids equivalent to the signal peptides of soluble secretory proteins that target them to the ER (31). Membrane proteins with a cleavable N-terminal signal peptide (e.g. VSV G protein; 32) contain at least one additional hydrophobic region which serves to anchor them in the membrane. This region is called a stop–transfer sequence because it prevents translocation of the downstream nascent chain (33, 34). The N-terminus of membrane proteins with a cleavable signal peptide is always located in the ER lumen (type I membrane protein, Fig. 2). In other membrane proteins, such as the transferrin receptor and glycophorin C (35, 36), signal peptide function and membrane anchor function are combined in a single hydrophobic region—the signal–anchor sequence (37). Signal–anchors can be located anywhere within the protein. There are examples for both signal–anchor proteins with an ER lumenal N-terminus (type I, glycophorin C; 36; Fig. 2) and signal–anchor proteins with an ER lumenal C-terminus (type II, transferrin receptor; 35; Fig. 2).

Exchanging the hydrophobic region of signal–anchor sequences for other transmembrane segments does not affect the topology of the resulting protein, regard-

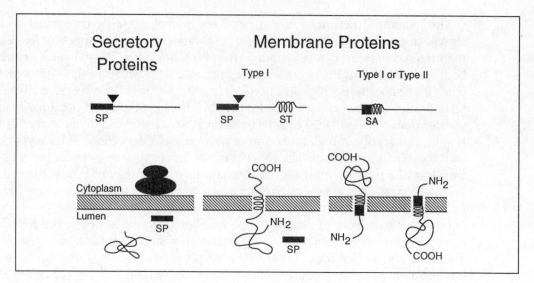

Fig. 2 Topology of secretory and membrane proteins in the ER. Top: linear diagrams of proteins destined for the ER (N-terminus left, C-terminus right). These proteins contain the following signals: SP, signal peptide; ST, stop–transfer sequence; SA, signal–anchor sequence. The arrowhead indicates the signal peptide cleavage site. Bottom: the topology of secretory and transmembrane proteins after translocation or membrane insertion.

less of the orientation these segments had in their original protein (38, 39). The conclusion from these experiments is that the hydrophilic flanking sequences rather than the hydrophobic portion of a signal–anchor sequence determine its orientation in the membrane. Flanking sequences on the cytosolic side of transmembrane regions frequently contain positively charged amino acids, which led von Heijne to propose the 'positive inside rule' (40). Hartmann *et al.* (41) performed a statistical analysis of the flanking regions of 91 eukaryotic signal–anchor proteins. The authors found that transmembrane orientation correlates closely with the charge difference between the 15 amino acids flanking the hydrophobic region on either side. A negative charge difference (C–N < 0) leads to a type II membrane orientation (N-terminus cytoplasmic), whereas a positive charge difference (C–N > 0) is found in type I proteins (N-terminus ER lumenal). A large body of experimental work supports the validity of these rules (18, 42, 43), although there are notable exceptions (44). In addition Beltzer *et al.* (42) found that the transmembrane orientation of a signal–anchor sequence is modulated by the domains to be translocated. If the polypeptide segment to be transported is not readily translocated, the signal–anchor may assume an orientation opposite that predicted from the flanking charges (42).

Both signal peptides and signal–anchor sequences can initiate translocation of a nascent chain across the ER membrane. Their functional difference is thought to result from the absence of an accessible signal peptidase cleavage site in the latter (37, 45). Sakaguchi *et al.* (18) showed that the functions of signal and signal–anchor sequences depend on the balance of the length of the hydrophobic segment and the charge at the N-terminus: long hydrophobic segments (12–15 leucine residues) function as signal peptides when positive charges are present at the N-terminus and as signal–anchors when the N-terminal charge is negative. Shorter hydrophobic segments (7–10 leucine residues) function as cleavable signal peptides regardless of the N-terminal charge.

In 1980, Blobel (33) proposed that polytopic membrane proteins achieve their integration into the ER membrane by using successive signal–anchor and stop–transfer sequences. This notion is now supported by a substantial amount of experimental data. As signal peptides can function as stop–transfer sequences and vice versa (38, 39, 46) the major determinant of the orientation of each individual transmembrane segment is the number of transmembrane segments preceding it (47). However, the hydrophilic stretches between the membrane-spanning regions can also influence transmembrane orientation (47). This is a concern if fusion proteins are used to determine the topology of multi-spanning membrane proteins (e.g. ref. 48).

5. Components involved in translocation into the ER

An essential step in the investigation of protein translocation into the ER was the development of cell-free systems which would faithfully reproduce the translocation of secretory proteins across the ER membrane *in vitro*. These translocation systems consist of a membrane-free cell lysate which serves as a translation system, purified or *in vitro* synthesized mRNA, and ER-derived rough microsomes. The experiments using these systems led to the discovery of cleavable signal peptides on secretory proteins (2, 3).

5.1 Co-translational translocation

5.1.1 Signal recognition particle (SRP)

If mammalian microsomes are washed with high salt-containing buffers, they lose their translocation competence *in vitro* (49). Adding back the salt extract to the *in vitro* assay restores translocation (49). The translocation promoting factor was purified (50) and its functions analysed. It specifically interacts with nascent secretory proteins via their signal peptide (4) and was therefore named signal recognition protein (SRP). Binding of SRP to the signal peptide causes an arrest of protein translation (5). Upon contact of the SRP–nascent chain–ribosome complex with the ER membrane, translation resumes and co-translational translocation into the ER begins (6). SRP is the equivalent of the 'binding factor' postulated in the signal hypothesis (13), a molecule which mediates the selective targeting of nascent secretory proteins to the ER.

Walter and Blobel (51, 52) analysed SRP on a molecular level and showed that it is a complex consisting of a 7S RNA and six proteins of 9, 14, 19, 54, 68, and 72 kDa molecular weight (SRP9, SRP14, etc.). Accordingly, the meaning of the acronym was changed to 'signal recognition particle'. SRP9/14 and SRP68/72 bind to the

RNA as heterodimers (52). SRP19 also binds to the 7S RNA (53) and its presence is required to mediate the association of SRP54 with SRP, but SRP54 and SRP19 do not interact in the absence of 7S RNA (54).

The components of SRP can be separated from each other under non-denaturing conditions, and by themselves are not able to promote translocation into the ER (52). However, 7S RNA and SRP proteins can be reconstituted *in vitro* into a functional particle (52). Alkylation of specific proteins prior to reconstitution and experiments with micrococcal nuclease-digested SRP, which lacks part of the 7S RNA and the SRP9/14 heterodimer, allowed analysis of the function of specific components. The SRP9/14 heterodimer was found to be responsible for the SRP-mediated translation arrest (54, 55). Alkylation or the absence of SRP9/14 from the particle also leads to a reduction of translocation efficiency, as nascent polypeptide chains are only translocation-competent up to a certain critical size in mammalian cells (55, 56). SRP–nascent chain–ribosome complexes containing alkylated SRP68/72 are unable to bind to microsomal membranes, indicating a role for this complex in targeting (54).

The interaction of SRP54 with secretory signal peptides has been shown directly by cross-linking to signals containing lysine residues modified by a UV-activatable cross-linker (57, 58). SRP54 consists of two domains:

- a 22 kDa C-terminal methionine-rich M domain
- a 33 kDa N-terminal G domain containing a GTP-binding site (59, 60).

The M domain is responsible for the association of SRP54 with SRP (61, 62) and is both necessary and sufficient to bind signal peptides (62–64). However, Zopf *et al.* (65) showed, using truncation experiments, that the presence of the G domain enhances the affinity of the M domain for signal peptides. Further evidence for communication between the two domains of SRP54 is derived from alkylation experiments in which cysteine residues of the protein were modified. SRP containing alkylated SRP54 can no longer bind to signal peptides (54), but the M domain itself does not contain any cysteine. Experiments using SRP containing a truncated SRP54 (65), showed that the G domain also plays a role in targeting to the ER membrane.

5.1.2 SRP receptor

The signal hypothesis postulates specific interaction of the signal peptide binding factor with the ER membrane as a prerequisite for translocation (13). Accordingly, the SRP receptor (SR) was identified by two different biochemical approaches. Meyer and Dobberstein (66) found that partial proteolysis of microsomal membranes abolishes their activity in a translocation assay. Translocation can be reconstituted if the supernatant from the protease treatment is added back to the membranes (67). The authors purified a 52 kDa polypeptide from the soluble fraction which they identified as a proteolytic fragment of a 69 kDa membrane-associated protein (docking protein; 68).

Gilmore and co-workers used affinity chromatography on immobilized SRP to purify the same protein from solubilized mammalian microsomes (SR; 69, 70). SR was later found to be a complex of the 69 kDa α-subunit (SRα) and a 30 kDa ER membrane protein (SRβ; 71). The C-terminal 33 kDa of SRα are homologous to the SRP54 G domain and contain a GTP-binding site (59, 60, 72). SRβ also binds GTP (73). Interaction of the SRP–nascent chain–ribosome complex with SR in the ER membrane causes release of SRP from the signal peptide and thus alleviates the SRP-imposed translation arrest (74). Concomitantly, translocation of the secretory protein across the ER membrane is initiated. Recently a large amount of experimental effort has been focused on clarifying the role of GTP in the process.

5.1.3 GTP and the targeting cycle

Three GTP-binding proteins are involved in the targeting of nascent secretory proteins to the ER membrane-SRP54, SRα, and SRβ (72, 73, 75). By themselves SRP and SR have very low intrinsic GTP hydrolysis rates, but interaction of the two complexes leads to a burst of GTP hydrolysis (73, 75). SRP54 has a low affinity for GTP and GDP and therefore in solution is most likely in equilibrium between the nucleotide-bound and the empty state (73). GTP is not necessary for signal peptide binding to SRP54 (64, 65), and SRP54 bound to a signal peptide is stabilized in the 'empty' state, as the presence of a signal peptide prevents binding of GTP to SRP54 (73). However, GTP or the non-hydrolysable analogue Gpp(NH)p is required for the SR-dependent displacement of SRP from the signal peptide (74) and GTP hydrolysis is necessary for the dissociation of SRP from SR (76). Furthermore, the stable interaction of SRP and SR requires a functional GTP binding site in SRα (72). Miller *et al.* (73) recently showed that interaction of SRP with SR increases the affinity of SRP54 for GTP and activates its GTPase activity.

Based on their data and the results described above, Miller *et al.* proposed the following model for SRP-mediated targeting to the ER membrane (Fig. 3). SRP binds the signal peptide with SRP54 in the empty conformation (73). Upon interaction with SR in the ER membrane, the conformation of SRP54 changes and the affinity of SRP54 for GTP increases resulting in a concomitant decrease in the affinity for the signal peptide and its release. This step requires the presence of a component other than SRP54/7S RNA or SR which has not yet been identified, but potentially could be the SRP68/72 complex or an additional ER membrane protein (73). After signal peptide release, the SRP–SR complex dissociates from the translocation complex in the ER membrane. SR induces the GTPase activity of SRP54 and the conformational change in SRP54 during hydrolysis leads to its release from SR. Due to the low affinity of SRP for the nucleotide, SRP and GDP dissociate, and 'empty' SRP is available for a new round of targeting.

Point mutations in the GTP-binding site of SRα abolish its stable interaction with SRP. Therefore, SRα is most likely to be in the nucleotide-bound state during the initial steps of the targeting cycle (72). It is unclear at which point SRα hydrolyses GTP. The role of the GTP-binding site in SRβ is also unknown, but an attractive hypothesis by Rapiejko and Gilmore proposes that it has a regulatory function in the association of SR with components of the translocation apparatus (72).

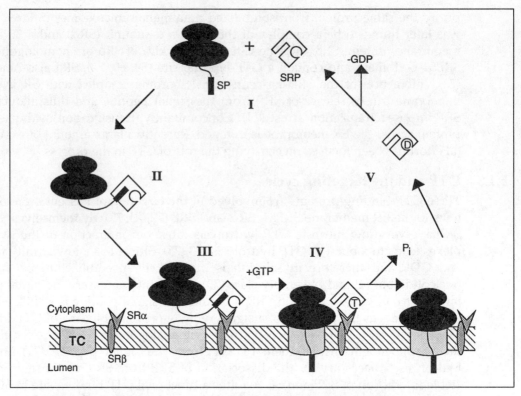

Fig. 3 GTP-dependent SRP targeting cycle in co-translational translocation. Cytoplasmic events during targeting of a nascent secretory protein to the translocation complex (TC) in the ER membrane are shown. In step I, the 'empty' (i.e. guanine nucleotide free) SRP binds to the signal peptide (SP) of a nascent secretory protein as it emerges from the ribosome. The resulting SRP–nascent chain–ribosome complex (II) is targeted to the ER membrane (III) where SRP interacts with the α-subunit of the SRP receptor (SRα). This interaction leads to an increase of the affinity of SRP54 for GTP and a concomitant decrease of its affinity for the SP. As SRP binds to the GTP (T in the diagram) the nascent chain is released and begins its transit across the ER membrane (IV). SR stimulates GTP hydrolysis by SRP54 to GDP (D in the diagram) which causes the dissociation of SRP from SR. GDP-bound (V) and 'empty' SRP are in equilibrium in solution (modelled after ref. 73).

5.2 Post-translational translocation

In mammalian cells SRP-mediated targeting of secretory proteins to the ER membrane ensures a tight coupling between translation and translocation (8). Protein translocation across the ER membrane of the yeast *Saccharomyces cerevisiae* can take place co- and post-translationally (77, 78). Using genetic and biochemical means, two members of the heat-shock protein family were identified as soluble factors necessary for post-translational translocation into the ER and into mitochondria, Ssa1p and Ssa2p (79, 80). These proteins probably act as molecular chaperones and prevent newly synthesized secretory proteins from folding into a compact conformation, thus keeping them in a translocation competent state. In addition to Ssa1p and Ssa2p, at least one more soluble factor is required to achieve post-translational translocation in yeast (81). At present, it is unclear whether post-translational translocation into the yeast ER also requires SRP; mammalian SRP binds to nascent chains of increasing length with progressively lower affinity which makes it seem unlikely that SRP is involved in post-translational targeting (82). In addition, post-translational translocation into the yeast ER *in vitro* does not require GTP (83). However, yeast cells depleted of Sec65p (the yeast homologue of SRP19), SRP54, or Srp101p (the yeast homologue of SRα) accumulate cytoplasmic precursors of both co-translationally and post-translationally translocated proteins (84–86).

5.3 SRP-independent translocation into the ER

In mammalian cells, some precursor proteins can be targeted to and translocated into the ER in an SRP-independent fashion (87–90). If the nascent chain is sufficiently short, for example prepromelittin, the SRP-imposed translation arrest is not necessary to keep the signal peptide exposed to the translocation apparatus in the ER membrane, and targeting can occur in the absence of SRP (88). Cytochrome b_5 does not contain a hydrophobic signal peptide but inserts into the ER via a hydrophobic C-terminal sequence in an SRP-independent fashion (87). Likewise, the SRα subunit does not contain a typical signal peptide and associates with the ER membrane independently of SRP (89, 90).

In yeast cells, depletion of SRP54 or SRα is not lethal, which indicates the existence of a bypass for SRP-dependent targeting to the ER (84, 86). Prolonged growth of yeast in the absence of SRα or SRP54 leads to a recovery from the secretion defects, which may indicate an upregulation of the expression or activity of the components involved in the bypass pathway (84, 86). The characterization of this process awaits further experimentation.

5.4 The translocation complex

5.4.1 The Sec61p complex

The gene for Sec61p was identified by a genetic selection for ER protein-translocation defects in yeast (91–93). Cross-linking studies with translocating secretory proteins demonstrated the close proximity of the nascent polypeptide to Sec61p in both yeast (94, 95) and mammalian ER (96, 97). Sec61p is a polytopic membrane protein with eight (yeast) or ten (mammalian) predicted transmembrane domains and cytosolically exposed N- and C-termini (93, 96). Sec61p shows homology to SecYp, a major component of the protein translocation apparatus in the *E. coli* plasma membrane (98), and is an essential protein in yeast (93). Mammalian Sec61p is complexed with two small proteins of 14 kDa and 8 kDa, which are named Sec61β and Sec61γ, respectively (22). A yeast homologue of Sec61γ has been identified by a genetic screen for suppressors of *sec61* (99). The corresponding protein, Sss1p, is essential for protein translocation into the yeast ER (99).

5.4.2 The Sec62p/Sec63p complex

Biochemical and genetic studies in yeast demonstrated the involvement of two other ER membrane proteins, Sec62p and Sec63p, in the ER translocation of secretory proteins (92–95, 100–103). Sec62p and Sec63p have no known mammalian homologues. Sec62p is a 30 kDa protein with two transmembrane domains, and both termini exposed to the cytoplasm (100). Sec63p is a 73 kDa protein which spans the bilayer three times (104). Part of the ER lumenal domain of Sec63p is homologous to DnaJ, an *E. coli* protein that interacts with the chaperone DnaK (104). Feldheim and colleagues (104) proposed that this domain is responsible for the association of the ER lumenal chaperone BiP (heavy-chain binding protein, Kar2p) with the translocation complex in yeast. Consistent with this idea, a point mutation in the DnaJ domain of Sec63p, *sec63-1*, causes a translocation defect *in vivo* and *in vitro* and a concomitant dissociation of BiP from Sec63p (102). Suppressors of *sec63-1* map in the gene for BiP, *KAR2*, further supporting the significance of Sec63p/BiP interaction for protein translocation (105).

Cross-linking and co-immunoprecipitation experiments with yeast ER membranes revealed that Sec61p, Sec62p and Sec63p form a multisubunit complex with an integral membrane glycoprotein of 31.5 kDa and a membrane-associated unglycosylated protein of 23 kDa (Fig. 4; 103). The genes for both proteins were cloned by reverse genetics and named *SEC66* and *SEC67* (102, 106; Feldheim and Schekman, in preparation). Complementation analysis revealed that alleles of these genes had been identified previously by a genetic screen for mutants defective in membrane protein integration into the ER (107). *SEC71* encodes the 31.5 kDa subunit and *SEC72* encodes the 23 kDa polypeptide. Null mutants of sec71 are viable at 30 °C, but not at 37 °C, and accumulate a variety of secretory protein precursors in the cytosol at the elevated temperature (106, 107). The absence of Sec71p also prevents a stable association of Sec72p with the translocation complex (106). A null mutation in *sec72* has no effect on cell growth, but leads to the cytosolic accumulation of a subset of secretory protein precursors (Feldheim and Schekman, in preparation). The *sec72*-mediated translocation defect is specific for the signal peptide rather than the mature region of the secretory protein (Feldheim and Schekman, in preparation).

The exact function of the Sec62p/63p complex is unclear. In yeast it has been demonstrated that action of Sec62p and Sec63p probably precedes that of Sec61p in translocation, as mutations in either gene prevent the association of the translocating chain with Sec61p (92). Based on this observation and the signal peptide specificity of the *sec72* defect, Feldheim and Schekman proposed a role for the Sec62/63p complex in signal peptide recognition (140).

5.4.3 Translocating chain associating membrane protein (TRAM)

In mammalian cells, crosslinking studies with nascent secretory protein chains at the ER membrane led to the discovery of another protein associated with translocation, translocating chain associating membrane (TRAM) protein (Fig. 4; 108–111). TRAM is an abundant 34 kDa polytopic protein that is predicted to span the ER membrane eight times with cytosolically exposed N- and C-termini (111). TRAM can be cross-linked only to short nascent chains, and is therefore thought to be involved at a step early in translocation where it interacts with amino acids preceding the hydrophobic core of the signal peptide, in contrast to Sec61p which contacts mostly the core and residues succeeding it (111, 112). Rapoport (113) proposed that the concerted action of Sec61p and TRAM leads to a specific orientation of the signal peptide in the membrane. Because TRAM is required for the translocation of some, but not all, nascent chains into the ER (22) it may function as a signal peptide receptor for proteins with weak signal peptides and enhance their chance for contact with Sec61p, similar to the role proposed for members of the Sec62p/Sec63p complex in yeast.

5.5 Definition of a minimal translocation apparatus

Secretory and membrane proteins can translocate into proteoliposomes reconstituted from ER membrane components (101, 114–116). Using this technique and purified components, Görlich and Rapoport recently showed that the only ER proteins required for the translocation of some mammalian secretory proteins are SRP receptor and the Sec61p complex (22; Fig. 4). Other proteins required the additional presence of TRAM which also stimulated translocation in cases in which it was not essential (22). Insertion of transmembrane proteins in either type I or type II orientation could also be achieved with this minimal translocation apparatus (22).

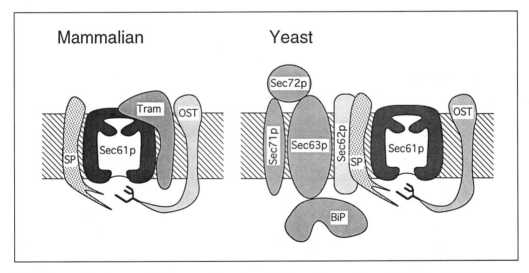

Fig. 4 The translocation complex in mammals and yeast. The protein complexes in the ER membrane involved in secretory protein translocation and co-translocational protein processing are shown. The mammalian and yeast translocation complexes have three proteins in common: Sec 61p, which constitutes the putative translocation pore, signal peptidase (SP), and oligosaccharyltransferase (OST). Translocating chain associating protein (Tram) is unique to the mammalian apparatus while Sec62p, Sec63p, Sec71p, Sec72p, and BiP have been found to be involved, so far, only in yeast translocation.

In contrast to these studies with mammalian ER, in yeast the soluble ER lumenal chaperone BiP is required for translocation of secretory proteins (95, 101, 102, 117; Fig. 4). The presence of functional BiP in the ER lumen is essential for both co- and post-translational translocation in yeast *in vivo* and *in vitro* (101, 102, 117, 141). It is possible that yeast and mammals differ with respect to their BiP requirement for translocation because the BiP association with Sec63p is essential for the integrity of the translocation apparatus (102). This function of BiP may have been lost in the course of evolution as no Sec63p homologue has been found in mammals to date. Reconstitution of translocation into proteoliposomes containing purified yeast components has not yet been achieved, but Brodsky and Schekman (102) showed that a complex containing Sec63p, BiP, Sec71p, and Sec72p is essential for transport into yeast-derived proteoliposomes (Fig. 4).

Savitz and Meyer demonstrated that depletion of a 180 kDa protein, previously characterized as a ribosome receptor, from mammalian ER proteoliposomes leads to a defect in protein translocation (118). However, Görlich and Rapoport (22) did not require the presence of this protein to reconstitute translocation. Instead, Görlich *et al.* demonstrated a ribosome binding capacity for Sec61p (96). Their results fulfil one of the predictions of the signal hypothesis, namely that the components constituting the protein translocation channel should have an affinity for the ribosome (3, 13). It is possible that the ER contains more than one ribosome receptor, but the controversy concerning the requirement for p180 in protein translocation has yet to be resolved.

6 The nature of the translocation site

The signal hypothesis postulates an aqueous channel in the ER membrane through which secretory proteins enter the ER (3, 13). Early indications that this notion is correct were derived from experiments by Gilmore and Blobel (119) who demonstrated that aqueous perturbants can extract short nascent secretory chains from the ER membrane. In addition, Simon and Blobel recently conducted electrophysiological experiments which led to the discovery of channels in the ER membrane that are large enough to allow the passage of proteins and which close upon the removal of ribosomes from the cytosolic face of the ER (120). The authors found similar channels in the plasma membrane of *E. coli* and showed that these channels open in the presence of a synthetic signal peptide (121).

A clear and direct demonstration of the aqueous environment of the nascent secretory chain early in translocation was recently provided by Crowley and colleagues (122). The authors incorporated fluorescent probes into the signal peptide of nascent secretory proteins. Fluorescence lifetime measurements of ribosome–nascent chain–membrane complexes revealed that the probes are in an aqueous milieu rather than buried in the non-polar core of the membrane. Because these membrane-bound probes are not susceptible to collisional quenching by iodide ions, the authors concluded that the space containing the signal peptide is sealed off from the cytoplasm by a tight ribosome–membrane junction. The fact that the nascent chain inside the ribosome also was not accessible to iodide was taken as an indication that the secretory polypeptide passes through a tunnel inside the large ribosomal subunit rather than through a groove on the surface of the ribosome.

The physical association with both the signal peptide and the mature region of translocating proteins and its absolute requirement for translocation make Sec61p a likely component of the actual translocating channel, but a pore-forming capacity for Sec61p remains to be experimentally proven. Likewise, it is unclear at present whether other proteins such as TRAM or the Sec62p/Sec63p complex contribute to the channel itself and perhaps modulate its properties, or whether they are simply located in close proximity (Fig. 4). Finally, it has been shown that the insertion of transmembrane proteins into the ER does not require protein components other than those essential for soluble secretory translocation (22). This implies that the translocation channel has a capacity to open laterally and release the signal–anchor or stop–transfer sequence of a membrane protein into the membrane itself.

7. Enzymes at the translocation site

7.1 Signal peptidase

The signal peptide of a secretory protein is cleaved off in the ER lumen (2, 3). The enzyme responsible has been purified from dog pancreas as a complex of five polypeptides of 12, 18, 21, 22/23, and 25 kDa (123). These proteins form a tight, protease-resistant signal peptidase complex in the ER membrane (124) that is stoichiometric with bound ribosomes, suggesting that this complex is part of each translocation site (Fig. 4). None of the subunits has signal peptidase activity on its own which made the assignment of enzymatic activity difficult (125). However, the 18 and 21 kDa polypeptides show homology to Sec11p (125), an essential yeast protein of 18 kDa (126) which co-purifies with signal peptidase activity from yeast and with three other proteins of 13, 20, and 25 kDa (127). Regions of homology exist between bacterial signal peptidases, which are active as a single polypeptide chain, and the Sec11p homologous group (128, 129). This finding implies that the catalytically active subunits of the eukaryotic complexes are Sec11p and the 18 and 21 kDa subunits of mammalian signal peptidase (128). Sequence similarities and results from site-directed mutagenesis experiments with the homologous *E. coli* enzyme (130) suggest that signal peptidases may be a new class of serine protease that does not require a histidine for enzymatic activity (16).

7.2 Oligosaccharyltransferase

Asparagine-linked glycosylation is a highly conserved protein modification in eukaryotes that occurs in the ER lumen (9). The central step is the co- or post-translocational transfer of dolichol-linked core oligosaccharides to the consensus acceptor sequence Asn–X–Ser/Thr catalysed by oligosaccharyltransferase (OST; 9). In yeast, two essential proteins are necessary for this activity (131–133); Wbp1p and

Swp1p. Wbp1p is a 45 kDa type I ER membrane protein with a C-terminal ER retention sequence (131). The gene for Swp1p was found as an allele-specific suppressor of a *wbp1* mutant, indicating physical interaction between the two proteins (133). *SWP1* encodes a 30 kDa type I transmembrane protein that can be chemically cross-linked to Wbp1p (133). Other proteins are found in the Wbp1p–Swp1p complex and overexpression of both *SWP1* and *WBP1* does not lead to increased OST activity, which implies that not all components of the enzymatically active complex have been identified yet (133).

The mammalian OST complex has been recently purified and found to contain three proteins; ribophorin I and II and a 48 kDa protein which is homologous to Wbp1p (134, 135). Ribophorins are abundant ER membrane glycoproteins which co-localize with ribosomes (136). Antibodies against the cytoplasmic tail of ribophorin I inhibit protein translocation into the ER (137), consistent with the notion that OST is located in close proximity to the translocation apparatus (Fig. 4). The membrane-spanning segment of ribophorin I contains a sequence motif that has been proposed to be a recognition site for dolichol and may therefore be involved in substrate binding of the OST complex (134, 138). Swp1p contains similar sequences, but is otherwise not homologous to ribophorin I (132). No yeast homologues of ribophorins have been identified so far, and likewise no mammalian homologues of Swp1p have been found.

Nilsson and von Heijne (139) recently demonstrated that the glycosylation acceptor site of translocating nascent transmembrane proteins must be at a distance of 12–14 amino acids from the membrane in order to be glycosylated. The catalytic mechanism for transfer of the oligosaccharyl moiety from dolichol to the asparagine of the acceptor site remains to be elucidated.

8. Conclusions

Since the signal hypothesis was first postulated in 1971, substantial progress has been made in identifying components involved in protein translocation into the ER. A molecular description of the SRP-mediated targeting cycle is now available, translocation into proteoliposomes containing a minimal set of ER proteins has been achieved, the catalytically active polypeptide of signal peptidase has been identified, and the elusive oligosaccharyltransferase complex has been purified. The translocation channel has been characterized as an aqueous pore in the ER membrane. After two decades of groundwork, the elucidation of a fundamental biological mechanism, the translocation of secretory proteins across the ER membrane, is now within reach.

Acknowledgements

We thank J. Brodsky, D. Feldheim, S. Lyman, and R. Schekman for critical reading of the manuscript. KR was supported by a fellowship from the BASF/German Scholarship Foundation.

References

1. Palade, G. (1975) Intracellular aspects of protein synthesis. *Science*, **189**, 347.
2. Milstein, C., Brownlee, G. G., Harrison, T. M., and Mathews, M. B. (1972) A possible precursor of immunoglobulin light chains. *Nature New Biol.*, **239**, 117.
3. Blobel, G. and Dobberstein, B. (1975) Transfer of proteins across membranes. I. Presence of proteolytically processed and unprocessed nascent immunoglobulin light chains on membrane-bound ribosomes of murine myeloma. *J. Cell Biol.*, **67**, 835.
4. Walter, P., Ibrahimi, I., and Blobel, G. (1981) Translocation of proteins across the endoplasmic reticulum. I. Signal recognition protein (SRP) binds to *in vitro* assembled polysomes synthesizing secretory protein. *J. Cell Biol.*, **91**, 545.
5. Walter, P. and Blobel, G. (1981) Translocation of proteins across the endoplasmic reticulum. III. Signal recognition protein (SRP) causes signal sequence-dependent and site-specific arrest of chain elongation that is released by microsomal membranes. *J. Cell Biol.*, **91**, 557.
6. Walter, P. and Blobel, G. (1981) Translocation of proteins across the endoplasmic reticulum. II. Signal recognition protein (SRP) mediates the selective binding to microsomal membranes of in-vitro-assembled polysomes synthesizing secretory protein. *J. Cell Biol.*, **91**, 551.
7. Gilmore, R., Walter, P., and Blobel, G. (1982) Protein translocation across the endoplasmic reticulum. II. Isolation and characterization of the signal recognition particle receptor. *J. Cell Biol.*, **95**, 470.
8. Walter, P. and Lingappa, V. R. (1986) Mechanism of protein translocation across the endoplasmic reticulum. *Annu. Rev. Cell Biol.*, **2**, 499.
9. Kornfeld, R. and Kornfeld, S. (1985) Assembly of asparagine-linked oligosaccharides. *Annu. Rev. Biochem.*, **54**, 631.
10. Rose, J. K. and Doms, R. W. (1988) Regulation of protein export from the endoplasmic reticulum. *Annu. Rev. Cell Biol.*, **4**, 257.
11. Hurtley, S. M. and Helenius, A. (1989) Protein oligomerization in the endoplasmic reticulum. *Annu. Rev. Cell Biol.*, **5**, 277.
12. Redman, C. M. (1969) Biosynthesis of serum proteins and ferritin by free and membrane-bound ribosomes of rat liver. *J. Biol. Chem.*, **244**, 4308.
13. Blobel, G. and Sabatini, D. D. (1971) Ribosome-membrane association in eukaryotic cells. In *Biomembranes*. Manson, L. A. (ed.). Plenum Press, New York, Vol. 2, p. 193.
14. von Heijne, G. (1988) Transcending the impenetrable: how proteins come to terms with membranes. *Biochim. Biophys. Acta*, **947**, 307.
15. Notwehr, S. F. and Gordon, J. I. (1990) Targeting of proteins into the eukaryotic secretory pathway: signal peptide structure/function relationships. *BioEssays*, **12**, 479.
16. Dalbey, R. E. and von Heijne, G. (1992) Signal peptidases in prokaryotes and eukaryotes —a new protease family. *Trends Biochem. Sci.*, **17**, 474.
17. Notwehr, S. F. and Gordon, J. I. (1989) Eukaryotic signal peptide structure/function relationships: identification of conformational features which influence the site and efficiency of co-translational proteolytic processing by site-directed mutagenesis of human pre(Δpro)apo A-II. *J. Biol. Chem.*, **264**, 3979.
18. Sakaguchi, M., Tomiyoshi, R., Kuroiwa, T., Mihara, K., and Omura, T. (1992) Functions of signal and signal–anchor sequences are determined by the balance between the hydrophobic segment and the N-terminal charge. *Proc. Natl Acad. Sci. USA*, **89**, 16.
19. Meek, R. L., Walsh, K. A., and Palmiter, R. D. (1982) The signal sequence of ovalbumin is located near the amino-terminus. *J. Biol. Chem.*, **257**, 12245.

20. Ye, R. D., Wun, T. C., and Sadler, J. E. (1988) Mammalian protein secretion without signal peptide removal. Biosynthesis of plasminogen-activator inhibitor 2 in U-937 cells. *J. Biol. Chem.*, **263**, 4869.

21. Talmadge, K., Stahl, S., and Gilbert, W. (1980) Eukaryotic signal sequence transports insulin antigen in *Escherichia coli*. *Proc. Natl Acad. Sci. USA*, **77**, 3369.

22. Görlich, D. and Rapoport, T. A. (1993) Protein translocation into proteoliposomes reconstituted from purified components of the endoplasmic reticulum membrane. *Cell*, **75**, 615.

23. von Heijne, G. and Abrahmsén, L. (1989) Species-specific variation in signal peptide design. Implications for protein secretion from foreign hosts. *FEBS Lett.*, **244**, 439.

24. Ngsee, J. K. and Smith, M. (1990) Changes in a mammalian signal sequence required for efficient protein secretion by yeasts. *Gene*, **86**, 251.

25. Bird, P., Gething, M.-J., and Sambrook, J. (1990) The functional efficiency of mammalian signal peptide is directly related to its hydrophobicity. *J. Biol. Chem.*, **265**, 8420.

26. Johansson, M., Nilsson, I., and von Heijne, G. (1993) Positively charged amino acids placed next to a signal sequence block protein translocation more efficiently in *Escherichia coli* than in mammalian microsomes. *Mol. Gen. Genet.*, **239**, 251.

27. Emr, S. D. and Silhavy, T. J. (1983) Importance of secondary structure in the signal sequence for protein secretion. *Proc. Natl Acad. Sci. USA*, **80**, 4599.

28. Kendall, D. A., Bock, S. C., and Kaiser, E. T. (1986) Idealization of the hydrophobic segment of the alkaline phosphatase signal peptide. *Nature*, **321**, 706.

29. Briggs, M. S. and Gierasch, L. M. (1984) Exploring the conformational roles of signal sequences: synthesis and conformational analysis of lambda receptor protein wild-type and mutant signal peptides. *Biochemistry*, **23**, 3111.

30. Caulfield, M. P., Park, K., Rosenblatt, M., and Fasman, G. D. (1991) Correlation of secondary structure with biological activity for a leader peptide: circular dichroism derived structure and *in vitro* biological activities of preproparathyroid hormone peptide and its analogs. *Arch. Biochem. Biophys.*, **289**, 208.

31. High, S. and Dobberstein, B. (1992) Mechanisms that determine transmembrane disposition of proteins. *Curr. Opin. Cell Biol.*, **4**, 581.

32. Lingappa, V. R., Katz, F. N., Lodish, H. F., and Blobel, G. (1978) A signal sequence for the insertion of a transmembrane glycoprotein. Similarities to the signals of secretory proteins in primary structure and function. *J. Biol. Chem.*, **253**, 8667.

33. Blobel, G. (1980) Intracellular protein topogenesis. *Proc. Natl Acad. Sci. USA*, **77**, 1496.

34. Kuroiwa, T., Sakaguchi, M., Mihara, K., and Omura, T. (1991) Systematic analysis of stop–transfer sequence for microsomal membrane. *J. Biol. Chem.*, **266**, 9251.

35. Zerial, M., Melancon, P., Schneider, C., and Garoff, H. (1986) The transmembrane segment of the human transferrin receptor functions as a signal peptide. *EMBO J.*, **5**, 1543.

36. High, S. and Tanner, J. A. (1987) Human erythrocyte membrane sialoglycoprotein β. The cDNA sequence suggests the absence of a cleaved N-terminal signal sequence. *Biochem. J.*, **243**, 277.

37. Lipp, J. and Dobberstein, B. (1986) The membrane spanning sequence of invariant chain (Iγ) contains a potentially cleavable signal sequence. *Cell*, **4**, 1103.

38. Audigier, Y., Friedlander, M., and Blobel, G. (1987) Multiple topogenic sequences in bovine opsin. *Proc. Natl Acad. Sci. USA*, **84**, 5783.

39. Zerial, M., Huylebroeck, D., and Garoff, H. (1987) Foreign transmembrane peptides replacing the internal signal sequence of transferrin receptor allow its translocation and membrane binding. *Cell*, **48**, 147.

40. von Heijne, G. (1986) The distribution of positively charged residues in bacterial inner membrane proteins correlates with the transmembrane topology. *EMBO J.*, **5**, 3021.

41. Hartmann, E., Rapoport, T. A., and Lodish, H. F. (1989) Predicting the orientation of eukaryotic membrane-spanning proteins. *Proc. Natl Acad. Sci. USA*, **86**, 5786.

42. Beltzer, J. P., Fiedler, K., Fuhrer, C., Geffen, I., Handschin, C., Wessels, H. P., and Spiess, M. (1991) Charged residues are major determinants of the transmembrane orientation of signal–anchor sequences. *J. Biol. Chem.*, **266**, 973.

43. Nilsson, I. and von Heijne, G. (1990) Fine-tuning the topology of a polytopic membrane protein: role of positively and negatively charged amino acids. *Cell*, **62**, 1135.

44. Andrews, D. W., Young, J. C., Mirels, L. F., and Csarnota, G. J. (1992) The role of the N region in signal sequence and signal–anchor function. *J. Biol. Chem.*, **267**, 7761.

45. Shaw, A. S., Rottier, P. J. M., and Rose, J. K. (1988) Evidence for the loop model of signal-sequence insertion into the endoplasmic reticulum. *Proc. Natl Acad. Sci. USA*, **85**, 7592.

46. Haeuptle, M.-T., Flint, N., Gough, N. M., and Dobberstein, B. (1989) A tripartite structure of the signals that determine protein insertion into the endoplasmic reticulum. *J. Cell Biol.*, **108**, 1227.

47. Lipp, J., Flint, N., Haeuptle, M.-T., and Dobberstein, B. (1989) Structural requirements for membrane assembly of proteins spanning the membrane several times. *J. Biol. Chem.*, **109**, 2013.

48. Calamia, J. and Manoil, C. (1990) lac permease of *Escherichia coli*: topology and sequence elements promoting membrane insertion. *Proc. Natl Acad. Sci. USA*, **87**, 4937.

49. Warren, G. and Dobberstein, B. (1978) Protein transfer across microsomal membranes reassembled from separate membrane components. *Nature*, **273**, 569.

50. Walter, P. and Blobel, G. (1980) Purification of a membrane associated complex required for protein translocation across the endoplasmic reticulum. *Proc. Natl Acad. Sci. USA*, **77**, 7112.

51. Walter, P. and Blobel, G. (1982) Signal recognition particle contains a 7S RNA essential for protein translocation across the microsomal membrane. *J. Cell Biol.*, **100**, 1913.

52. Walter, P. and Blobel, G. (1983) Disassembly and reconstitution of signal recognition particle. *Cell*, **34**, 525.

53. Siegel, V. and Walter, P. (1988) Binding sites of the 19 kDa and 68/72 kDa signal recognition particle (SRP) proteins on SRP RNA as determined by protein–RNA-'footprinting'. *Proc. Natl Acad. Sci. USA*, **85**, 1801.

54. Siegel, V. and Walter, P. (1988) Each of the activities of signal recognition particle (SRP) is contained within a distinct domain: analysis of biochemical mutants of SRP. *Cell*, **52**, 39.

55. Siegel, V. and Walter, P. (1985) Elongation arrest is not a prerequisite for secretory protein translocation across the microsomal membrane. *J. Cell Biol.*, **100**, 1913.

56. Perara, E., Rothman, R. E., and Lingappa, V. R. (1986) Uncoupling translocation from translation: implications for transport of proteins across membranes. *Science*, **232**, 348.

57. Krieg, U. C., Walter, P., and Johnson, A. E. (1986) Photocrosslinking of signal sequence of nascent preprolactin to the 54–kilodalton polypeptide of the signal recognition particle. *Proc. Natl Acad. Sci. USA*, **83**, 8604.

58. Kurzchalia, T. V., Wiedmann, M., Girshovich, A. S., Bochkareva, E. S., Bielka, H., and Rapoport, T. A. (1986) The signal sequence of nascent preprolactin interacts with the 54 K polypeptide of signal recognition particle. *Nature*, **320**, 634.

59. Bernstein, H. D., Poritz, M. A., Strub, K., Hoben, P. J., Brenner, S., and Walter, P. (1989)

Model for signal-sequence recognition from amino acid sequence of 54 K subunit of signal recognition particle. *Nature*, **340,** 482.

60. Römisch, K., Webb, J., Herz, J., Prehn, S., Frank, R., Vingron, M., and Dobberstein, B. (1989) Homology of 54 K protein of signal recognition particle, docking protein and two *E. coli* proteins with putative GTP-binding domains. *Nature*, **340,** 478.

61. Römisch, K., Webb, J., Lingelbach, K., Gausepohl, H., and Dobberstein, B. (1990) The 54 kDa protein of signal recognition particle contains a methionine-rich RNA binding domain. *J. Cell Biol.*, **111,** 1793.

62. Zopf, D., Bernstein, H. D., Johnson, A. E., and Walter, P. (1990) The methionine-rich domain of the 54 kD protein subunit of the signal recognition particle contains an RNA binding site and can be crosslinked to signal sequences. *EMBO J.*, **9,** 4511.

63. High, S. and Dobberstein, B. (1991) The signal sequence interacts with the methionine-rich domain of the 54 kD protein of signal recognition particle. *J. Cell Biol.*, **113,** 229.

64. Lütcke, H., High, S., Römisch, K., Ashford, A. J., and Dobberstein, B. (1992) The methionine-rich domain of the 54 kDa subunit of signal recognition particle is sufficient for the interaction with signal sequence. *EMBO J.*, **11,** 1543.

65. Zopf, D., Bernstein, H. D., and Walter, P. (1993) GTPase domain of the 54 kD subunit of mammalian signal recognition particle is required for protein translocation but not for signal sequence binding. *J. Cell Biol.*, **120,** 1113.

66. Meyer, D. I. and Dobberstein, B. (1980) A membrane component essential for vectorial translocation of nascent proteins across the endoplasmic reticulum: Requirements for its extraction and reassociation with the membrane. *J. Cell Biol.*, **87,** 498.

67. Meyer, D. I. and Dobberstein, B. (1980) Identification and characterization of a membrane component essential for the translocation of nascent secretory proteins across the membrane of the endoplasmic reticulum. *J. Cell Biol.*, **87,** 503.

68. Meyer, D. I., Krause, E., and Dobberstein, B. (1982) Secretory protein translocation across membranes — the role of 'docking protein'. *Nature*, **297,** 647.

69. Gilmore, R., Blobel, G., and Walter, P. (1982) Protein translocation across the membrane of the endoplasmic reticulum. I. Detection of the membrane receptor for the signal recognition particle. *J. Cell Biol.*, **95,** 463.

70. Gilmore, R., Blobel, G., and Walter, P. (1982) Protein translocation across the endoplasmic reticulum. II. Isolation and characterization of the signal recognition particle receptor. *J. Cell Biol.*, **95,** 470.

71. Tajima, S., Lauffer, L., Rath, V. L., and Walter, P. (1986) The signal recognition particle receptor is a complex that contains two distinct polypeptide chains. *J. Cell Biol.*, **103,** 1167.

72. Rapiejko, P. J. and Gilmore, R. (1992) Protein translocation across the endoplasmic reticulum requires a functional GTP-binding site in the α subunit of the signal recognition particle receptor. *J. Cell Biol.*, **117,** 493.

73. Miller, J. D., Wilhelm, H., Gierasch, L., Gilmore, R., and Walter, P. (1993) GTP binding and hydrolysis by the signal recognition particle during initiation of protein translocation. *Nature*, **366,** 351.

74. Connolly, T. and Gilmore, R. (1989) The signal recognition particle receptor mediates the GTP-dependent displacement of SRP from the signal sequence of the nascent polypeptide. *Cell*, **57,** 599.

75. Connolly, T. and Gilmore, R. (1993) GTP hydrolysis by complexes of the signal recognition particle and the single recognition particle receptor. *J. Cell Biol.*, **123,** 799.

76. Connolly, T., Rapiejko, P., and Gilmore, R. (1991) Requirement of GTP hydrolysis for dissociation of the signal recognition particle from its receptor. *Science*, **252,** 1171.

77. Waters, M. G., Chirico, W. J., and Blobel, G. (1986) Protein translocation across the yeast microsomal membrane is stimulated by a soluble factor. *J. Cell Biol.*, **103**, 2629.

78. Hansen, W., Garcia, P. D., and Walter, P. (1986) *In vitro* protein translocation across the yeast endoplasmic reticulum: ATP-dependent posttranslational translocation of the pre-pro-alpha-factor. *Cell*, **45**, 397.

79. Chirico, W. J., Water, M. G., and Blobel, G. (1988) 70K heat shock related proteins stimulate protein translocation into microsomes. *Nature*, **332**, 805.

80. Deshaies, R. J., Koch, B. D., Werner-Washburne, M., Craig, E. A., and Schekman, R. (1988) A subfamily of stress proteins facilitates translocation of secretory and mitochondrial precursor polypeptides. *Nature*, **332**, 800.

81. Deshaies, R. J., Koch, B. D., and Schekman, R. (1988) The role of stress proteins in membrane biogenesis. *Trends Biochem. Sci.*, **13**, 384.

82. Wiedmann, M., Kurzchalia, T. V., Bielka, H., and Rapoport, T. A. (1987) Direct probing of the interaction between the signal sequence of nascent preprolactin and the signal recognition particle by specific cross-linking. *J. Cell Biol.*, **104**, 201.

83. Waters, M. G. and Blobel, G. (1986) Secretory protein translocation into a yeast cell-free system can occur posttranslationally and requires ATP hydrolysis. *J. Cell Biol.*, **102**, 1543.

84. Hann, B. C. and Walter, P. (1991) The signal recognition particle in *S. cerevisiae*. *Cell*, **67**, 131.

85. Stirling, C. J. and Hewitt, E. W. (1992) The *S. cerevisiae SEC65* gene encodes a component of yeast signal recognition particle with homology to human SRP19. *Nature*, **356**, 534.

86. Ogg, S. C., Poritz, M. A., and Walter, P. (1992) Signal recognition particle receptor is important for cell growth and protein secretion in *Saccharomyces cerevisiae*. *Mol. Biol. Cell*, **3**, 895.

87. Sabatini, D. D., Kreibich, G., Morimoto, T., and Adesnick, M. (1982) Mechanisms for the incorporation of proteins into membranes and organelles. *J. Cell Biol.*, **92**, 1.

88. Müller, G. and Zimmermann, R. (1987) Import of honeybee prepromelittin into the endoplasmic reticulum: structural basis for independence of SRP and docking protein. *EMBO J.*, **6**, 2099.

89. Hortsch, M. and Meyer, D. I. (1988) The human docking protein does not associate with the membrane of the rough endoplasmic reticulum via a signal or insertion sequence mediated mechanism. *Biochem. Biophys. Res. Commun.*, **150**, 111.

90. Andrews, D. W., Lauffer, L., Walter, P., and Lingappa, V. R. (1989) Evidence for a two-step mechanism in assembly of functional signal recognition particle receptor. *J. Cell Biol.*, **108**, 797.

91. Deshaies, R. J. and Schekman, R. (1987) A yeast mutant defective at an early stage in import of secretory protein precursors into the endoplasmic reticulum. *J. Cell Biol.*, **105**, 633.

92. Rothblatt, J. A., Deshaies, R. J., Sanders, S. L., Daum, G., and Schekman, R. (1989) Multiple genes are required for proper insertion of secretory proteins into the endoplasmic reticulum in yeast. *J. Cell Biol.*, **109**, 2641.

93. Stirling, C. S., Rothblatt, J., Hosobuchi, M., Deshaies, R., and Schekman, R. (1992) Protein translocation mutants defective in the insertion of integral membrane proteins into the endoplasmic reticulum. *Mol. Biol. Cell*, **3**, 129.

94. Müsch, A., Wiedmann, M., and Rapoport, T. A. (1992) Yeast Sec proteins interact with polypeptides traversing the endoplasmic reticulum membrane. *Cell*, **69**, 343.

95. Sanders, S. L., Whitfield, K. M., Vogel, J. P., Rose, M. D., and Schekman, R. (1992) Sec61p and BiP directly facilitate polypeptide translocation into the ER. *Cell*, **69**, 353.

96. Görlich, D., Prehn, S., Hartmann, E., Kalies, K.-U., and Rapoport, T. A. (1992) A mammalian homolog of Sec61p and SecYp is associated with ribosomes and nascent polypeptides during translocation. *Cell*, **71**, 489.

97. High, S., Andersen, S. S., Görlich, D., Hartmann, E., Prehn, S., Rapoport, T. A., and Dobberstein, B. (1993) Sec61p is adjacent to type I and type II signal–anchor proteins during their membrane insertion. *J. Cell Biol.*, **121**, 743.

98. Akiyama, Y. and Ito, K. (1987) Topology analysis of the SecY protein, and integral membrane protein involved in protein export in *E. coli*. *EMBO J.*, **6**, 3645.

99. Esnault, Y., Blondel, M.-O., Deshaies, R. J., Schekman, R., and Képes, F. (1993) The yeast *SSS1* genes is essential for secretory protein translocation and encodes a conserved protein of the endoplasmic reticulum. *EMBO J.*, **12**, 4083.

100. Deshaies, R. J. and Schekman, R. (1989) *SEC62* encodes a putative membrane protein required for protein translocation into the yeast endoplasmic reticulum. *J. Cell Biol.*, **109**, 2653.

101. Brodsky, J. L., Hamamoto, S., Feldheim, D., and Schekman, R. (1993) Reconstitution of protein translocation from solubilized yeast membranes reveals topologically distinct roles for BiP and cytosolic Hsc70. *J. Cell Biol.*, **120**, 95.

102. Brodsky, J. L. and Schekman, R. (1993) A Sec63–BiP complex is required for protein translocation in a reconstituted proteoliposome. *J. Cell Biol.*, **123**, 1355.

103. Deshaies, R. J., Sanders, S. L., Feldheim, D. A., and Schekman, R. (1991) Assembly of yeast Sec proteins involved in translocation into the endoplasmic reticulum into a membrane-bound multisubunit complex. *Nature*, **349**, 806.

104. Feldheim, D., Rothblatt, J., and Schekman, R. (1992) Topology and functional domains of Sec63p, an endoplasmic reticulum membrane protein required for secretory protein translocation. *J. Cell Biol.*, **12**, 3288.

105. Scidmore, M., Okamura, H. H., and Rose, M. D. (1993) Genetic interactions between *KAR2* and *SEC63*, encoding eukaryotic homologs of *DnaK* and *DnaJ* in the endoplasmic reticulum. *J. Cell Biol.*, **4**, 1145.

106. Feldheim, D., Yoshimura, K., Admon, A., and Schekman, R. (1993) Structural and functional characterization of Sec66p, a new subunit of the polypeptide translocation apparatus in the yeast endoplasmic reticulum. *Mol. Biol. Cell*, **4**, 931.

107. Green, N., Fang, H., and Walter, P. (1992) Mutants in three novel complementation groups inhibit membrane protein insertion into and soluble protein translocation across the endoplasmic reticulum membrane of *Saccharomyces cerevisiae*. *J. Cell Biol.*, **116**, 597.

108. Wiedmann, M., Kurzchalia, T. V., Hartmann, E., and Rapoport, T. A. (1987) A signal sequence receptor in the endoplasmic reticulum membrane. *Nature*, **328**, 830.

109. Krieg, U. C., Johnson, A. E., and Walter, P. (1989) Protein translocation across the endoplasmic reticulum membrane: identification by photo-crosslinking of a 39 kD integral membrane glycoprotein as part of a putative translocation tunnel. *J. Cell Biol.*, **109**, 2033.

110. Wiedmann, M., Görlich, D., Hartmann, E., Kurzchalia, T. V., and Rapoport, T. A. (1989) Photocrosslinking demonstrates proximity of a 34 kDa membrane protein to different portions of preprolactin during translocation through the endoplasmic reticulum. *FEBS Lett.*, **257**, 263.

111. Görlich, D., Hartmann, E., Prehn, S., and Rapoport, T. A. (1992) A protein of the endoplasmic reticulum involved early in polypeptide translocation. *Nature*, **357**, 47.

112. High, S., Martoglio, B., Görlich, D., Andersen, S. S. L., Ashford, A. J., Giner, A., Hartmann, E., Prehn, S., Rapoport, T. A., Dobberstein, B., and Brunner, J. (1993) Site-specific photocross-linking reveals that Sec61p and TRAM contact different regions of a membrane-inserted signal sequence. *J. Cell Biol.*, **268**, 26745.

113. Rapoport, T. A. (1992) Transport of proteins across the endoplasmic reticulum membrane. *Science*, **258**, 931.

114. Yu, Y., Zhang, Y., Sabatini, D. D., and Kreibich, G. (1989) Reconstitution of translocation competent membrane vesicles from detergent solubilized dog pancreas rough microsomes. *Proc. Natl Acad. Sci. USA*, **86**, 9931.

115. Nicchitta, C. V. and Blobel, G. (1990) Assembly of translocation-competent proteoliposomes from detergent-solubilized rough microsomes. *Cell*, **60**, 259.

116. Zimmermann, D. L. and Walter, P. (1990) Reconstitution of protein translocation activity from partially solubilized microsomal vesicles. *J. Biol. Chem.*, **265**, 4048.

117. Vogel, J., Misra, L. M., and Rose, M. D. (1990) Loss of BiP/GRP78 function blocks translocation of secretory proteins in yeast. *J. Cell Biol.*, **110**, 1885.

118. Savitz, A. J. and Meyer, D. (1993) 180 kD ribosome receptor is essential for both ribosome binding and protein translocation. *J. Cell Biol.*, **120**, 853.

119. Gilmore, R. and Blobel, G. (1985) Translocation of secretory proteins across the microsomal membrane occurs through an environment accessible to aqueous perturbants. *Cell*, **42**, 497.

120. Simon, S. M. and Blobel, G. (1991) A protein-conducting channel in the endoplasmic reticulum. *Cell*, **65**, 371.

121. Simon, S. M. and Blobel, G. (1992) Signal peptides open protein conducting channels in *E. coli. Cell*, **69**, 677.

122. Crowley, K. S., Reinhart, G. D., and Johnson, A. E. (1993) The signal sequence moves through a ribosomal tunnel into a noncytoplasmic aqueous environment at the ER membrane early in translocation. *Cell*, **73**, 1101.

123. Evans, E. A., Gilmore, R., and Blobel, G. (1986) Purification of microsomal signal peptidase as a complex. *Proc. Natl Acad. Sci. USA*, **83**, 581.

124. Shelness, G. S., Lin, L., and Nicchitta, C. V. (1993) Membrane topology and biogenesis of eukaryotic signal peptidase. *J. Cell Biol.*, **268**, 5201.

125. Shelness, G. S. and Blobel, G. (1990) Two subunits of the canine signal peptidase complex are homologous to yeast *SEC11* protein. *J. Biol. Chem.*, **265**, 9512.

126. Böhni, P. C., Deshaies, R. J., and Schekman, R. (1988) *SEC11* is required for signal peptide processing and yeast cell growth. *J. Cell Biol.*, **106**, 1035.

127. YaDeau, J. T., Klein, C., and Blobel, G. (1991) Yeast signal peptidase contains a glycoprotein and the Sec11 gene product. *Proc. Natl Acad. Sci. USA*, **88**, 517.

128. van Dijl, J. M., de Jong, A., Vehmaanperä, J., Venema, G., and Bron, S. (1992) Signal peptidase I of *Bacillus subtilis*: patterns of conserved amino acids in prokaryotic and eukaryotic type I signal peptidase. *EMBO J.*, **11**, 2819.

129. Nunnari, J., Fox, T. D., and Walter, P. (1993) A mitochondrial protease with two catalytic subunits of non-overlapping specificities. *Science*, **262**, 1997.

130. Sung, M. and Dalbey, R. (1992) Identification of a potential active site in the *Escherichia coli* leader peptidase. *J. Biol. Chem.*, **267**, 13154.

131. te Heesen, S., Rauhut, R., Aebersold, R., Abelson, J., Aebi, M., and Clark, M. W. (1991) An essential 45 kDa yeast transmembrane protein reacts with anti-nuclear pore antibodies: purification of the protein, immunolocalization and cloning of the gene. *Eur. J. Cell Biol.*, **56**, 8.

132. te Heesen, S., Janetzky, B., Lehle, L., and Aebi, M. (1992) The yeast *WBP1* gene is essential for oligosaccharyl transferase activity *in vivo* and *in vitro*. *EMBO J.*, **11**, 2071.

133. te Heesen, S., Knauer, R., Lehle, L., and Aebi, M. (1993) Yeast Wbp1p and Swp1p form a complex essential for oligosaccharyl transferase activity. *EMBO J.*, **12**, 279.

134. Kelleher, D. J., Kreibich, G., and Gilmore, R. (1992) Oligosaccharyltransferase activity is associated with a protein complex composed of ribophorins I and II and a 48 kD protein. *Cell*, **69**, 55.

135. Silberstein, S., Kelleher, D. J., and Gilmore, R. (1992) The 48-kDa subunit of the mammalian oligosaccharyltransferase complex is homologous to the essential yeast protein WBP1. *J. Biol Chem.*, **267**, 23658.

136. Marcantonio, E. E., Amar-Costesec, A., and Kreibich, G. (1984) Segregation of the polypeptide translocation apparatus to regions of the endoplasmic reticulum containing ribophorins and ribosomes. II. Rat liver microsomal subfractions contain equimolar amounts of ribophorins and ribosomes. *J. Cell Biol.*, **99**, 2254.

137. Yu, Y., Sabatini, D. D., and Kreibich, G. (1990) Antiribophorin antibodies inhibit the targeting to the ER membrane of ribosomes containing nascent secretory polypeptides. *J. Cell Biol.*, **111**, 1335.

138. Albright, C. F., Orlean, P., and Robbins, P. W. (1989) A 13-amino acid peptide in three yeast glycosyltransferases may be involved in dolichol recognition. *Proc. Natl Acad. Sci. USA*, **86**, 7366.

139. Nilsson, I. and von Heijne, G. (1993) Determination of the distance between the oligosaccharyltransferase active site and the endoplasmic reticulum membrane. *J. Biol. Chem.*, **268**, 5798.

140. Feldheim, D. and Schekman, R. (1994) Sec72p contributes to the selective recognition of signal peptides by the secretory polypeptide translocation complex. *J. Cell Biol.*, **126**, 935.

141. Brodsky, J. L., Goeckeler, J., and Schekman, R. (1995) BiP and Sec63p are required for both co- and posttranslational protein translocation into the yeast endoplasmic reticulum. *Proc. Natl Acad. Sci. USA*, **92**, 9643.

5 | ER–Golgi membrane traffic and protein targeting

CAROLYN E. MACHAMER

1. Introduction

In eukaryotic cells, the vast majority of proteins destined for secretion or insertion at the plasma membrane follow a common pathway through the cell (1). They are synthesized on ribosomes bound to the endoplasmic reticulum (ER) membrane, co-translationally translocated completely or partially into the ER lumen (see Chapter 4), and then transported via multiple budding and fusion steps through the Golgi complex to the cell surface. Many co- and post-translational modifications occur to newly synthesized proteins as they move through this exocytic pathway. In the ER, polypeptides must be folded properly: formation of disulfide bonds, core glycosylation, and oligomerization contribute to this process. In the Golgi complex, oligosaccharides are processed and proteolytic processing can occur. The enzymes involved in these modifications are resident proteins of the ER or Golgi complex, and must be retained in the appropriate compartment in the face of substantial membrane flow.

2. ER–Golgi and intra-Golgi transport

2.1 Vesicular transport

How are newly synthesized proteins moved from the ER through multiple membrane-bounded compartments to their final destination? How is the vectorial nature of transport and the precise order of steps ensured? Transport vesicles that bud from one compartment and fuse with the next are the carriers. The mechanisms and components involved are subjects of active investigation.

2.1.1 Topology is always conserved in multiple vesicular budding and fusion events

When a transport vesicle buds from the donor compartment, it removes a portion of the lipid bilayer with embedded membrane proteins as well as a sample of the lumenal contents from that compartment. Membrane topology is always conserved during vesicular transport (Fig. 1). Even though a newly synthesized secreted or

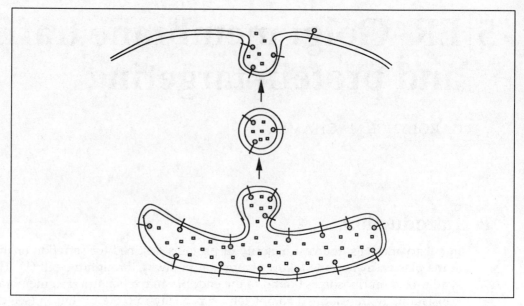

Fig. 1 Membrane topology is always conserved during vesicular transport. When a transport vesicle buds from the donor membrane, it removes a portion of the lipid bilayer and the contents of the lumen. After fusion with the acceptor compartment, membrane and lumenal contents are delivered to that compartment. If the acceptor compartment is the plasma membrane (as shown), the lumenal contents are secreted outside the cell, and the lumenal portions of membrane proteins now face the extracellular space.

membrane protein moves through multiple compartments on its way to the cell surface, it is only translocated across one membrane (that of the ER). Since the cytosolic side of the membrane-bounded compartment is the cytosolic side of the transport vesicle, the lumen of both the compartment and of the vesicle is the topological equivalent of outside the cell.

2.1.2 Mechanism of vesicular transport

Substantial progress has been made in recent years by Rothman and colleagues in defining the components involved in vesicle budding and fusion. Using *in vitro* reconstitution of intra-Golgi transport, biochemical dissection of the process has been achieved (reviewed in 2, 3). The assay follows transfer of a marker protein, vesicular stomatitis virus (VSV) G protein, between one set of Golgi membranes to another by measuring a specific glycosylation step. This purified system requires the addition of cytosol, which can be manipulated and fractionated easily. ER to Golgi traffic has been reconstituted in semi-intact cells (4), where cytosol can also be removed and manipulated, while the organelles remain in place. Using yeast as a model system, Schekman and colleagues have performed both genetic analysis of mutants defective in secretion and *in vitro* reconstitution of ER to Golgi transport (5). The results have frequently converged with the studies in mammalian cells, turning up homologues of mammalian proteins and highlighting the remarkable conservation of components required for intracellular vesicular transport.

Each stage of vesicular transport can be divided into three steps: budding, targeting, and fusion. Most of the proteins known to be involved in these steps were identified in the intra-Golgi transport assay. However, many of them appear to be used in ER to Golgi traffic and other stages of transport as well. Some of the differences between ER to Golgi and intra-Golgi transport will be discussed in Section 2.2.

Budding

Budding transport vesicles are coated with a complex of polypeptides (coat proteins or COPs), that exist as an oligomeric coatomer complex in the cytosol when not membrane-associated. Identification of COPs was facilitated by the finding that transport vesicles formed in the presence of GTPγS remained stably coated (6). The coat proteins include α-, β-, γ-, δ, ε, and ζ-COPs (7). β-COP has been studied most thoroughly, since it was previously identified as a peripheral Golgi membrane protein using a monoclonal antibody (8). β-COP has homology to β-adaptin (9, 10), a protein important for receptor recruitment into clathrin-coated vesicles. β'-COP, a newly described COP with a size close that of β-COP, has homology to β-subunits of heterotrimeric G proteins (11, 12). Recently, the idea that the coatomer exists as a preassembled complex in the cytosol that binds membranes *en bloc* was confirmed (13).

What regulates coatomer binding to membranes? Another peripheral Golgi protein, ADP-ribosylation factor (ARF), must bind Golgi membranes before coatomer can bind (reviewed in ref. 14). ARF is a GTP-binding protein that is *N*-myristoylated. It binds tightly to membranes in its GTP-bound form, and triggers coatomer binding (Fig. 2). Membrane proteins that function as the ARF receptor and that regulate nucleotide exchange (perhaps the same protein) are still unidentified. There is indirect evidence to suggest that heterotrimeric GTP-binding proteins may be involved at the stage where ARF binds membranes (15). GTP hydrolysis is required for the release of ARF and subsequent uncoating of the transport vesicle. Uncoating is required before fusion with the target membrane can occur. Since ARF is a weak GTPase, an ARF-specific GTPase activating protein (GAP) must be involved. The

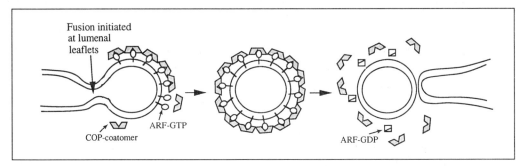

Fig. 2 Protein coats mediate vesicle budding. ARF-GTP binds to membranes and recruits coatomer to bind. Fusion initiated at the lumenal leaflets of the bilayer is required for release of the vesicle from the donor compartment. Hydrolysis of GTP by ARF (stimulated by an ARF-GAP, not shown) releases ARF-GDP from the vesicle, thereby releasing the COP coat.

GDP-bound form of ARF is soluble, and thus can be efficiently recycled for further cycles of membrane binding and release.

What drives membrane deformation during budding of vesicles? One idea is that assembly of the coat on to the bilayers is the driving force for budding (2), as hypothesized for the formation of clathrin-coated vesicles. Another possibility is that the membrane itself is responsible. Deformation of the bilayer could be induced by localized lipid asymmetry (16). The recent finding that ARF stimulates phospholipase D activity (17) raises the possibility that changes in phospholipid composition across the bilayer could be involved in membrane budding.

The release of the vesicle from the donor compartment requires fusion of the lumenal membrane leaflets at the site of vesicle fission (Fig. 2). Although some information is available regarding proteins required for fusion of transport vesicles with the acceptor membrane (where the *cytoplasmic* sides of the two membranes must fuse, see below), there is as yet no information on the components required to initiate the lumenal leaflet fusion required for release of the vesicle. In a recent study, vesicle formation and release could be separated into a two-stage reaction, where only palmitoyl CoA was required in the second stage to release the transport vesicles (18). One possibility is that a protein must be palmitoylated for fission of the vesicle from the donor compartment. Now that the steps of vesicle formation and fission can be dissociated *in vitro*, it is likely that any proteins required for the release of budded vesicles will soon be isolated.

Targeting

Once a transport vesicle has formed, it must find the correct acceptor membrane with which to fuse. This specificity is instrumental to the vectorial and precise nature of intracellular transport. Cytoskeletal elements are not required for vesicular transport, since the process can be reconstituted *in vitro* in the absence of microtubules and actin filament networks. However, it is likely that the cytoskeleton enhances the efficiency of transport in intact cells. The cytosol is a relatively dense matrix, and transport vesicles may not be capable of diffusing efficiently. Motor proteins capable of moving vesicles along microtubules or actin filaments have been identified.

Small GTP-binding proteins called Rab proteins are localized at different membranes throughout the exocytic and endocytic pathways, and are thought to be involved in vesicle targeting (reviewed in 19, 20). Rab proteins are homologous to Ras, and are isoprenylated (usually with geranylgeranyl) at the C-terminus, a modification required for binding to membranes. Rab proteins are probably targeted to specific membranes by a hypervariable region found near the C-terminus. The restricted localizations of these GTP-binding proteins suggested that they could be specific vesicular targeting molecules. Specificity could be achieved by coupling GTP hydrolysis to vesicle docking at the proper acceptor membrane, which would initiate the fusion event. However, recent analysis of chimeric Rab proteins in yeast suggests that Rab proteins are not the sole targeting components (21, 22). Identification of membrane proteins on both the vesicle and the target membrane called v-

SNAREs and t-SNAREs (23) raise the possibility that these molecules are the primary targeting components, with Rab proteins perhaps regulating their activity (see below).

Fusion

Once a transport vesicle is docked at the appropriate acceptor membrane, fusion initiated at the cytoplasmic leaflets of the two bilayers must occur. One protein required at this late step in vesicular transport is NSF (N-ethylmaleimide-sensitive factor), the first component to be identified when intra-Golgi vesicular transport was reconstituted (24). Uncoated vesicles accumulated when N-ethylmaleimide (NEM)-treated cytosol was used in the *in vitro* assay. The step at which NEM blocked transport was later than that blocked by GTPγS, and it was hypothesized that the NEM-sensitive factor was involved in membrane fusion. NSF is a tetramer that binds ATP, and ATP hydrolysis is required for NSF activity (25). The yeast *SEC18* gene encodes the yeast homologue of NSF (26, 27). Additional cytosolic proteins called SNAPs (soluble NSF attachment proteins) are required for NSF binding to membranes. Three SNAPs (α, β, γ) have been identified, although β-SNAP is a brain-specific isoform (28, 29).

Membrane proteins recognized by SNAPs (SNAP receptors) were purified from detergent-solubilized brain membranes by their ability to bind immobilized NSF–SNAP complexes (23). These SNAP receptors, or SNAREs, turned out to be previously characterized proteins with a role in synaptic vesicle-mediated secretion. The three proteins identified as SNAREs were known to exist on synaptic vesicles (VAMP/synaptobrevin), or on the presynaptic plasma membrane (syntaxin and SNAP-25). Each of these proteins is specifically proteolysed by different neurotoxins that rapidly block neurotransmitter secretion, implying that they play an instrumental role in the process (30). Since the predominant type of membrane traffic in brain tissue is synaptic transmission, it is not surprising that synaptic vesicle and presynaptic membrane proteins were isolated as SNAP receptors.

It was hypothesized that at each stage of vesicular transport, both the vesicle and the target membrane would have its own unique SNARE (v-SNAREs and t-SNAREs) required for targeting and fusion (23). This 'SNARE hypothesis' also states that fusion could occur only after the complete assembly of SNAPs and NSF on to the v-SNARE–t-SNARE complex (Fig. 3). In regulated secretory events, complete assembly of the fusion complex may occur without fusion. Additional components (acting as a 'fusion clamp') would block fusion until a signal for secretion caused the clamp to be removed. Synaptotagmin was suggested to play such a role in synaptic vesicle docking and fusion (31).

Yeast proteins homologous to syntaxin or VAMP have been identified, and are required for different steps of secretion (32–36). Existing evidence regarding the functions of these proteins is consistent with their potential roles as v-SNAREs or t-SNAREs (reviewed in ref. 37). All of these proteins (including VAMP and syntaxin) have a C-terminal hydrophobic sequence that anchors the protein in the membrane, with the bulk of the protein facing the cytosol. It is not known if proteins belonging

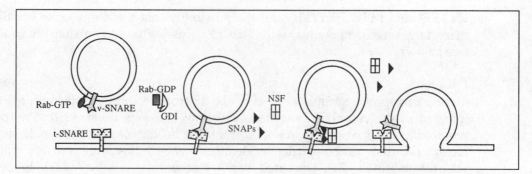

Fig. 3 SNAREs and Rab proteins are required for vesicle targeting. This model indicates that specific membrane proteins in the vesicle (v-SNARE) and in the target compartment (t-SNARE) interact to allow fusion via NSF and SNAPs that are subsequently recruited. GTP hydrolysis by a Rab protein in the vesicle may activate the v-SNARE; however, the Rab protein is released prior to or during vesicle docking. Once fusion has occurred, the v-SNARE must be recycled to the original compartment.

to this class actually span the membrane, or if they traverse the normal exocytic pathway after synthesis (38).

The actual mechanism of membrane fusion is unknown. Presumably, the v-SNARE: t-SNARE complex must be dissociated, and *in vitro*, the ATPase activity of NSF promotes this dissociation (31). By analogy with certain viral fusion proteins (39), it might be predicted that exposure of hydrophobic sequences in one of the assembled proteins (NSF?) allows perturbation of the lipid bilayer to initiate the fusion reaction.

One intriguing problem inherent in the SNARE hypothesis is how v-SNAREs are removed from the acceptor membrane where they reside after fusion has occurred (fig. 3), and recycled back to the donor compartment. Specificity would rapidly break down unless active v-SNAREs were returned to their original compartment before being incorporated into new transport vesicles. Perhaps the v-SNAREs are maintained in an inactive form most of the time. As mentioned above, hydrolysis of bound GTP by Rab proteins could activate a neighbouring v-SNARE in a transport vesicle prior to fusion of the vesicle with the target membrane. Rab proteins are dissociated from the membrane after GTP hydrolysis, and kept soluble by association with a protein that both masks the hydrophobic isoprenyl group at the C-terminus and blocks dissociation of GDP (GDP-dissociation inhibitor, or GDI; 40). However, the v-SNAREs are integral membrane proteins that would presumably have to be recycled by vesicular or some other form of membrane transport in the reverse (retrograde) direction (see Section 2.3).

2.1.3 Incorporation of cargo into transport vesicles

Is the cargo in transport vesicles selected or incorporated by default? Several types of experiment have suggested that transport by default might occur. In the Golgi complex, no concentration of an itinerant membrane protein (VSV G) was observed in transport vesicles and buds relative to its concentration in surrounding cisternae (41). In addition, a tripeptide marker assumed to lack transport signals was secreted from cells as fast or faster than any known secreted protein (42). The idea of transport by default with the bulk flow of lipid has gained wide support (43). However, qualitative observations suggested that several viral glycoproteins (44, 45) and albumin (46) are concentrated before arrival in the Golgi stacks. A recent study using quantitative immunoelectron microscopy showed that the VSV G protein is concentrated 5- to 10-fold as it exists the ER, but that no further concentration occurs as it moves through the Golgi stacks (47). Thus, signals may be required for efficient exit from the ER. For the VSV G protein, it has long been known that the cytoplasmic tail is required for efficient exit from the ER. Truncated proteins lacking part or all of the cytoplasmic tail leave the ER 6- to 8-fold more slowly than the full-length protein, even though they fold and oligomerize normally (48, 49). This implies that there is a signal in the VSV G tail for efficient incorporation into transport vesicles exiting the ER. It will be of interest to localize the signal in the VSV G cytoplasmic tail and to identify the components with which it interacts. It will also be important to determine if other proteins (both secreted and membrane-bound) are concentrated during export from the ER.

Specific localization signals in resident proteins of the ER and Golgi complex have been identified (see below). These studies have shown that deletion or mutation of these signals results in transport of the resident protein to the cell surface. Thus, it seems likely that a bulk flow default pathway does exist (although the rate of this membrane flow is likely to be slower than previous estimates, 42). However, specific signals probably exist on rapidly transported proteins to enhance their rate of transport relative to the bulk flow.

2.2 Similarities and differences between ER–Golgi and intra-Golgi transport

Many of the proteins identified in the reconstituted intra-Golgi transport assay are also required for ER to Golgi transport. For example, NSF is required at a late step in the ER–Golgi transport assay (50), as well as in intra-Golgi transport. In yeast, mutants with a temperature-sensitive version of the NSF homologue (Sec18p) are blocked in ER to Golgi transport at the non-permissive temperature, and transport vesicles accumulate (51). Pulse–chase temperature shift experiments showed that Sec18p was also required for multiple transport steps through the Golgi (52). The yeast homologue of α-SNAP (Sec17p) is required for ER to Golgi transport (53). ARF is also required for ER to Golgi transport in yeast and animal cells (54, 55). Thus, many of the components used for vesicular transport through the Golgi complex appear to be used for ER to Golgi transport as well.

However, some differences do exist. As noted above, concentration of cargo in vesicles has been recently demonstrated for the ER to Golgi step, whereas no concentration has been detected for proteins moving through the Golgi complex. Are additional components required for ER to Golgi transport? Potential candidates are additional small GTP-binding proteins like Sar1p (56). This protein does not appear to be required for intra-Golgi transport (57). Calcium was shown to be required for ER to Golgi transport (58), but does not seem essential for intra-Golgi transport.

Although some evidence suggests that COPs are involved in ER to Golgi transport, other evidence suggests they may not be. ER to Golgi transport is blocked in yeast cells with a defective *SEC21* gene, which codes for the yeast homologue of γ-COP (59). However, coatomer complexes containing Sec21p were not required for formation of active transport vesicles in a reconstituted yeast system (60). Microinjected antibodies to β-COP inhibited transport of the VSV G protein from the ER to the Golgi complex in mammalian cells (61). However, the inhibition in *sec21* cells and that produced by anti-β-COP antibodies could be a downstream effect of blocking intra-Golgi transport or another membrane traffic step.

2.3 Retrograde traffic

Membrane traffic in the opposite direction to that discussed for the transport of newly synthesized proteins has been hypothesized to occur. This 'retrograde' traffic would be required to return v-SNAREs to their appropriate compartment, as well as to return escaped resident proteins to their steady-state locations (see below). Although the transport of lipid and proteins in this direction is believed to occur, the mechanism and components involved are still, for the most part, unknown.

Studies with the drug brefeldin A (BFA) suggested that there might be a normal Golgi to ER membrane traffic pathway (reviewed in ref. 62). Shortly after addition of BFA, ARF binding to membranes is inhibited because exchange of GDP for GTP on ARF is blocked. Coats are thus unable to assemble on to membranes. Uncoated Golgi membranes tubulate and fuse with each other and with the ER. The eventual result is a complete mixing of the Golgi with the ER, with disruption of the normal Golgi morphology. One interpretation is that BFA blocks 'anterograde' transport while retrograde traffic is unaffected. Membrane tubules were hypothesized to return membrane to the ER via a normal retrograde pathway that is accentuated by BFA (63). Such tubules can be visualized under certain conditions in non-BFA treated cells (63, 64). However, the massive tubulation seen in BFA-treated cells suggests that disruption of the normal coated structures in the ER and Golgi may completely change the dynamics of these organelles. Vesicles generated by some of the same mechanisms discussed for anterograde traffic may be the carriers for normal (non-drug-induced) retrograde transport. Different targeting proteins would be required, however. In the case of retrograde transport from the Golgi complex, carriers could return directly to the previous compartment, or to the ER where proteins would be transported by anterograde traffic to their appropriate location.

3. Specific localization of ER- and Golgi-resident proteins

3.1 How are these dynamic compartments defined?

In recent years, the actual definition of the number of compartments that exist in the ER–Golgi pathway has been the subject of much debate. What has become increas-

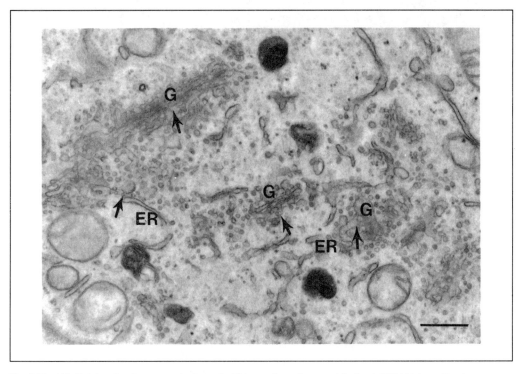

Fig. 4 The ER–Golgi region is extremely dynamic. Thin sections from rapidly fixed (151) HeLa cells show numerous coated and uncoated vesicles surrounding Golgi stacks (G), and close apposition of ER to the Golgi cisternae (arrows). It is difficult to define the boundary between the ER and the Golgi complex. Bar, 500 mm. Micrograph kindly provided by Dr E. B. Cluett.

ingly clear is that both the ER and Golgi complex are extremely dynamic, and there is much intercommunication. Morphologically, it is difficult to define the boundary between the ER and the Golgi complex. In electron micrographs of rapidly fixed cells, the dynamic nature of the ER–Golgi interface can be appreciated (Fig. 4). A number of vesicles are observed, including those that appear to line up along tracks connecting the Golgi stacks. Close apposition of ER and Golgi membranes is also evident.

Biochemical definition of the number of vesicular transport steps between the ER and Golgi complex is also problematic. Some researchers support the idea that only one vesicular step is required, whereas others invoke an 'intermediate' compartment between the ER and Golgi, with a vesicular transport step on each side of this compartment (Fig. 5). Even those that favour the idea of only one transport step between the ER and Golgi differ on where this step occurs. For example, in mouse L cells infected with murine hepatitis virus, the 'intermediate compartment' is apparently directly connected to the ER, and only one vesicular transport step is required to move newly synthesized proteins to the Golgi complex (65). Other proposals suggest that transport vesicles that leave the ER congregate and fuse into structures that will form the early face of the Golgi complex (66). At the other extreme is the

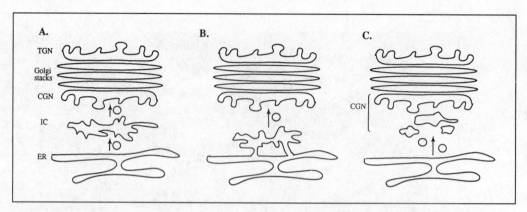

Fig. 5 Vesicular transport steps between the ER and Golgi. Various models predict that individual vesicular transport steps occur on both sides of an 'intermediate compartment', IC (A), or between the intermediate compartment and the Golgi only (B), or between the ER and the *cis* Golgi network (CGN) only (C).

idea that the intermediate compartment is a stable and separate compartment, with a composition intermediate between ER and Golgi membranes (67). Vesicular transport steps would be required for both entry into and exit from this compartment. It is not yet clear if some of the differences reported by different investigators are due to cell type variation or experimental systems used (such as virus infection).

In addition to the controversy regarding the ER–Golgi boundary, the number of distinct compartments in the Golgi complex is unclear. In the traditional view, the Golgi complex is organized into functional subcompartments, called *cis*, *medial*, and *trans* (68). Resident proteins such as glycosyltransferases and glycosidases were believed to be enriched in different subcompartments in the order in which they function (69). More recently, two tubular-reticular networks on either side of the stacks, the *cis* Golgi network (CGN) and the *trans* Golgi network (TGN), have been proposed to function in protein and lipid sorting (70, 71). The subcompartmentalization idea is being questioned with recent demonstration of significant overlap in cisternal localization of two glycosyltransferases (72) and of two glycosidases (73) as determined by double-label immunoelectron microscopy. Mellman and Simons (74) proposed that the Golgi stacks could correspond to the 'medial' subcompartment, where oligosaccharide processing occurs. The TGN and CGN would correspond to the other functional compartments of the Golgi. Mellman and Simons point out that glycosidases and glycosyltransferases need not be compartmentalized, since they act only in a precise order that depends on the presence of the substrate processed by the previous enzyme. Again, the debate regarding the number of Golgi compartments is complicated by the analysis of different cell types, including comparison of tissues and cell lines.

3.2 ER- and Golgi-resident proteins

Given all the confusion regarding the definition of the ER and Golgi complex, how are resident proteins of these organelles defined? For the purposes of this discussion, a steady-state distribution of a protein in the ER or Golgi determines its residence status. For example, even though some lumenal ER proteins cycle through the early Golgi region, they are considered ER residents since this is their steady-state distribution. ER residents include the proteins involved in ER translocation, such as the signal recognition particle receptor and membrane proteins of the translocation pore. Other examples include signal peptidase, oligosaccharyltransferase, protein disulfide isomerase, and GRP78/BiP. Resident proteins of the Golgi complex include enzymes involved in glycosylation (glycosidases, glycosyltransferases, and sugar nucleotide transporters), sulfotransferases, and prohormone processing proteases. Interestingly, all Golgi-resident proteins identified to date are membrane-bound; no lumenal (non membrane-bound) proteins are known.

The localization of ER- and Golgi-resident proteins is the second stage of sorting for proteins that enter the exocytic pathway. The first stage of sorting separates proteins destined for the plasma membrane and intermediate points along the exocytic pathway from all other proteins (those that will remain in the cytoplasm as well as those destined for the nucleus, mitochondria, or peroxisomes). The first sorting step occurs via signal-sequence mediated translocation into the ER (Chapter 4). The third stage of sorting occurs after retention of ER and Golgi residents, and involves sorting in the TGN for diversion to several possible destinations (Chapter 6).

How are resident proteins of the ER and Golgi complex retained in the appropriate compartment in the face of substantial anterograde and retrograde membrane traffic? Localization signals present in a resident protein's structure are responsible. The term 'localization' signal will be used here instead of the more commonly used term 'retention' signal. This reflects the finding that some such signals actually function via a retrieval mechanism, instead of retention *per se*. Some examples of ER and Golgi localization signals will be discussed below.

Much progress has been made in the past years in defining localization signals and the mechanisms by which they function. The basic strategy is to perform site-directed mutagenesis on the gene encoding a resident protein, and assess the effects of the mutations in transfected cells. Loss of localization should result in transport further into the exocytic pathway: the plasma membrane for membrane proteins, or secretion for lumenal proteins. After identification of a sequence that is required for localization of a given protein, it must be shown to be sufficient for localization when transplanted into a reporter protein.

3.3 General mechanisms for localization: retention versus retrieval

There are two general mechanisms by which a localization signal might function to maintain steady-state distribution of a resident protein (reviewed in ref. 75). True retention implies that a newly synthesized resident protein stops once it reaches the appropriate compartment. Such immobilization could occur by interaction with stationary receptors, with stable lipid microdomains (for membrane proteins), or

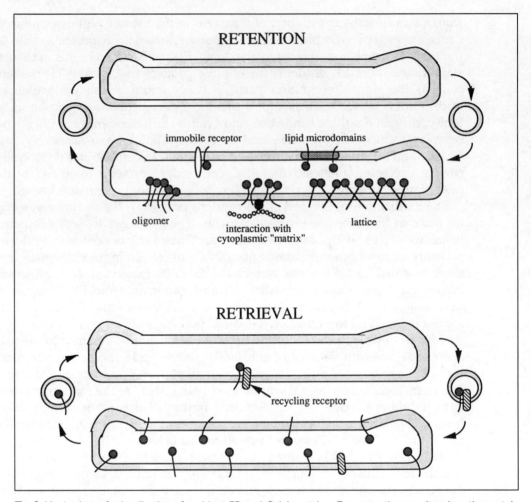

Fig. 6 Mechanisms for localization of resident ER and Golgi proteins. True retention results when the protein fails to move further once it reaches the appropriate compartment. Various ways in which this could occur include formation of large oligomers or lattices that fail to enter transport vesicles for steric reasons, or stable association with lipid microdomains, cytosolic matrix components, or immobile receptors. Retrieval is a dynamic process that returns escaped resident proteins to the appropriate compartment. A recycling receptor returns the escaped protein via the retrograde membrane traffic pathway. It should be noted that these two general mechanisms are not mutually exclusive: one protein could use both for most efficient localization. Such proteins would most likely have two different localization signals.

via formation of large oligomers that are prevented by steric reasons from entering transport vesicles. The other type of mechanism would be a more dynamic one, where active retrieval of escaped molecules by a cycling receptor would occur. These general mechanisms are illustrated in Fig. 6. Both require some type of microenvironment difference between relevant compartments. For retrieval, the ligand must be bound in one compartment, but released in another. For retention, the structural change or specific interaction must not occur until the protein reaches the appropriate compartment in the exocytic pathway. It should be noted that the two mechanisms are not mutually exclusive, one protein might use both for efficient localization. As will be discussed below, evidence for both retention and retrieval of ER- and Golgi-resident proteins has been obtained.

3.4 ER residents

3.4.1 KDEL proteins

The best characterized localization signal is that found on many lumenal (non-membrane-bound) ER proteins. A C-terminal sequence of -Lys-Asp-Glu-Leu (-KDEL) or closely related sequence (-RDEL, -KEEL, etc.) is present on many soluble ER-resident proteins (76). Munro and Pelham (77) first noticed that protein disulfide isomerase, GRP78/BiP, and GRP94 terminated in this sequence. They showed that the KDEL sequence was required for ER localization of BiP and could retain lysozyme (normally secreted) in the ER (77). A loss of ER retention was observed if the KDEL sequence was not at the extreme C-terminus, or if it was mutated to KDES. Although the KDEL sequence retained lysozyme very efficiently, the secretion of other reporter proteins (growth hormone and the α-subunit of chorionic gonadotropin) was delayed but not prevented (78). Thus, it is likely that other structural features of KDEL proteins contribute to efficient ER localization.

The presence of a simple tetrapeptide sequence at the C-terminus of these proteins suggested that a receptor might recognize the motif. A retrieval mechanism was favoured because KDEL proteins are very abundant and there was no candidate protein of sufficient abundance in the ER to serve as a receptor. It was hypothesized that perhaps a receptor might act catalytically to return escaped KDEL proteins from a post-ER compartment (77). Additional experiments where the KDEL sequence was appended to a lysosomal hydrolase showed that although the steady-state distribution of the tagged hydrolase was the ER, it received carbohydrate modifications consistent with exposure to cis Golgi enzymes (79).

In the search for the KDEL receptor, standard affinity chromatography techniques failed. In retrospect, this is not surprising, since the microenvironment differences that allow the receptor to bind its ligand in one compartment and release it in another are not known. Pelham and colleagues initiated a genetic approach for identification of the receptor using the budding yeast, Saccharomyces cerevisiae. They found that in S. cerevisiae, the sequence at the C-terminus of lumenal ER residents like GRP78/BiP (Kar2p) contained His instead of Lys (-HDEL). They expressed a fusion protein consisting of the secreted form of yeast invertase with a C-terminal HDEL sequence (separated by a spacer sequence), and found it was retained in the ER (80). They then screened for mutant cells that secreted the modified invertase and found mutations that fell into two complementation groups. The gene defective in the first group, ERD1 (ER retention defective), seems to be involved in Golgi function (81). Isolation of the gene defective in the second complementation group, ERD2, led to the identification of the putative HDEL receptor. The protein encoded by this gene (Erd2p) is a membrane protein with seven potential membrane-spanning domains (82).

Erd2p determines the specificity of ligand binding, since expression of the homologous protein from another yeast, *Kluyveromyces lactis* (which recognizes both HDEL and KKEL) in *S. cerevisiae* resulted in efficient retention of proteins containing both signals (83). A human homologue of Erd2p was isolated, and its localization is consistent with its function (84). It resides in the Golgi region and/or CGN, but redistributes to the ER when KDEL containing ligands are overexpressed (85). Evidence that the human Erd2 protein was actually the KDEL receptor (and not just a protein involved in regulating the actual receptor) was recently obtained in *in vitro* binding experiments. Interestingly, the pH optimum for binding of peptides terminating in -KDEL to membranes containing the KDEL receptor was pH 5.0 (86). This is much lower than the pH of the *cis* Golgi/CGN region, but Wilson *et al.* (86) hypothesized that weak binding at neutral pH may be instrumental to the proper functioning of the receptor. Ligands must be released in the ER where the concentration of KDEL proteins is estimated to be millimolar.

Extensive mutagenesis of the Erd2 protein has identified hydrophilic residues in the first lumenal loop and in the first, second, fifth, and six transmembrane domains that are important for ligand binding (87, 88). Interestingly, mutations in the cytoplasmic loops had no effect on retrograde movement from the Golgi to the ER, but an aspartic acid residue in the seventh transmembrane domain was essential (88). It was suggested that ligand binding could induce a conformational change (somehow dependent on the aspartate) that would allow the receptor to cluster and enter the retrograde pathway.

In addition to its presence on lumenal ER proteins, the HDEL motif has been found at the C-terminus of a yeast ER *membrane* protein, Sec20p (89). The topology of this type II membrane protein places the HDEL sequence in the ER lumen. It was shown that the HDEL sequence was required for intracellular retention of Sec20p, and that this retention was mediated by *ERD2* (89). Thus, the motif does not need to be on a soluble protein to be recognized by the receptor.

Is retrieval of escaped KDEL proteins back to the ER the primary function of the Erd2 protein? The number of KDEL proteins leaving the ER at any given time could serve as a monitor of the amount of membrane traffic leaving the ER. The return of KDEL proteins via vesicular or other type of retrograde transport to the ER would replenish lipid as well, which could be an important function of the receptor. Consistent with this idea, overexpression of a human ERD-like protein (ELP-1) caused disruption of the Golgi complex, with a phenotype similar to that seen in BFA-treated cells (90). In yeast, the original *erd2* mutants that secreted HDEL proteins were viable, although deletion of the *ERD2* gene was lethal (82). Similarly, the HDEL sequence on Kar2p/BiP was not required for cell viability, presumably because these proteins are secreted slowly when they lack the signal (82). All of these observations suggest that the retrieval of HDEL proteins to the ER is not the most important function of the receptor.

With the idea that one important function of the KDEL receptor could be to monitor the flow of lipid from the ER, it is interesting to note that, in yeast, phospholipid composition of Golgi membranes appears to be critical for normal secretion. The

SEC14 gene, originally identified in a mutant where membrane traffic from the Golgi is blocked, encodes a phosphatidylinositol (PI)/phosphatidylcholine (PC) transfer protein that preferentially binds PI *in vitro* (91). One hypothesis is that the PI/PC ratio needs to be elevated in Golgi membranes relative to bulk membranes for active secretion (92). One mutation that bypassed the requirement for Sec14p inactivated the CDP-choline pathway of PC synthesis (93), one of the two PC synthesis pathways in yeast. It was recently demonstrated that Golgi membranes from *sec14* mutant cells have increased PC content and unchanged PI content compared to wild-type cells (94). In light of these results, Sec14p was proposed to be a phospholipid sensor protein that maintains a level of PC compatible with Golgi secretory function (94). This would be accomplished by regulating PC biosynthesis locally in Golgi membranes via the CDP-choline pathway.

3.4.2 The double lysine motif on ER membrane proteins

A localization signal for ER membrane proteins was first identified on the cytoplasmic tail of an adenovirus protein, E19k (95, 96). The motif consists of a Lys residue at the −3 position, and another Lys at the −4 or −5 position in the cytoplasmic tail (-KKxx or -KxKxx, where x is any amino acid). This motif seems to be sufficient to retain several reporter proteins in the ER, although its distance from the membrane may be critical for efficient function. The motif is present on several endogenous ER membrane proteins, including UDP-glucuronosyltransferase, ribophorins I and II, and HMG-CoA reductase (see ref. 96). In the E19k protein and several endogenous proteins, the -KKxx sequence is at the C-terminus. A similar sequence (but containing arginines) functions at the N-terminus in type II membrane proteins such as the Iip33 form of the invariant chain (97).

The oligosaccharide processing of a CD8 reporter containing the -KKxx signal suggested that its steady-state ER distribution is maintained by retrieval from a post-ER compartment (98). In recent experiments designed to identify receptors for this ER localization signal, Cosson and Letourner (99) made the surprising observation that COP-coatomers bound specifically to peptides terminating in -KKxx. α-COP and β'-COP seemed to bind most tightly. This raises the interesting possibility that COP-coated vesicles are involved in retrograde traffic from the Golgi to the ER.

Interestingly, the double lysine motif was recently reported on an 'intermediate compartment' protein, ERGIC-53 (100). This protein has a steady-state localization intermediate to the ER and Golgi, but has been shown to cycle between these organelles (101). It is possible that proteins with the double lysine motif and a steady-state ER localization (like the adenovirus E19k protein) have additional signals for efficient retention and use the di-lysine signal as a back-up mechanism to retrieve escaped proteins. If ERGIC-53 lacks retention information of any sort, it might be continuously subjected to retrieval.

3.4.3 Other ER localization signals

Other types of ER localization signals have been identified, although the sequence requirements and mechanisms by which they function have not yet been deciphered. Cytochrome b_5 is targeted to the ER by a C-terminal sequence that is near the membrane anchor (102). Cytochrome P_{450} requires its N-terminal 29 amino acids for ER localization (103). Rotavirus VP7 also requires N-terminal sequences for ER localization, including residues within the hydrophobic domain that presumably functions as a signal peptide, and three specific residues downstream of the signal cleavage site (104). It will be of interest to determine if any of these signals function by true retention rather than by retrieval from a post-ER compartment. Another type of signal consisting of unpaired charged residues was identified within the transmembrane domain of unassembled oligomeric proteins that seemed specifically to target such proteins for retention and degradation in the ER (reviewed in 105). Such proteins may not be well anchored in the membrane, however (106), and misfold when extruded completely into the ER lumen. Misfolded or unassembled proteins are not exported from the ER, perhaps because they remain bound to ER-resident chaperones or are degraded (107).

3.5 Golgi-resident proteins

3.5.1 Transmembrane domain signals

The first Golgi localization signal to be described was discovered in a coronavirus membrane protein called M (formerly called E1). The M protein is thought to be important for assembly of coronaviruses at intracellular membranes (108). The M protein from the avian infectious bronchitis virus is targeted to the *cis* Golgi/CGN when expressed from cDNA in the absence of the other coronavirus proteins (109). The M protein has three membrane-spanning domains, and the first of these was found to be required for Golgi localization of the avian coronavirus M protein (110). It was subsequently shown that this transmembrane domain could retain a reporter protein in the early Golgi region (111). Transmembrane residues critical for retention of the reporter were polar uncharged residues (Asn, Thr, Gln) that line up on one face of a predicted α-helix (112).

Endogenous glycosyltransferases have also been examined for localization signals. Studies with β-1,4-galactosyltransferase (β-1,4-GT), α-2,6-sialyltransferase (α-2,6-ST), and *N*-acetylglucosaminyltransferase I (NAGT-I) all showed that the transmembrane domains of these proteins contained information critical for Golgi localization (reviewed in refs 113, 114). In several cases, sequences flanking the transmembrane domain were required for efficient localization. There is no primary sequence homology in the transmembrane domains of these transferases or of the coronavirus M protein.

The transmembrane domain of β-1,4-GT was sufficient for localization of ovalbumin to the Golgi region (115), and the invariant chain (a subunit of class II major histocompatibility antigens) to the *trans* Golgi (116). In the latter study, a chimeric protein containing only the lumenal half of the β-1,4-GT transmembrane domain was properly localized. In another study using a different reporter protein (chorionic gonadotropin fused to the VSV G membrane anchor), the β-1,4-GT transmembrane domain was also sufficient for Golgi region localization (117). In contrast to the study cited above, this group found that a cysteine residue and a histidine residue in the cytoplasmic half of the transmembrane domain were critical. A fourth group reported that the length of the β1,4-GT transmembrane domain was an important factor (118).

The transmembrane domain of NAGT-I was sufficient for localization of dipeptidylpeptide IV (DPPIV) in the Golgi region of MDCK cells, although flanking sequence from the lumenal domain enhanced localization (119). Burke *et al.* (120) showed that the short cytoplasmic tail and transmembrane domain of NAGT-I retained ovalbumin in the *medial* Golgi of mouse L cells using immunoelectron microscopy.

Of the three transferases that have been studied, α2,6-ST seems to contain the most complicated targeting information. It may contain a localization signal within a region called the stem (near the transmembrane domain on the lumenal side of the membrane) in addition to that within the transmembrane domain. Wong *et al.* (121) showed that the transmembrane domain of α-2,6-ST was sufficient to retain DPPIV in the Golgi complex of MDCK cells. However, Munro found that flanking sequences on the lumenal side of the membrane were required, in addition to the transmembrane domain, to retain lysozyme in the Golgi region of COS cells (122). He also showed that efficient Golgi localization was obtained when the transmembrane domain of α-2,6-ST was replaced with 17 leucines (the length of the actual transmembrane domain). However, replacing the transmembrane domain with 23 leucines eliminated Golgi localization. In contrast, Dahdal and Colley (123) found that replacing the sialyltransferase transmembrane domain with the 29 amino acid transmembrane domain of influenza neuraminidase did not interfere with Golgi localization in COS cells, and concluded that the transmembrane domain was neither necessary nor sufficient for Golgi targeting. However, although the experiments suggest that the stem region of α-2,6-ST is required for Golgi targeting, they do not show that it is sufficient. Localization of sialyltransferase in both the *trans* Golgi and TGN in some cell types, as well as more broad distribution in cells of certain tissues, may explain the apparent multiplicity of targeting signals. None of the studies cited above have determined the precise localizations of the expressed proteins by electron microscopy, which could resolve some of the confusion.

Mechanisms by which transmembrane domain targeting signals function

To date, the available evidence suggests that transmembrane domain localization signals in Golgi resident proteins may act by true retention. In general, overexpression of Golgi membrane proteins does not result in transport to the cell surface (as might be expected if saturation of a retrieval receptor occurred), but rather in 'backing up' of the protein into the ER (111, 116, 122, 124, 125). Some evidence has been obtained for large oligomers that could prevent transport for steric reasons. A chimeric VSV G protein containing the Golgi-targeting transmembrane domain from the coronavirus M protein forms large, detergent-resistant oligomers as it arrives in the early Golgi (126). Single amino acid substitutions within the transmembrane domain that inactivate the localization signal prevented formation of these oligomers, and the mutant proteins were transported to the cell surface. Endogenous proteins are probably included in the oligomers, but have not yet been identified. Although the coronavirus M transmembrane domain in the chimeric VSV G protein is required for Golgi localization and oligomerization, the cytoplasmic tail is involved in stabilizing the oligomer during detergent solubilization (126). It is possible that interaction of this tail with a cytosolic 'matrix' serves to retain absolutely the protein once trans-membrane domain-mediated oligomerization has occurred.

Nilsson *et al.* (127) have recently provided evidence that two *medial* Golgi enzymes (mannosidase II and *N*-AGT-I) form hetero-oligomers. They showed that expression of a modified form of mannosidase II containing an ER localization signal caused endogenous NAGT-I to be retained in the ER. In the reciprocal experiment, ER-retained NAGT-I caused redistribution of endogenous mannosidase II to the ER. Retention of the *trans* Golgi enzyme β-1,4-GT in the ER with the same ER localization signal did not cause redistribution of either of the two *medial* Golgi enzymes, and replacing the transmembrane domain and part of the stem domain of the ER-retained form of NAGT-I with that of β-1,4-GT prevented the redistribution of mannosidase II. Interestingly, with increasing expression levels of the ER-retained form of NAGT-I, the Golgi complex became progressively smaller and eventually completely vesiculated. Thus, complete redistribution of *medial* Golgi enzymes to the ER caused a breakdown of the normal stacked cisternal arrangement of the Golgi complex. Nilsson *et al.* (127, 128) proposed a model for Golgi protein retention where after dimerization in the ER, proteins move to their resident cisterna, where their transmembrane domains interact with other 'kin' dimers (those residing in the same compartment), adding on to pre-existing hetero-oligomers. Thus, 'kin recognition' would specify the cisterna in which a newly synthesized Golgi protein would remain. The model also suggests how overlapping distribution of Golgi enzymes could occur: interactions of the cytoplasmic tails of kin oligomers in the intercisternal space would result in a greater spread in the enzyme's distribution. The recent isolation of a detergent-insoluble 'matrix' from Golgi membranes that binds the cytoplasmic tails of mannosidase II and NAGT-I supports this idea (129). The kin recognition model makes some predictions that can be tested. For example, expression of Golgi proteins that can dimerize but cannot add on to kin oligomers (non-membrane-bound forms or those with altered transmembrane domains) should 'cap' the ends of the oligomer and eventually disrupt the Golgi retention mechanism.

Are Golgi resident proteins found in large oligomers?

Several early Golgi enzymes have been reported to be relatively insoluble in non-ionic detergent or to form large oligomers (130, 131). However, available evidence suggests that many of the *medial* and *trans* enzymes exist as dimers (128). In addition, analysis of target size by radiation inactivation has shown that the active forms of Golgi enzymes are dimers (sialyltransferase, galactosyltransferase, and UDPase, 132), or monomers (*N*-deacetylase/*N*-sulfotransferase, 133). Although these data do not preclude the presence of large oligomers, they indicate that such structures are not essential for enzymatic activity. The *medial* Golgi hetero-oligomers described by Nilsson *et al.* are not stable in non-ionic detergent (127). Therefore, there is no evidence that large oligomers of Golgi membrane proteins exist within the lipid bilayer. Altering the lipid environment by solubilization with detergent may induce aggregation of some Golgi membrane proteins, especially if they exist as an extended lattice in the membrane. Biophysical measurements such as fluorescence photobleaching and recovery to assess the mobility of Golgi proteins in native membranes would be very informative.

Does the lipid bilayer contribute to localization?

The kin recognition model described above depends on pre-existing oligomers and is not dependent on the local microenvironment in a given cisterna. Alternative suggestions have been proposed for specific Golgi protein localization mediated by transmembrane domains. Bretscher and Munro (134) observed that Golgi proteins appear to have shorter transmembrane domains than those of plasma membrane proteins. They speculated that the lipid composition of Golgi membranes may play a major role in localization of resident proteins. There is indirect evidence that a gradient of cholesterol (with the highest concentration on the *trans* side) exists in the Golgi complex (135). Since cholesterol can influence both the permeability and the thickness of biological membranes, Bretscher and Munro (134) hypothesized that membrane proteins with shorter transmembrane segments might be retained by exclusion from cholesterol-rich regions of Golgi membranes *en route* for the plasma membrane. This idea can be tested in cells with altered levels of cholesterol. It seems unlikely that membrane thickness could be the sole determinant of specific localization however, since newly synthesized Golgi proteins would prefer to remain in the ER where the cholesterol concentration is lowest. Also, several notable exceptions of Golgi membrane proteins with long transmembrane domains exist (136–138).

A combination of both protein–protein interactions (transmembrane domains as well as lumenal or cytoplasmic domains) and lipid–protein interactions is likely to retain Golgi membrane proteins. The lipid composition of a given membrane could be expected to influence the potential protein–protein interactions within that bilayer. Differences in lipid composition or the presence of specific lipid microdomains would thus be major factors in specific localization of Golgi membrane proteins. Like cholesterol, sphingolipids could have a major influence on Golgi membrane microenvironments. Unlike phospholipids and cholesterol, sphingolipids are synthesized in the Golgi complex, where a newly synthesized membrane protein would meet this class of lipids for the first time (reviewed in ref. 139). In this regard, glucosylceramide could be a particularly important sphingolipid, since it is synthesized on the cytoplasmic leaflet of early Golgi membranes and then presumably flipped to the lumenal leaflet (140). The early Golgi membranes, where glucosylceramide is synthesized, would thus be the only place that newly synthesized membrane proteins encounter glucosylceramide in both leaflets of the membrane. In addition, certain glycosphingolipids can self-associate *in vitro* (141), which could contribute to the formation of microenvironments *in vivo*. More information regarding the specific lipid composition across Golgi stacks will be required before such speculations can be evaluated.

3.5.2 Cytoplasmic tail signals in TGN proteins

The cytoplasmic tail of TGN38, a protein of unknown function that resides in the TGN, is required for TGN localization (142). A tyrosine residue (in the sequence Tyr-Gln-Arg-Leu) in the cytoplasmic tail is critical (143, 144), since mutation of this residue resulted in accumulation of the protein at the cell surface. This sequence motif is similar to sequences in the cytoplasmic tails of endocytic receptors that mediate their internalization into clathrin-coated vesicles (145). TGN38 cycles rapidly through the plasma membrane (144), so the sequence in the tail may actually mediate retrieval from the plasma membrane to the TGN. Alternatively, the sequence may interact with clathrin in the TGN to prevent further transport. Consistent with this, one mutation in the cytoplasmic tail of TGN38 prevented its localization in the TGN, but did not block its internalization from the plasma membrane (143). The furin processing protease also constitutively cycles through the plasma membrane, and requires sequences in the cytoplasmic tail for TGN localization (146).

In yeast, the processing proteases Kex2p, Kex1p, and dipeptidylaminopeptidase A (DPAP A) are believed to reside in the latest functional Golgi compartment, the TGN or its equivalent. These proteins also require their cytoplasmic tails for Golgi localization (reviewed in 147). Kex2p has a critical tyrosine in its cytoplasmic tail and DPAP A requires a phenylalanine-rich region of its tail. Clathrin is required for localization of these enzymes as well, since a temperature-sensitive mutation in clathrin heavy chain caused missorting to the cell surface at the non-permissive temperature (148).

4. Conclusions

Transport of newly synthesized proteins from the ER through the Golgi complex in a vectorial manner requires many components and precise regulation. Proteins required for the process were discovered in assays that reconstitute vesicular transport *in vitro*, and their roles confirmed genetically in yeast. Transport of newly synthesized proteins to the cell surface can occur by default, although signals may enhance the rate of exit from the ER.

Appropriate localization of ER and Golgi resident proteins requires specific retention or retrieval signals. Such signals have been identified in endogenous and viral proteins that are targeted to the ER or Golgi complex. The localization signals discussed here are those identified to date. More undoubtedly await discovery. In addition, the mechanism by which these localization signals function are not well understood. The examples discussed suggest that the ER and TGN may favour retrieval, while the Golgi stacks use retention for localizing their resident proteins.

However, it is likely that some ER residents will be efficiently retained and not require retrieval. Membrane proteins that make up the translocation pore would be likely candidates. It is also likely that some Golgi resident proteins are not efficiently retained, and must be continuously retrieved to maintain their steady-state distribution. Several possibilities are GIMPc and MG160 (*cis* and *medial* residents, respectively); both receive late Golgi oligosaccharide modifications (149, 150).

Many questions remain. Do transport vesicles form only in specialized regions of the ER and Golgi? If transport vesicles form only in specialized regions of the ER, how are newly synthesized proteins moved to these locations? If transport vesicles bud and fuse only at the rims of the stacked Golgi cisternae, how are itinerant proteins processed by resident enzymes, thought to be concentrated in the flattened, closely apposed region of the stacks? How are the ER and Golgi complex disassembled at mitosis, and then reassembled afterwards? Are there alternative pathways besides vesicular transport for movement of newly synthesized (and remodelled) lipids to the plasma membrane? Understanding the mechanism of intracellular transport and the localization signals for retention of resident proteins is just the first step in understanding organelle maintenance and dynamics. Interesting new ideas have been generated by current work in this area, and future experiments will continue to challenge our way of thinking.

References

1. Palade, G. (1975) Intracellular aspects of the process of protein synthesis. *Science*, **189**, 347.
2. Rothman, J. E. and Orci, L. (1992) Molecular dissection of the secretory pathway. *Nature*, **355**, 33.
3. Pryer, N. K., Wuestehube, L. J., and Schekman, R. (1992) Vesicle-mediated protein sorting. *Annu. Rev. Biochem.*, **61**, 471.
4. Schwanginger, R., Plutner, H., Davidson, H. W., Pind, S., and Balch, W. E. (1992) Transport of protein between endoplasmic reticulum and Golgi compartments in semiintact cells. *Methods Enzymol.*, **219**, 110.
5. Schekman, R. (1992) Genetic and biochemical analysis of vesicular traffic in yeast. *Curr. Opin. Cell Biol.*, **4**, 587.
6. Malhotra, V., Serafini, T., Orci, L., Shepherd, J. C., and Rothman, J. E. (1989) Purification of a novel class of coated vesicles mediating biosynthetic protein transport through the Golgi stack. *Cell*, **58**, 329.
7. Waters, M. G., Serafini, T., and Rothman, J. E. (1991) Coatomer: a cytosolic protein complex containing subunits of non-clathrin-coated Golgi transport vesicles. *Nature*, **349**, 248.
8. Alan, V. J. and Kreis, T. E. (1986) A microtubule-binding protein associated with membranes of the Golgi apparatus. *J. Cell Biol.*, **103**, 2229.
9. Serafini, T., Stenbeck, G., Brecht, A., Lottspeich, F., Orci, L., Rothman, J. E., and Wieland, F. T. (1991) A coat subunit of Golgi-derived non-clathrin-coated vesicles with homology to the clathrin-coated vesicle coat protein beta-adaptin. *Nature*, **349**, 215.
10. Duden, R., Griffiths, G., Frank, R., Argos, P., and Kreis, T. E. (1991) β-COP, a 110 kd protein associated with non-clathrin-coated vesicles and the Golgi complex, shows homology to β-adaptin. *Cell*, **64**, 649.

11. Stenbeck, G., Harter, C., Brecht, A., Herrmann, D., Lottspeich, F., Orci, L., and Wieland, F. T. (1993) β'-COP, a novel subunit of coatomer. *EMBO J.*, **12**, 2841.
12. Harrison-Lavoie, K. J., Lewis, V. A., Hynes, G. M., Collison, K. S., Nutland, E., and Willison, K. R. (1993) A 102 kDa subunit of a Golgi-associated particle has homology to β subunits of trimeric G proteins. *EMBO J.*, **12**, 2847.
13. Hara-Kuge, S., Kuge, O., Orci, L., Amherdt, M., Ravazzola, M., Wieland, F. T., and Rothman, J. E. (1994) *En bloc* incorporation of coatomer subunits during the assembly of COP-coated vesicles. *J. Cell Biol.*, **124**, 883.
14. Kahn, R. A., Yucel, J. K., and Malhotra, V. (1993) ARF signaling: a potential role for phospholipase D in membrane traffic. *Cell*, **75**, 1045.
15. Barr, F. A., Leyte, A., and Huttner, W. B. (1992) Trimeric G proteins and vesicle formation. *Trends Cell Biol.*, **2**, 91.
16. Sheetz, M. P. and Singer, S. J. (1974) Biological membranes as bilayer couples. A molecular mechanism of drug–erythrocyte interactions. *Proc. Natl Acad. Sci. USA*, **71**, 4457.
17. Brown, H. A., Gutowski, S., Moomaw, C. R., Slaughter, C., and Sternweis, P. C. (1993) ADP-ribosylation factor, a small GTP-dependent regulatory protein, stimulates phospholipase D activity. *Cell*, **75**, 1137.
18. Ostermann, J., Orci, L., Tani, K., Amherdt, M., Ravazzola, M., and Rothman, J. E. (1993) Stepwise assembly of functionally active transport vesicles. *Cell*, **75**, 1015.
19. Zerial, M. and Stenmark, H. (1993) Rab GTPases in vesicular transport. *Curr. Opin. Cell Biol.*, **5**, 613.
20. Novick, P. and Brennwald, P. (1993) Friends and family: the role of Rab GTPases in vesicular traffic. *Cell*, **75**, 597.
21. Brennwald, P. and Novick, P. (1993) Interactions of three domains distinguishing the Ras-related GTP-binding proteins Ypt1 and Sec4. *Nature*, **362**, 560.
22. Dunn, B., Stearns, T., and Botstein, D. (1993) Specificity domains distinguish the Ras-related GTPases Ypt1 and Sec4. *Nature*, **362**, 563.
23. Sollner, T., Whiteheart, S. W., Brunner, M., Erdjument-Bromage, H., Geromanos, S., Tempst, P., and Rothman, J. E. (1993) SNAP receptors implicated in vesicle targeting and fusion. *Nature*, **362**, 318.
24. Block, M. R., Glick, B. S., Wilcox, C. A., Wieland, F. T., and Rothman, J. E. (1988) Purification of an N-ethylmaleimide-sensitive protein catalyzing vesicular transport. *Proc. Natl Acad. Sci. USA*, **85**, 7852.
25. Wilson, D. W., Whiteheart, S. W., Wiedmann, M., Brunner, M., and Rothman, J. E. (1992) A multisubunit particle implicated in membrane fusion. *J. Cell Biol.*, **117**, 531.
26. Wilson, D. W., Wilcox, C. A., Flynn, G. C., Chen, E., Kuang, W.-J., Henzel, W. J., Block, M. R., Ullrich, A., and Rothman, J. E. (1989) A fusion protein required for vesicle-mediated transport in both mammalian cells and yeast. *Nature*, **339**, 355.
27. Eakle, K. A., Bernstein, M., and Emr, S. D. (1988) Characterization of a component of the yeast secretion machinery: identification of the *SEC18* gene product. *Mol. Cell Biol.*, **8**, 4098.
28. Clary, D. O., Griff, I. C., and Rothman, J. E. (1990) SNAPs, a family of NSF attachment proteins involved in intracellular membrane fusion in animals and yeast. *Cell*, **61**, 709.
29. Whiteheart, S. W., Griff, I. C., Brunner, M., Clary, D. O., Mayer, T., Buhrow, S. A., and Rothman, J. E. (1993) SNAP family of NSF attachment proteins includes a brain isoform. *Nature*, **362**, 353.
30. Sudhof, T. C., DeCamilli, P., Niemann, H., and Jahn, R. (1993) Membrane fusion machinery: insights from synaptic proteins. *Cell*, **75**, 1.

31. Sollner, T., Bennett, M. K., Whiteheart, S. W., Scheller, R. H., and Rothman, J. E. (1993) A protein assembly–disassembly pathway *in vitro* that may correspond to sequential steps of synaptic vesicle docking, activation, and fusion. *Cell*, **75**, 409.

32. Lian, J. P. and Ferro-Novick, S. (1993) Bos1p, an integral membrane protein of the endoplasmic reticulum to Golgi transport vesicles, is required for their fusion competence. *Cell*, **73**, 735.

33. Hardwick, K. G. and Pelham, H. R. B. (1992) SED5 encodes a 39-kD integral membrane protein required for vesicular transport between the ER and the Golgi complex. *J. Cell Biol.*, **119**, 513.

34. Gerst, J. E., Rodgers, L., Riggs, M., and Wigler, M. (1992) *SNC1*, a yeast homolog of the synaptic-vesicle-associated membrane protein/synaptobrevin gene family: genetic interactions with the RAS and CAP genes. *Proc. Natl Acad. Sci. USA*, **89**, 4338.

35. Protopopov, V., Govindan, B., Novick, P., and Gerst, J. E. (1993) Homologs of the synaptobrevin/VAMP family of synaptic vesicle proteins function on the late secretory pathway in *S. cerevisiae. Cell*, **74**, 855.

36. Aalto, M. K., Ronne, H., and Keranen, S. (1993) Yeast syntaxins Sso1p and Sso2p belong to a family of related membrane proteins that function in vesicular transport. *EMBO J.*, **12**, 4095.

37. Rothman, J. E. and Warren, G. (1994) Implications of the SNARE hypothesis for intracellular membrane topology and dynamics. *Curr. Biol.*, **4**, 220.

38. Kutay, U., Hartmann, E., and Rapoport, T. A. (1993) A class of membrane proteins with a C-terminal anchor. *Trends Cell Biol.*, **3**, 72.

39. White, J. M. (1992) Membrane fusion. *Science*, **258**, 917.

40. Ullrich, O., Stenmark, H., Alexandrov, K., Huber, L. A., Kaibuchi, K., Sasaki, T., Takai, Y., and Zerial, M. (1993) Rab GDI as a general regulator for the membrane association of Rab proteins. *J. Biol. Chem.*, **268**, 18143.

41. Orci, L., Glick, B. S., and Rothman, J. E. (1986) A new type of coated vesicular carrier that appears not to contain clathrin: its possible role in protein transport within the Golgi stack. *Cell*, **46**, 171.

42. Wieland, F. T., Gleason, M. L., Serafini, T. A., and Rothman, J. E. (1987) The rate of bulk flow from the endoplasmic reticulum to the cell surface. *Cell*, **50**, 289.

43. Pfeffer, S. R. and Rothman, J. E. (1987) Biosynthetic protein transport and sorting by the endoplasmic reticulum and Golgi. *Annu. Rev. Biochem.*, **56**, 829.

44. Copeland, C. S., Zimmer, K.-P., Wagner, K. R., Healy, G. A., Mellman, I., and Helenius, A. (1988) Folding, trimerization, and transport are sequential events in the biogenesis of influenza virus hemagglutinin. *Cell*, **53**, 197.

45. Hobman, T. C., Woodward, L., and Farquhar, M. G. (1992) The rubella virus E1 glycoprotein is arrested in a novel, post-ER-pre-Golgi compartment. *J. Cell Biol.*, **118**, 795.

46. Mizuno, M. and Singer, S. J. (1993) A soluble secretory protein is first concentrated in the endoplasmic reticulum before transfer to the Golgi apparatus. *Proc. Natl Acad. Sci. USA*, **90**, 5732.

47. Balch, W. E., McCaffery, J. M., Plutner, H., and Farquhar, M. G. (1994) Vesicular stomatitis virus glycoprotein is sorted and concentrated during export from the endoplasmic reticulum. *Cell*, **76**, 841.

48. Rose, J. K. and Bergmann, J. E. (1983) Altered cytoplasmic domains affect intracellular transport of the vesicular stomatitis virus glycoprotein. *Cell*, **34**, 513.

49. Doms, R. W., Ruusala, A., Machamer, C., Helenius, J., Helenius, A., and Rose, J. K. (1988) Differential effects of mutations in three domains on folding, quaternary

structure, and intracellular transport of vesicular stomatitis virus G protein. *J. Cell Biol.*, **107,** 89.

50. Beckers, C. J., Block, M. R., Glick, B. S., Rothman, J. E., and Balch, W. E. (1989) Vesicular transport between the endoplasmic reticulum and the Golgi stack requires the NEM-sensitive fusion protein. *Nature*, **339,** 397.

51. Kaiser, C. A. and Schekman, R. (1990) Distinct sets of *SEC* genes govern vesicle formation and fusion early in the secretory pathway. *Cell*, **61,** 723.

52. Graham, T. R. and Emr, S. D. (1991) Compartmental organization of Golgi-specific protein modification and vacuolar protein sorting events defined in a yeast *sec18* (NSF) mutant. *J. Cell Biol.*, **114,** 207.

53. Griff, I. C., Schekman, R., Rothman, J. E., and Kaiser, C. A. (1992) The yeast *SEC17* gene product is functionally equivalent to mammalian alpha-SNAP. *J. Biol. Chem.*, **267,** 12106.

54. Stearns, T., Kahn, R. A., Botstein, D., and Hoyt, M. A. (1990) ADP-ribosylation factor is an essential protein in *Saccharomyces cerevisiae* and is encoded by two genes. *Mol. Cell. Biol.*, **10,** 6690.

55. Balch, W. E., Kahn, R. A., and Schwaninger, R. (1992) ADP-ribosylation factor is required for vesicular trafficking between the endoplasmic reticulum and the *cis*-Golgi compartment. *J. Biol. Chem.*, **267,** 13053.

56. Nakano, A. and Muramatsu, M. (1989) A novel GTP-binding protein, Sar1p, is involved in transport from the endoplasmic reticulum to the Golgi apparatus. *J. Cell Biol.*, **109,** 2677.

57. Kuge, O., Dascher, C., Orci, L., Amherdt, M., Rowe, T., Plutner, H., Ravazzola, M., Tanigawa, G., Rothman, J. E., and Balch, W. E. (1994) Sar1 promotes vesicle budding from the endoplasmic reticulum but not Golgi compartments. *J. Cell Biol.*, **125,** 51.

58. Beckers, C. J. and Balch, W. E. (1989) Calcium and GTP: essential components in vesicular trafficking between the endoplasmic reticulum and Golgi apparatus. *J. Cell Biol.*, **108,** 1245.

59. Hosobuchi, M., Kreis, T., and Schekman, R. (1992) *SEC21* is a gene required for ER to Golgi protein transport that encodes a subunit of a yeast coatomer. *Nature*, **360,** 603.

60. Salama, N. R., Yeung, T., and Schekman, R. (1993) The Sec13p complex and reconstitution of vesicle budding from the ER with purified cytosolic proteins. *EMBO J.*, **12,** 4073.

61. Pepperkok, R., Scheel, J., Horstmann, H., Hauri, H. P., Griffiths, G., and Kreis, T. E. (1993) β-COP is essential for biosynthetic membrane transport from the endoplasmic reticulum to the Golgi complex *in vivo. Cell*, **74,** 71.

62. Klausner, R. D., Donaldson, J. G., and Lippincott-Schwartz, J. (1992) Brefeldin A: insights into the control of membrane traffic and organelle structure. *J. Cell Biol.*, **116,** 1071.

63. Lippincott-Schwartz, J., Donaldson, J. G., Schweizer, A., Berger, E. G., Hauri, H. P., Yuan, L. C., and Klausner, R. D. (1990) Microtubule-dependent retrograde transport of proteins into the ER in the presence of brefeldin A suggests an ER recycling pathway. *Cell*, **60,** 821.

64. Cluett, E. B., Wood, S. A., Banta, M., and Brown, W. J. (1993) Tubulation of Golgi membranes *in vivo* and *in vitro* in the absence of brefeldin A. *J. Cell Biol.*, **120,** 15.

65. Krinjse-Locker, J., Ericsson, M., Rottier, P. J. M., and Griffiths, G. (1994) Characterization of the budding compartment of mouse hepatitis virus: evidence that transport from the RER to the Golgi complex requires only one vesicular transport step. *J. Cell Biol.*, **124,** 55.

66. Saraste, J. and Kuismanen, E. (1992) Pathways of protein sorting and membrane traffic between the rough endoplasmic reticulum and the Golgi complex. *Semin. Cell Biol.*, **3,** 343.

67. Schweizer, A., Matter, K., Ketcham, C. A., and Hauri, H. P. (1991) The isolated ER–Golgi intermediate compartment exhibits properties that are different from ER and *cis*-Golgi. *J. Cell Biol.*, **113,** 45.

68. Farquhar, M. G. (1985) Progress in unraveling pathways of Golgi traffic. *Annu. Rev. Cell Biol.*, **1,** 477.

69. Dunphy, W. G. and Rothman, J. E. (1985) Compartmental organization of the Golgi stack. *Cell*, **42,** 13.

70. Hsu, V. W., Yuan, L. C., Nuchtern, J. G., Lippincott-Schwartz, J., Hammerling, G. J., and Klausner, R. D. (1991) A recycling pathway between the endoplasmic reticulum and the Golgi apparatus for retention of unassembled MHC class I molecules. *Nature*, **352,** 441.

71. Griffiths, G. and Simons, K. (1986) The *trans* Golgi network: sorting at the exit site of the Golgi complex. *Science*, **234,** 438.

72. Nilsson, T., Pypaert, M., Hoe, M. H., Slusarewicz, P., Berger, E. G., and Warren, G. (1993) Overlapping distribution of two glycosyltransferases in the Golgi apparatus of HeLa cells. *J. Cell Biol.*, **120,** 5.

73. Velasco, A., Hendricks, L., Moremen, K. W., Tulsiani, D. R. P., Touster, O., and Farquhar, M. G. (1993) Cell type-dependent variations in the subcellular distribution of α-mannosidase I and II. *J. Cell Biol.*, **122,** 39.

74. Mellman, I. and Simons, K. (1992) The Golgi complex: *in vitro veritas*? *Cell*, **68,** 829.

75. Machamer, C. E. (1991) Golgi retention signals: do membranes hold key? *Trends Cell Biol.*, **1,** 144.

76. Pelham, H. R. B. (1989) Control of protein exit from the endoplasmic reticulum. *Annu. Rev. Cell Biol.*, **5,** 1.

77. Munro, S. and Pelham, H. R. B. (1987) A C-terminal signal prevents secretion of luminal ER proteins. *Cell*, **48,** 899.

78. Zagouras, P. and Rose, J. K. (1989) Carboxy-terminal SEKDEL sequences retard but do not retain two secretory proteins in the endoplasmic reticulum. *J. Cell Biol.*, **109,** 2633.

79. Pelham, H. R. B. (1988) Evidence that luminal ER proteins are sorted from secreted proteins in a post-ER compartment. *EMBO J.*, **7,** 913.

80. Pelham, H. R. B., Hardwick, K. G., and Lewis, M. J. (1988) Sorting of soluble ER proteins in yeast. *EMBO J.*, **7,** 1757.

81. Hardwick, K. G., Lewis, M. J., Semenza, J., Dean, N., and Pelham, H. R. B. (1990) *ERD1*, a yeast gene required for the retention of luminal endoplasmic reticulum proteins, affects glycoprotein processing in the Golgi apparatus. *EMBO J.*, **9,** 623.

82. Semenza, J., Hardwick, K. G., Dean, N., and Pelham, H. R. B. (1990) *ERD2*, a yeast gene required for the receptor-mediated retrieval of luminal ER proteins from the secretory pathway. *Cell*, **61,** 1349.

83. Lewis, M. J., Sweet, D. J., and Pelham, H. R. B. (1990) The *ERD2* gene determines the specificity of the luminal ER protein retention system. *Cell*, **61,** 1359.

84. Lewis, M. J. and Pelham, H. R. B. (1990) A human homologue of the yeast HDEL receptor. *Nature*, **348,** 162.

85. Lewis, M. J. and Pelham, H. R. B. (1992) Ligand-induced redistribution of a human KDEL receptor from the Golgi complex to the endoplasmic reticulum. *Cell*, **68,** 353.

86. Wilson, D. W., Lewis, M. J., and Pelham, H. R. B. (1993) pH-dependent binding of KDEL to its receptor *in vitro*. *J. Biol. Chem.*, **268,** 7465.

87. Semenza, J. C. and Pelham, H. R. B. (1992) Changing the specificity of the sorting receptor for luminal endoplasmic reticulum proteins. *J. Mol. Biol.*, **224**, 1.
88. Townsley, F. M., Wilson, D. W., and Pelham, H. R. B. (1993) Mutational analysis of the human KDEL receptor: distinct structural requirements for Golgi retention, ligand binding and retrograde transport. *EMBO J.*, **12**, 2821.
89. Sweet, D. J. and Pelham, H. R. B. (1992) The *Saccharomyces cerevisiae SEC20* gene encodes a membrane glycoprotein which is sorted by the HDEL retrieval system. *EMBO J.*, **11**, 423.
90. Hsu, V. W., Shah, N., and Klausner, R. D. (1992) A brefeldin A-like phenotype is induced by the overexpression of a human ERD-2–like protein, ELP-1. *Cell*, **69**, 625.
91. Bankaitis, V. A., Aitken, J. F., Cleves, A. E., and Dowhan, W. (1990) An essential role for a phospholipid transfer protein in yeast Golgi function. *Nature*, **347**, 561.
92. Cleves, A. E., McGee, T. P., and Bankaitis, V. A. (1991) Phospholipid transfer proteins: a biological debut. *Trends Cell Biol.*, **1**, 30.
93. Cleves, A. E., McGee, T. P., Whitters, E. A., Champion, K. M., Aitken, J. R., Dowhan, W., Goebl, M., and Bankaitis, V. A. (1991) Mutations in the CDP-choline pathway for phospholipid biosynthesis bypass the requirement for an essential phospholipid transfer protein. *Cell*, **64**, 789.
94. McGee, T. P., Skinner, H. B., Whitters, E. A., Henry, S. A., and Bankaitis, V. A. (1994) A phosphatidylinositol transfer protein controls the phosphatidylcholine content of yeast Golgi membranes. *J. Cell Biol.*, **124**, 273.
95. Nilsson, T., Jackson, M. R., and Peterson, P. A. (1989) Short cytoplasmic sequences serve as retention signals for transmembrane proteins in the endoplasmic reticulum. *Cell*, **58**, 707.
96. Jackson, M. R., Nilsson, T., and Peterson, P. A. (1990) Identification of a consensus motif for retention of transmembrane proteins in the endoplasmic reticulum. *EMBO J.*, **9**, 3153.
97. Lotteau, V., Teyton, L., Peleraux, A., Nilsson, T., Karlsson, L., Schmid, S. L., Quaranta, V., and Peterson, P. A. (1990) Intracellular transport of class II MHC molecules directed by invariant chain. *Nature*, **348**, 600.
98. Jackson, M. J., Nilsson, T., and Peterson, P. A. (1993) Retrieval of transmembrane proteins to the endoplasmic reticulum. *J. Cell Biol.*, **121**, 317.
99. Cosson, P. and Letourner, F. (1994) Coatomer interaction with di-lysine endoplasmic reticulum retention motifs. *Science*, **263**, 1629.
100. Schindler, R., Itin, C., Zerial, M., Lottspeich, F., and Hauri, H. P. (1993) ERGIC-53, a membrane protein of the ER–Golgi intermediate compartment, carries an ER retention motif. *Eur. J. Cell Biol.*, **61**, 1.
101. Hauri, H. P. and Schweizer, A. (1992) The endoplasmic reticulum–Golgi intermediate compartment. *Curr. Opin. Cell Biol.*, **4**, 600.
102. Mitoma, J. and Ito, A. (1992) The carboxy-terminal 10 amino acid residues of the cytochrome b5 are necessary for its targeting to the endoplasmic reticulum. *EMBO J.*, **11**, 4197.
103. Ahn, K., Szczesna-Skorupa, E., and Kemper, B. (1993) The amino-terminal 29 amino acids of cytochrome P450 2C1 are sufficient for retention in the endoplasmic reticulum. *J. Biol. Chem.*, **268**, 18726.
104. Maass, D. R. and Atkinson, P. H. (1994) Retention by the endoplasmic reticulum of rotavirus VP7 is controlled by three adjacent amino-terminal residues. *J. Virol.*, **68**, 366.
105. Bonifacino, J. S. and Lippincott-Schwartz, J. (1991) Degradation of proteins within the endoplasmic reticulum. *Curr. Opin. Cell Biol.*, **3**, 592.

106. Shin, J., Lee, S., and Strominger, J. L. (1993) Translocation of TCRα chains into the lumen of the endoplasmic reticulum and their degradation. *Science*, **259**, 1901.

107. Hurtley, S. M. and Helenius, A. (1989) Protein oligomerization in the endoplasmic reticulum. *Annu. Rev. Cell Biol.*, **5**, 277.

108. Griffiths, G. and Rottier, P. J. M. (1992) Cell biology of viruses that assemble along the biosynthetic pathway. *Semin. Cell Biol.*, **3**, 367.

109. Machamer, C. E., Mentone, S. A., Rose, J. K., and Farquhar, M. G. (1990) The E1 glycoprotein of an avian coronavirus is targeted to the *cis* Golgi complex. *Proc. Natl Acad. Sci. USA*, **87**, 6944.

110. Machamer, C. E. and Rose, J. K. (1987) A specific transmembrane domain of a coronavirus E1 glycoprotein is required for its retention in the Golgi region. *J. Cell Biol.*, **105**, 1205.

111. Swift, A. M. and Machamer, C. E. (1991) A Golgi retention signal in a membrane-spanning domain of coronavirus E1 protein. *J. Cell Biol.*, **115**, 19.

112. Machamer, C. E., Grim, M. G., Esquela, A., Chung, S. W., Rolls, M., Ryan, K., and Swift, A. M. (1993) Retention of a *cis* Golgi protein requires polar residues on one face of a predicted α-helix in the transmembrane domain. *Mol. Biol. Cell*, **4**, 695.

113. Shaper, J. H. and Shaper, N. L. (1992) Enzymes associated with glycosylation. *Curr. Opin Struct. Biol.*, **2**, 701.

114. Machamer, C. E. (1993) Targeting and retention of Golgi membrane proteins. *Curr. Opin. Cell Biol.*, **5**, 606.

115. Teasdale, R. D., D'Agnostaro, G., and Gleeson, P. A. (1992) The signal for Golgi retention of bovine β1,4-galactosyltransferase is in the transmembrane domain. *J. Biol. Chem.*, **267**, 9241.

116. Nilsson, T., Lucocq, J. M., Mackay, D., and Warren, G. (1991) The membrane spanning domain of β1,4-galactosyltransferase specifies *trans* Golgi localization. *EMBO J.*, **10**, 1367.

117. Aoki, D., Lee, N., Yamaguchi, N., Dubois, C., and Fukuda, M. N. (1992) Golgi retention of a *trans*-Golgi membrane protein, galactosyltransferase, requires cysteine and histidine residues within the membrane-anchoring domain. *Proc. Natl Acad. Sci. USA*, **89**, 4319.

118. Masibay, A. S., Balaji, P. V., Boeggeman, E. E., and Qasba, P. K. (1993) Mutational analysis of the Golgi retention signal of bovine β-1,4-galactosyltransferase. *J. Biol. Chem.*, **268**, 9908.

119. Tang, B. L., Wong, S. H., Low, S. H., and Hong, W. (1992) The transmembrane domain of N-glucosaminyltransferase I contains a Golgi retention signal. *J. Biol. Chem.*, **267**, 10122.

120. Burke, J., Pettitt, J. M., Schachter, H., Sakar, M., and Gleeson, P. A. (1992) The transmembrane and flanking sequences of β1,2-N-acetylglucosaminyltransferase I specify *medial* Golgi localization. *J. Biol. Chem.*, **267**, 24433.

121. Wong, S. H., Low, S. H., and Hong, W. (1992) The 17 residue transmembrane domain of β-galactoside α2,6 sialyltransferase is sufficient for Golgi retention. *J. Cell Biol.*, **117**, 245.

122. Munro, S. (1991) Sequences within and adjacent to the transmembrane segment of α2,6-sialyltransferase specify Golgi retention. *EMBO J.*, **10**, 3577.

123. Dahdal, R. Y. and Colley, K. J. (1993) Specific sequences in the signal anchor of the β-galactoside α2,6-sialyltransferase are not essential for Golgi localization. *J. Biol. Chem.*, **268**, 26310.

124. Russo, R. N., Shaper, N. L., Taatjes, D. J., and Shaper, J. H. (1992) β1,4-Galactosyltransferase:

a short NH2-terminal fragment that includes the cytoplasmic and transmembrane domain is sufficient for Golgi retention. *J. Biol. Chem.*, **267**, 9241.

125. Colley, K. J., Lee, E. U., and Paulson, J. C. (1992) The signal anchor and stem regions of the β-galactoside α2,6–sialyltransferase may each act to localize the enzyme to the Golgi apparatus. *J. Biol. Chem.*, **267**, 7784.

126. Weisz, O. A., Swift, A. M., and Machamer, C. E. (1993) Oligomerization of a membrane protein correlates with its retention in the Golgi complex. *J. Cell Biol.*, **122**, 1185.

127. Nilsson, T., Hoe, M. H., Slusarewicz, P., Rabouille, C., Watson, R., Hunte, F., Watzele, G., Berger, E. G., and Warren, G. (1994) Kin recognition between medial Golgi enzymes in HeLa cells. *EMBO J.*, **13**, 562.

128. Nilsson, T., Slusarewicz, P., Hoe, M. H., and Warren, G. (1993) Kin recognition. A model for the retention of Golgi enzymes. *FEBS Lett.*, **330**, 1.

129. Slusarewicz, P., Nilsson, T., Hui, N., Watson, R., and Warren, G. (1994) Isolation of a matrix that binds *medial* Golgi enzymes. *J. Cell Biol.*, **124**, 405.

130. Tulsiani, D. R. P. and Touster, O. (1988) The purification and characterization of mannosidase IA from rat liver Golgi membranes. *J. Biol. Chem.*, **263**, 5408.

131. Hiraizumi, S., Spohr, U., and Spiro, R. G. (1994) Ligand affinity chromatographic purification of rat liver Golgi endomannosidase. *J. Biol. Chem.*, **269**, 4697.

132. Fleischer, B., McIntyre, J. O., and Kempner, E. S. (1993) Target sizes of galactosyltransferase, sialyltransferase, and uridine diphosphatase in Golgi apparatus of rat liver. *Biochemistry*, **32**, 2076.

133. Mandon, E., Kempner, E. S., Ishihara, M., and Hirschberg, C. B. (1994) A monomeric protein in the Golgi membrane catalyzes both *N*-deacetylation and *N*-sulfation of heparan sulfate. *J. Biol. Chem.*, **269**, 11729.

134. Bretscher, M. S. and Munro, S. (1993) Cholesterol and the Golgi apparatus. *Science*, **261**, 1280.

135. Orci, L., Montesano, R., Meda, P., Malaisse-Lagae, F., Brown, D., Perrelet, A., and Vassalli, P. (1981) Heterogeneous distribution of filipin–cholesterol complexes across the cisternae of the Golgi apparatus. *Proc. Natl Acad. Sci. USA*, **78**, 293.

136. Kumar, R., Yang, J., Larsen, R. D., and Stanley, P. (1990) Cloning and expression of *N*-acetylglucosaminyltransferase I, the *medial* Golgi transferase that initiates complex *N*-linked carbohydrate formation. *Proc. Natl Acad. Sci. USA*, **87**, 9948.

137. Yamamoto, F., Marken, J., Tsuji, T., White, T., Clausen, H., and Hakomori, S. (1990) Cloning and characterization of DNA complementary to human UDP-GalNAc:Fucα1,2Galα1,3GalNAc transferase (histo-blood group A transferase) mRNA. *J. Biol. Chem.*, **265**, 1146.

138. Hashimoto, Y., Orellana, A., Gil, G., and Hirschberg, C. B. (1992) Molecular cloning and expression of rat liver *N*-heparan sulfate sulfotransferase. *J. Biol. Chem.*, **267**, 15744.

139. Allan, D. and Kallen, K. J. (1993) Transport of lipids to the plasma membrane in animal cells. *Prog. Lipid Res.*, **32**, 195.

140. Futerman, A. H. and Pagano, R. E. (1991) Determination of the intracellular sites and topology of glucosylceramide synthesis in rat liver. *Biochem. J.*, **280**, 295.

141. Thompson, T. E. and Tillack, T. W. (1985) Organization of glycosphingolipids in bilayers and plasma membranes of mammalian cells. *Annu. Rev. Biophys. Biophys. Chem.*, **14**, 361.

142. Luzio, J. P., Brake, B., Banting, G., Howell, K. E., Braghetta, P., and Stanley, K. K. (1990) Identification, sequencing and expression of an integral membrane protein of the *trans*-Golgi network (TGN38). *Biochem. J.*, **270**, 97.

143. Humphrey, J. S., Peters, P. J., Yuan, L. C., and Bonifacino, J. S. (1993) Localization of TGN38 to the *trans*-Golgi network: involvement of a cytoplasmic tyrosine-containing motif. *J. Cell Biol.*, **120**, 1123.

144. Stanley, K. K. and Howell, K. E. (1993) TGN38/41: a molecule on the move. *Trends Cell Biol.*, **3**, 252.

145. Trowbridge, I. S. (1991) Endocytosis and signals for internalization. *Curr. Opin. Cell Biol.*, **3**, 634.

146. Molloy, S. S., Thomas, L., VanSlyke, J. K., Stenberg, P. E., and Thomas, G. (1994) Intracellular trafficking and activation of the furin proprotein convertase: localization to the TGN and recycling to the cell surface. *EMBO J.*, **13**, 18.

147. Wilsbach, K. and Payne, G. S. (1993) Dynamic retention of TGN membrane proteins in *Saccharomyces cerevisiae*. *Trends Cell Biol.*, **3**, 426.

148. Seeger, M. and Payne, G. S. (1992) Selective and immediate effects of clathrin heavy chain mutations on Golgi membrane protein retention in *Saccharomyces cerevisiae*. *J. Cell Biol.*, **118**, 531.

149. Yuan, L., Barriocanal, J. G., Bonifacino, J. S., and Sandoval, I. V. (1987) Two integral membrane proteins located in the *cis*-middle and *trans*-part of the Golgi system acquire sialylated N-linked carbohydrates and display different turnovers and sensitivity to cAMP-dependent phosphorylation. *J. Cell Biol.*, **105**, 215.

150. Gonatas, J. O., Mezitis, S. G. E., Stieber, A., Fleischer, B., and Gonatas, N. K. (1989) MG160, a novel sialoglycoprotein of the *medial* cisternae of the Golgi apparatus. *J. Biol. Chem.*, **264**, 646.

151. Login, G. R. and Dvorak, A. M. (1985) Microwave energy fixation for electron microscopy. *Am. J. Pathol.*, **120**, 230.

6 | Biogenesis of constitutive secretory vesicles, secretory granules, and synaptic vesicles—facts and concepts

MASATO OHASHI, RUDOLF BAUERFEIND, and
WIELAND B. HUTTNER

1. Introduction

Vesicle-mediated secretion from animal cells utilizes three types of secretory vesicles (1, 2) (Fig. 1). The first type is the constitutive secretory vesicle (CSV), which is thought to be common to all eukaryotic cells and which mediates the continuous secretion of newly synthesized proteins. The second type is the secretory granule, in neurones also referred to as the large dense core vesicle, which occurs only in cells capable of regulated protein secretion and which mediates the stimulus-dependent release of stored proteins and biogenic amines. The third type is the synaptic vesicle of neurones, which mediates the release of neurotransmitters but lacks secretory proteins, and its counterpart in certain endocrine cells, the synaptic-like microvesicle (SLMV). In this chapter, we will summarize current knowledge, and describe concepts concerning the molecular machinery involved in the biogenesis of these vesicles, its regulation, and the protein sorting processes that occur in the course of vesicle formation.

2. Biogenesis of constitutive secretory vesicles

CSVs are formed in the *trans* Golgi network (TGN) and deliver proteins to the plasma membrane (3). In cells which lack a polar distribution of plasma membrane proteins, only one class of CSVs is thought to exist. In contrast, polarized cells such as certain epithelial cells and neurones form two classes of CSVs in the TGN, which deliver proteins directly to the apical/axonal and basolateral/dendritic domains of the plasma membrane. Certain other epithelial cells such as hepatocytes are believed to lack apical CSVs, and in these cells both apical and basolateral membrane proteins are thought to leave the TGN in basolateral CSVs followed by transcytosis of

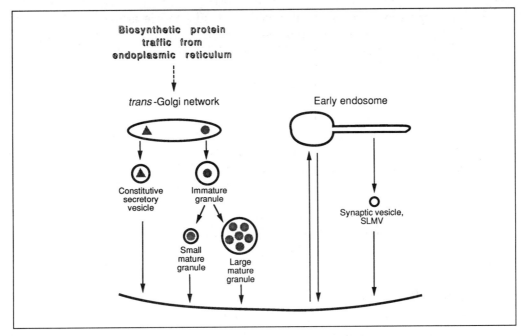

Fig. 1 Biogenesis of three types of secretory vesicles. (i) Constitutive secretory vesicles mediate the secretion of constitutive secretory proteins (shaded triangles). (ii) Secretory granules mediate the secretion of regulated secretory proteins (shaded circles). Secretory granules originate as vesicular intermediates, the immature secretory granules, and mature into either small mature or large mature secretory granules. (iii) Synaptic vesicles of neurones and their endocrine counterpart, the synaptic-like microvesicles (SLMVs), mediate the secretion of classical neurotransmitters.

the apical proteins from the basolateral to the apical plasma membrane. The latter route is taken by some, but not all, apical proteins in certain epithelial cells which form both apical and basolateral CSVs. The subject of cell polarity and the sorting of apical and basolateral proteins has been extensively reviewed (4–6), and will not be dealt with further.

Besides CSVs, at least two other classes of vesicles are formed in the TGN. One is the immature secretory granule (ISG) which originates from the TGN in cells containing the regulated pathway of protein secretion and whose biogenesis will be discussed in Section 3. The other is the clathrin-coated vesicle involved in the delivery of lysosomal enzymes to lysosomes. Membrane traffic to lysosomes and the sorting of lysosomal proteins have also been reviewed in depth (7–9) and will not be discussed further, except for the destination of the TGN-derived, clathrin-coated, lysosomal enzyme-containing vesicle, a topic that may be relevant for constitutive protein secretion.

Two destinations of this vesicle have been discussed, i.e. the early and the late endosome (7–9). If the acceptor compartment of the clathrin-coated, lysosomal enzyme-containing vesicle derived from the TGN should indeed turn out to be the early endosome, this vesicle may contribute to constitutive protein secretion.

Although the membrane composition of this vesicle is highly selective, reflecting the clustering of the mannose-6-phosphate receptor in type 1 adaptor/clathrin-coated pits of the TGN (10), this vesicle is likely to include, in the course of its formation, some of the fluid phase of the TGN, which contains constitutive secretory proteins. Upon delivery to early endosomes, at least part of these soluble proteins would be expected to be secreted by bulk flow via the constitutively recycling vesicles which return endocytosed receptors such as the transferrin receptor from the early endosome to the cell surface (11). The efficiency with which constitutive secretory proteins would be delivered from the TGN to the cell surface via this route is unclear. However, it should be borne in mind that a vesicular traffic pathway from the TGN to early endosomes could mediate some constitutive protein secretion, whereas this is unlikely to be the case if the clathrin-coated vesicles derived from the TGN would be destined to late endosomes.

In the following, we will discuss the basic aspects of formation of CSVs, irrespective of the specific features of polarized cells.

2.1 Soluble proteins

The proteinaceous material packaged into the CSV lumen is thought to be any protein that is soluble in the lumen of the TGN (1, 12, 13). Entry of these proteins into the forming CSV is believed to occur without the need for any specific protein–protein interaction, i.e. by 'bulk flow' (14) or 'default'. This implies that soluble proteins are packaged into CSVs at the same concentration at which they are present in the fluid phase of the lumen of the TGN. Proteins that would be soluble in the TGN include:

- Constitutive secretory proteins (under normal circumstances).
- Secretory proteins which are normally packaged into secretory granules (the regulated secretory proteins, see below), if they fail to aggregate in the TGN or fail to interact with the granule-forming zone of the TGN membrane (13).
- Lysosomal enzymes that have escaped the mannose-6-phosphate receptor-mediated delivery to endosomes (7–9).
- Soluble proteins which normally reside in the endoplasmic reticulum (ER), if they have escaped the ER retention mechanism (15).
- Membrane-associated proteins which normally reside in the ER and Golgi complex, if they lose their membrane association, e.g. by proteolytic cleavage of their membrane anchor (16).

The bulk flow concept of explaining the lumenal content of CSVs does not imply that different constitutive secretory proteins exit from the TGN with the same kinetics, nor that the same constitutive secretory protein exits with the same kinetics irrespective of its post-translational modifications. The TGN membrane, like the ER, presumably has 'stationary' zones, i.e. membrane zones containing TGN-resident proteins mediating TGN-specific functions such as certain post-translational modifications, which are distinct from 'exit' zones, i.e. zones of the TGN membrane at which vesicles bud (see Fig. 2 for details). The longer any given constitutive secretory protein transiently interacts with these stationary zones of the TGN membrane (this interaction being not necessarily with the modifying enzymes but, for example, with lipids), the slower its kinetics of exit will be. It is conceivable that different constitutive secretory proteins, or the same constitutive secretory protein in different states of post-translational modification (17), exhibit different degrees of transient interaction with the TGN stationary zones, but none the less exit from the TGN by bulk flow. The bulk flow concept also does not imply that the same constitutive secretory protein exits from the TGN necessarily with the same kinetics in different states of the cell. As will be discussed below, the formation of CSVs is subject to both positive and negative regulation, and different kinetics of exit from the TGN of the same constitutive secretory protein in different conditions may simply reflect an acceleration or deceleration of CSV-mediated bulk flow out of the TGN.

2.2 Membrane proteins

Do membrane proteins enter the forming CSV also by bulk flow, i.e. without the need for any specific protein–lipid or protein–protein interaction? Before addressing this question, let us first consider some principles of bulk flow of membrane constituents. A frequently held, but none the less erroneous, view of bulk flow along the secretory pathway is that there is no concentration of the cargo. The following three considerations, which take into account characteristic features of the secretory pathway, make the point that in bulk flow along the secretory pathway, there may well be concentration of cargo.

The first of these considerations concerns the organization of the membrane of a compartment from which a vesicle buds (see Fig. 3 for details). The membrane of compartments like the ER and the TGN are presumably characterized by the existence of at least two types of zones. One would be the above-mentioned stationary zones which harbour the proteins mediating compartment-specific functions (for example, the protein translocation machinery in the case of the ER, or the machinery for certain post-translational modifications in the case of the TGN). The other would be zones characterized by 'mobile' membrane constituents such as exit zones where vesicles bud and entry zones where vesicles fuse. Membrane constituents that are not resident in a given compartment are likely to be largely excluded from the stationary zones. From this it follows that if these membrane constituents are distributed randomly over the membrane area that is actually available in this compartment, they will be more concentrated in the mobile zones than in membrane domains rich in stationary zones. For example, a newly synthesized membrane protein destined for the plasma membrane will be more concentrated in those areas of the ER membrane which are not engaged in protein translocation, such as the ER exit zones, than in areas of the ER membrane that are, simply because a large proportion of the latter zone of the ER membrane is occupied by the protein translocation machinery itself. In this case, the concentration factor would reflect the ratio of

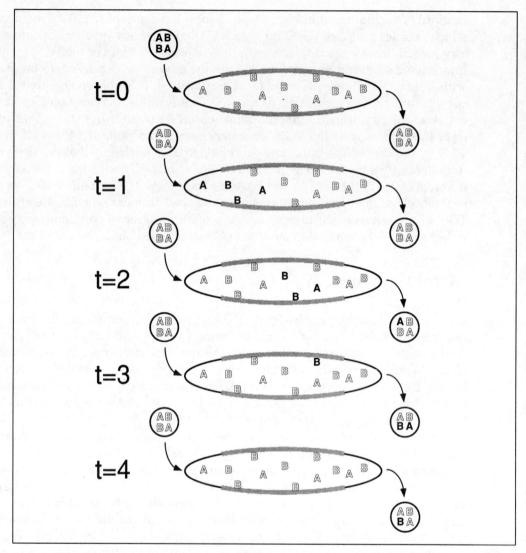

Fig. 2 Differential kinetics of the movement of two constitutive secretory proteins (A and B) through the TGN by bulk flow, as revealed in a pulse–chase experiment at five different chase times ($t = 0$ to $t = 4$). Labelled molecules of A and B are shown in black, unlabelled ones in white. $t = 0$, a vesicle destined to fuse with the TGN contains labelled A and B. $t = 1$, labelled A and B are in the TGN; A has no tendency for a transient interaction with the TGN stationary zone (thick, shaded membrane areas), whereas B does. This results in (i) half the B molecules interacting with the TGN stationary zone in steady state, (ii) the pool size of B molecules in the TGN being twice as large as that of A molecules, and (iii) the kinetics of exit of labelled B molecules being twice as slow as that of labelled A molecules ($t = 2$ to $t = 4$).

ER exit zones to ER stationary zones. (The recently reported observation (18) that the VSV-G protein is concentrated at the exit zones of the ER relative to the cisternal parts of the ER, most of which are engaged in protein translocation, signal peptide cleavage, N-glycosylation, etc., is therefore not evidence against bulk flow; rather, this is precisely what bulk flow would predict.)

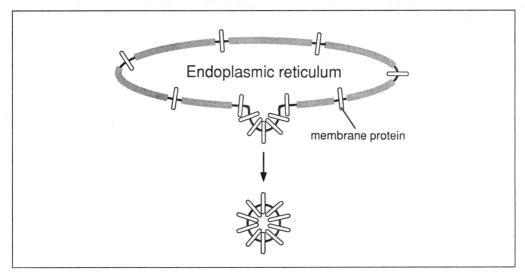

Fig. 3 Exit of a membrane protein from the endoplasmic reticulum (ER) by bulk flow. The membrane protein does not diffuse laterally into the ER stationary zones (thick, shaded membrane areas) because of steric hindrance. This results in its concentration being higher at the exit zone of the ER and in the transport vesicle than in the areas of the ER membrane rich in stationary zones.

The second consideration takes into account the recycling of membrane constituents (for details see Fig. 4, left). A vesicle mediating traffic between two compartments, say from the ER to the Golgi complex, contains as membrane constituents not only cargo to be delivered to the Golgi complex for further forward transport, but also membrane constituents to be recycled from the Golgi complex to the ER, such as the KDEL receptor (15). The sorting of the latter membrane constituents into a recycling vesicle, for example by a specific interaction with a cytosolic sorting machinery, will automatically cause the concentration of those membrane constituents that are not recycled. In this case, the concentration factor would reflect the ratio between recycled and forward moving membrane.

The third consideration (for details see Fig. 4, right) concerns the fact that a compartment such as the TGN distributes the mixture of membrane constituents it receives from the proximal compartments of the secretory pathway into at least two post-TGN vesicles, the clathrin-coated vesicle destined to endosomes and the CSV destined to the plasma membrane (3). The fact that certain newly synthesized membrane constituents are sorted into the clathrin-coated vesicle implies that those membrane constituents which are not sorted will be more concentrated in the CSV leaving from the TGN than in the transport vesicle that delivered them to the TGN. This is a variation of the second theme discussed earlier, i.e. that segregation of a mixture of membrane constituents by the selective removal of one results in the concentration of those remaining.

From these three considerations, it follows that membrane constituents may become more concentrated as they travel by bulk flow from the ER to the Golgi complex and beyond. In other words, a concentration of membrane cargo along the

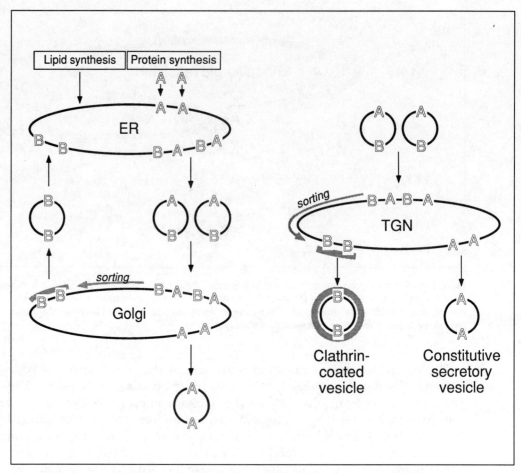

Fig. 4 Movement of a membrane protein (A) by bulk flow from the ER to the plasma membrane, and its concentration resulting from the sorting and removal of another membrane protein (B). *Left*, selective recycling of the ER-resident membrane protein B. Let us assume that (i) the ER has a net production of two molecules of A and an amount of lipid bilayer corresponding to one vesicle per unit time, (ii) the ER has no net production of the ER-resident membrane protein B, and (iii) the concentrations of A and B in the exit zones of the ER membrane are equal. If two vesicles, each containing a mixture of one molecule of A and one molecule of B, leave the ER, one vesicle containing two molecules of B has to recycle from the Golgi to the ER for the ER to be in steady state. Formation of the latter vesicle from the Golgi involves the sorting of B into this vesicle via a cytoplasmic machinery (curved shaded bar). For the Golgi to be in steady state, another vesicle containing two molecules of A has to leave the Golgi. Hence, although the movement of A is by bulk flow, its concentration in the vesicle leaving the Golgi will be twice as high as that in the vesicles transporting A from the ER to the Golgi. *Right*, sorting of B upon exit from the TGN. Let us assume that per unit of time, two vesicles, each containing a mixture of one molecule of A and one molecule of B, enter the TGN and two vesicles leave the TGN. If the two molecules B delivered into the TGN are sorted into one vesicle leaving the TGN via a clathrin-involving process (shaded coat), the two molecules A delivered into the TGN will exit in the other vesicle in steady state. Again, although the movement of A is by bulk flow, its concentration in the vesicle leaving the TGN will be twice as high as that in the vesicles delivering A into the TGN.

secretory pathway does not necessarily argue against bulk flow but can be consistent with it. Bearing this in mind, the key question with respect to the formation of CSV from the TGN is whether all, or only some (19), membrane constituents enter CSVs by bulk flow. Having outlined above that studying cargo concentration may not always be an appropriate way to distinguish between these two possibilities, the answer to this question requires the investigation of whether or not cargo molecules need to undergo a specific protein–lipid or protein–protein interaction while being packaged into CSVs.

2.3 Molecular machinery

What about the cytoplasmic machinery mediating the formation of CSVs? The formation of CSVs from the TGN has been reconstituted in various systems (20–25), and the molecular machinery involved in CSV formation has begun to be elucidated using these systems. CSV formation seems to involve an ADP-ribosylation factor (ARF) protein since it is stimulated by myristoylated peptides corresponding to the N-terminal domain of ARF1 and inhibited by the corresponding non-myristoylated peptides of ARF1 and ARF4 (F. A. Barr and W. B. Huttner, submitted). ARF is a small GTP-binding protein (26, 27) that in the GTP-bound state binds to membranes (28, 29) and mediates the recruitment of cytosolic coat proteins of both the clathrin (30, 31) and non-clathrin (coatomer) (28, 32) type to Golgi membranes (see discussion in ref. 33). The involvement of ARF in CSV formation explains why brefeldin A inhibits this process (34, 35); brefeldin A blocks ARF binding to membranes by inhibiting the guanine nucleotide exchange on ARF (36, 37). CSV formation also requires cytosol (F. A. Barr and W. B. Huttner, submitted; M. Ohashi and W. B. Huttner, unpublished observations). If the requirement for cytosol reflects the need for coat proteins, they may be distinct from the previously characterized coatomer (38) since coatomer-depleted cytosol still supports CSV formation (F. A. Barr and W. B. Huttner, submitted).

2.4 Regulation

The use of a cell-free system (21) also revealed that the machinery mediating CSV formation includes at least two types of regulatory elements. One type are the heterotrimeric G-proteins (Fig. 5). At least four distinct trimeric G-proteins have been found to be associated with the TGN membrane, i.e. G_{i2}, G_{i3}, G_{o2}, and G_s (39, 40). Various agents known to affect trimeric G-proteins have been found to regulate CSV formation. The inhibition of CSV formation is thought to be mediated by activation of TGN-associated $G_{i/o}$, whereas the stimulation of CSV formation appears to involve activation of TGN-associated G_s (39, 40).

The other type of regulatory element is an, as yet, unidentified cytosolic phosphoprotein (83). Phosphorylation of this protein on serine/threonine antagonizes the inhibition of CSV formation observed upon activation of $G_{i/o}$. This phenomenon could be explained by the phosphoprotein either augmenting the G_s-mediated

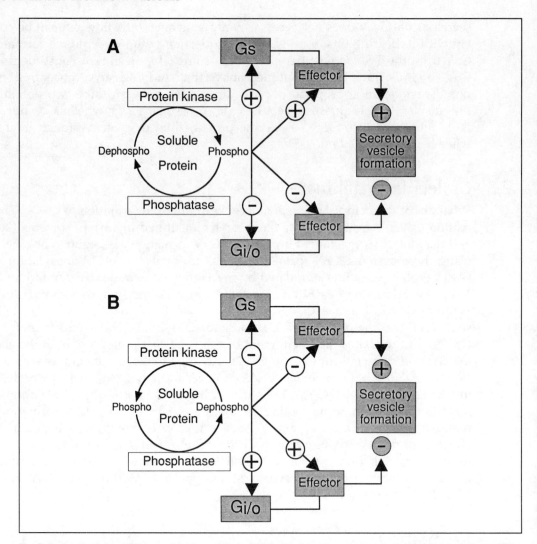

Fig. 5 Regulation of secretory vesicle formation by TGN-associated heterotrimeric G_s and $G_{i/o}$ proteins and the phospho- (top) or dephospho- (bottom) form a cytosolic protein.

effects or inhibiting the $G_{i/o}$-mediated effects (Fig. 5). In the absence of G-protein activators, the inhibition of neither protein kinases nor serine/threonine protein phosphatases significantly inhibits CSV formation, showing that phosphorylation–dephosphorylation of this cytosolic protein indeed exerts a regulatory rather than an obligatory role in this process (83). These findings differ from those reported by Howell and colleagues (41), who showed that (i) a cytosolic protein (p62) forms a complex with the small GTP-binding protein rab6, (ii) this complex interacts with TGN-38 in the course of CSV formation, (iii) this complex is required for CSV formation in a cell-free system, (iv) p62 is a phosphoprotein, and (v) phosphorylation of p62 results in its dissociation from the membrane. The reason for these divergent observations, i.e. phosphorylation of a cytosolic protein stimulating CSV formation (83) vs. phosphorylation of p62 being expected to inhibit CSV formation (41) is presently unknown, but may reflect differences in the cell-free systems used (PC12 cells vs. liver).

3. Biogenesis of secretory granules

The biogenesis of secretory granules has recently been reviewed (13). It comprises two major steps: (i) the formation of a vesicular intermediate, the immature secretory granule (ISG), from the TGN, and (ii) the maturation of the ISG to the mature secretory granule (MSG) which is stored in the cell (Fig. 1).

3.1 Immature secretory granules

Both morphological and biochemical evidence has shown that ISGs are formed in the TGN. ISGs and CSVs are formed concomitantly, and thus their formation entails the differential distribution of soluble and membrane proteins between these two classes of post-Golgi secretory vesicles (1).

3.1.1 Soluble proteins

The soluble proteins typically found in ISGs largely comprise regulated secretory proteins. It is now widely accepted that the selective aggregation of regulated secretory proteins is part of the mechanism which ensures their efficient segregation from constitutive secretory proteins (13). The biochemical evidence in support of this concept has largely come from studies on the granins (chromogranins/secretogranins), a family of regulated secretory proteins sorted to secretory granules in a wide variety of endocrine cells and neurones (42). In the case of the granins, it has been shown that their selective aggregation in the TGN is induced by the specific lumenal milieu of this compartment, in particular the low pH and the elevated calcium concentration (43). We find it likely that milieu-induced aggregation also occurs for other regulated secretory proteins.

To generate a properly assembled secretory granule, the aggregated regulated secretory proteins must somehow interact with a membrane component. The nature of the structure on regulated secretory proteins that mediates this interaction, referred to as an S-M (Secretory protein-Membrane recognition) signal, remains to be established. Recent work on chromogranin B, a member of the granin family (42), has suggested that such an S-M signal is associated with the polypeptide loop formed by the single intramolecular disulfide bond in this protein (44). This loop has long been suspected to have a role in sorting as it is highly homologous between chromogranin B and chromogranin A (45). Reduction of this disulfide bond (caused by addition of dithiothreitol to intact cells) results in the selective misrouting of chromogranin B to CSVs, whereas the packaging into ISGs of the related regulated secretory protein secretogranin II, another member of the granin family which lacks disulfide bonds (46), was unaffected. The reduced chromogranin B still exhibited low pH/calcium-induced aggregation (44). These observations suggest that in the case of chromogranin B, the conformational stabilization of the loop by the disulfide bond is not required for aggregation but rather for its interaction with the membrane. Our conclusion that the 20 amino acid residues between the two cysteines of chromogranin B contain an S-M signal is fully consistent with a recent report showing the binding to secretory granule membranes of a synthetic peptide corresponding to the related loop of chromogranin A (47).

The observation that reduction of the disulfide bond of chromogranin B is sufficient to inactivate the putative S-M signal is significant, given that the sequence comparison of a wide variety of regulated secretory proteins has not revealed a consensus sequence that qualifies as an S-M signal. Perhaps the S-M signal in regulated secretory proteins is very degenerate or consists of only a few amino acid residues, and it is their correct three-dimensional configuration which is crucial for membrane recognition. Peng Loh and colleagues (48) have reported that the N-terminal 24 amino acids of pro-opiomelanocortin (POMC), which form a loop stabilized by two intramolecular disulfide bonds, are able to direct a reporter protein to the regulated pathway. There is no extensive sequence homology between the chromogranin B and the POMC loops except, perhaps, for an L-S/T-X-X-S motif. This leads us to propose that the hallmark of the S-M signal of regulated secretory proteins may be a loop structure containing hydroxylamino acids in a degenerate sequence but receptor-recognizable three-dimensional configuration. It should be realized that aggregates of regulated secretory proteins would constitute multivalent ligands, and thus efficient binding of these aggregates to the putative membrane receptor would occur even if the affinity of the individual regulated secretory proteins for this receptor is relatively low.

3.1.2 Membrane proteins

In contrast to the traffic of secretory proteins, which has been studied both from the TGN to ISGs and from ISGs to MSGs, the traffic of membrane proteins from the TGN to secretory granules has not been investigated in comparable detail. Studies on the sorting of membrane proteins are largely confined to describing their presence in MSGs (49), without analysis of their exact traffic route. However, although direct evidence is lacking, it is probably reasonable to assume that membrane proteins characteristic of MSGs are segregated from plasma membrane proteins at the level of the TGN, i.e. that they, like secretory proteins, take the TGN→ISG→MSG route. We shall therefore discuss the sorting of secretory granule membrane proteins in this section, which addresses the formation of ISGs.

There are three main possibilities as to how membrane proteins could interact with the aggregated regulated secretory proteins to generate a properly assembled secretory granule (see Fig. 6 for details). The first possibility is that every granule membrane protein co-aggregates, via its lumenal domains, with the regulated secretory proteins (Fig. 6a). In this case, aggregation alone would be sufficient for the proper assembly of secretory granules. Second, some, but not all, granule mem-

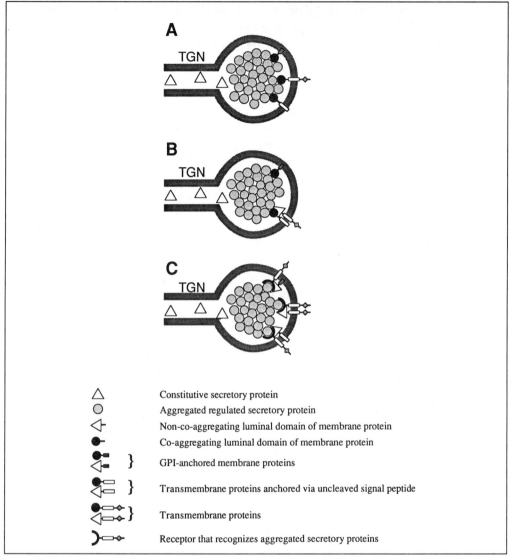

Fig. 6 Diagrams illustrating the three main possibilities as to how aggregation of regulated secretory proteins (grey circles) can direct membrane proteins into secretory granules. (A) Co-aggregation of all membrane proteins via their lumenal domains (black circles). (B) Co-aggregation of some membrane proteins which in turn interact with other membrane proteins that do not co-aggregate via their lumenal domains (attached triangles). (C) Recognition of the aggregate by a receptor which in turn interacts with membrane proteins that do not co-aggregate via their lumenal domains (attached triangles).

brane proteins co-aggregate with the regulated secretory proteins (Fig. 6b). To generate a properly assembled secretory granule, the granule membrane proteins that co-aggregate with the regulated secretory proteins would also have to interact with those granule membrane proteins which do not co-aggregate. This 'lateral' interaction could occur either directly, via granule membrane lipids, or via cytosolic coat proteins assembled on to the cytoplasmic tails of the granule membrane proteins. Third, no granule membrane protein co-aggregates with the regulated secretory proteins (Fig. 6c). In this case, one would have to postulate the existence of a receptor that on the one hand recognizes the aggregated regulated secretory proteins and on the other hand interacts with the granule membrane proteins, via the mechanisms mentioned above. The second and third possibilities are similar, in that interactions between membrane proteins are required in addition to the binding of membrane proteins to secretory proteins. The third possibility, however, differs from the second in the existence of a membrane protein (the receptor) which has no function other than (i) providing the link between regulated secretory proteins and the granule membrane in the course of granule formation and (ii) being part of a signal transduction machinery from the TGN lumen to the cytoplasm (see below). The observations summarized below suggest that co-aggregation of membrane proteins with regulated secretory proteins is a major mechanism in their segregation from plasma membrane proteins, although lateral interactions may also contribute.

Several types of secretory granule membrane proteins can be distinguished, (i) transmembrane proteins with a cytoplasmic domain, (ii) lumenal proteins with a glycosylphosphatidylinositol (GPI) anchor, and (iii) lumenal proteins which exhibit a tight, usually poorly understood, association with the membrane (Fig. 6). Several studies have examined the cytoplasmic domain in the type I (i.e. N-terminus lumenal) transmembrane proteins for the possible presence of signals for sorting to secretory granules. In the case of P-selectin, a membrane protein localized in the Weibel–Palade bodies of endothelial cells, Disdier et al. (50), using transfection of P-selectin mutants into AtT-20 cells, identified a 23 amino acid C-terminal sequence in the cytoplasmic tail necessary and sufficient for the sorting of this protein to secretory granules. As might be expected, the, as yet unidentified, cytoplasmic machinery recognizing this signal appears to operate only in regulated secretory cells since wild-type P-selectin expressed in constitutive cells resides at the plasma membrane (50, 51). Also in the case of peptidylglycine α-amidating monooxygenase (PAM), a processing enzyme found in secretory granules of neuroendocrine cells (52), the C-terminal cytoplasmic domain has been shown to contain structural information for sorting to secretory granules (53). Interestingly, however, Milgram et al. (54) found that whereas PAM lacking most of its cytoplasmic domain was no longer sorted, soluble PAM lacking both the transmembrane and cytoplasmic domains was sorted to secretory granules. One possible explanation for these observations is that soluble PAM was sorted via co-aggregation with endogenous regulated secretory proteins, whereas PAM containing the transmembrane anchor was not, perhaps because its membrane association prevented efficient co-aggregation. On the other hand, the presence of a membrane anchor does not necessarily preclude sorting of a granule membrane protein by co-aggregation. GP-2 of zymogen granules, which exists in a soluble and a GPI-anchored form, is thought to be sorted via low pH/calcium-induced co-aggregation (55). The same may be true for dopamine β-hydroxylase of chromaffin granules, which also exists in soluble and membrane-bound form (the latter anchored to the membrane by an uncleaved signal peptide) (56). In fact, the picture emerges that the difference between soluble proteins and membrane proteins of secretory granules is more a quantitative than a qualitative one, because not only do 'membrane' proteins of secretory granules, as just outlined, also exist in soluble form, but 'soluble' proteins, such as the granins (57) and certain processing enzymes (58), also exist in membrane-associated form. This raises the intriguing possibility that in aggregation-mediated sorting, the membrane-associated form of a lumenal protein mediates the membrane enveloping of the aggregate in the course of ISG formation (57).

3.1.3 Molecular machinery

Present knowledge concerning the cytoplasmic machinery mediating the formation of ISGs, as revealed by the use of a cell-free assay (21), indicates substantial similarities to that mediating the formation of CSVs. The formation of ISGs, too, requires cytosol, seems to involve ARF (F. A. Barr and W. B. Huttner, submitted), and is inhibited by brefeldin A (34, 35). Coatomer-depleted cytosol still supports secretory granule formation (F. A. Barr and W. B. Huttner, submitted). Given that clathrin patches are observed on the membrane of ISGs budding from the TGN (13), the cytosolic proteins recruited on to this membrane by ARF may well be clathrin and the appropriate adaptors.

3.1.4 Regulation

The regulation of ISG formation is similar to that of CSV formation. It involves both TGN-associated heterotrimeric G-proteins of the G_i, G_o, and G_s class (39, 40) and protein phosphorylation–dephosphorylation (83) (Fig. 5). With respect to the regulation of ISG formation by TGN-associated heterotrimeric G-proteins, an unanswered question is how the GDP–GTP exchange on these proteins is controlled. One attractive, though at present speculative, possibility is that a putative receptor recognizing S-M signals on regulated secretory proteins (see above) acts as a guanine nucleotide exchange-promoting factor for the TGN-associated trimeric G-proteins involved in ISG formation.

3.2 Granule maturation

ISGs are a short-lived intermediate in the biogenesis of MSGs (59). There are at least two routes of maturation of the ISG (Fig. 1). ISGs may mature into large MSGs by self-fusion (59) or into small MSGs by condensation of the contents (60). Both routes of maturation entail the removal of the excess membrane, presumably by a vesicle budding from the ISG (Fig. 7). In the following, we will address the content of this ISG-derived vesicle and the fate of its membrane.

3.2.1 Soluble proteins

Given the fact that the aggregated regulated secretory proteins are further condensed as the ISG matures into an MSG (59), the content of the ISG-derived vesicle will include the proteinaceous material that is soluble in the ISG, i.e. excluded from the aggregate (33) (Fig. 7). One such component could be residual constitutive secretory proteins (25). Although it is clear that most of the constitutive secretory proteins are segregated from regulated secretory proteins upon exit from the TGN (21), a small amount of constitutive secretory protein would be included in the forming ISG if the fluid phase of the TGN is not completely extruded in the course of membrane enveloping of the aggregate. This residual constitutive secretory protein would presumably remain soluble as the aggregate of regulated secretory proteins is further condensed in the course of granule maturation, and thus would be expected to enter the ISG-derived vesicle by default.

A second component of the content of the ISG-derived vesicle would be soluble proteinaceous material that has been generated in the ISG. The soluble products of the proteolytic processing of regulated secretory proteins are obvious candidates. Kuliawat and Arvan (61) have proposed that the C-peptide cleaved from proinsulin is removed from the ISG by the ISG-derived vesicle formed in the course of granule maturation. Again, the C-peptide would be expected to enter the ISG-derived vesicle by default as the insulin aggregate is further condensed and remains with the MSG. The C-peptide-containing, ISG-derived vesicle is thought to mediate the 'constitutive-like' secretion of the C-peptide (61). An open issue is the acceptor membrane to which the ISG-derived vesicle is targeted. Constitutive-like secretion would occur irrespective of whether this vesicle fuses with the TGN, the early endosome, or directly with the plasma membrane.

3.2.2 Membrane proteins

Little is known about the membrane constituents that may be removed from the ISG in the course of maturation and thus would be expected to be present in the ISG-derived vesicle. On theoretical grounds, these membrane constituents should essentially be those which do not (any longer) interact, either directly or indirectly, with the aggregate of regulated secretory proteins in the ISG (Fig. 7). Candidate proteins include (i) the putative receptor recognizing the S-M signal on regulated secretory proteins and (ii) mannose-6-phosphate receptors. With respect to the putative receptor recognizing the S-M signal on regulated secretory proteins, its removal from the ISG would allow its recycling to the TGN (33) which otherwise would be dependent on the exocytosis the MSG. It is conceivable that the more acidic pH of the ISG relative to the TGN (estimated to be pH $\approx$ 5.5 vs. pH $\approx$ 6.5) causes such a receptor to dissociate from the aggregate of regulated secretory proteins.

With respect to the possible presence of mannose-6-phosphate receptors in the ISG-derived vesicle, it should be realized that, so far, there is little evidence for the concept of three different exit routes out of the TGN (62), i.e. to the plasma membrane via CSVs, to secretory granules, and to lysosomes via endosomes. This concept has apparently resulted from a theoretical superimposition of the TGN of a constitutive secretory cell (CSVs, lysosomes) with that of a regulated secretory cell (CSVs, secretory granules). However, on the basis of the available evidence, it is also feasible that the biosynthetic pathway to lysosomes diverges from the secretory pathway at the level of the ISG rather than the TGN. In other words, the possibility

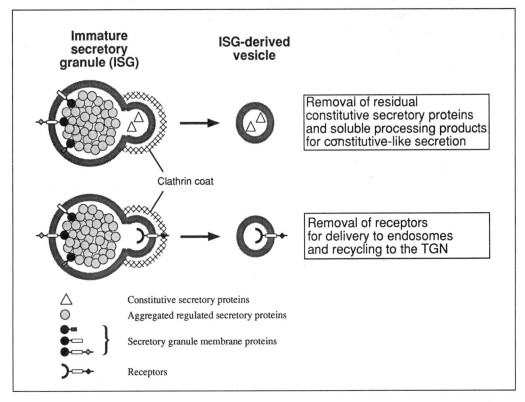

Immature
secretory
granule (ISG)

ISG-derived
vesicle

Removal of residual
constitutive secretory proteins
and soluble processing products
for constitutive-like secretion

Clathrin coat

Removal of receptors
for delivery to endosomes
and recycling to the TGN

△   Constitutive secretory proteins

◎   Aggregated regulated secretory proteins

}   Secretory granule membrane proteins

)⊸   Receptors

Fig. 7 Implications of the budding of a vesicle from the immature secretory granule (ISG) in the course of granule maturation. Top, soluble proteins such as residual constitutive secretory proteins and peptides resulting from the proteolytic processing of prohormones are excluded from the aggregate of regulated secretory proteins and enter the ISG-derived vesicle by bulk flow. Bottom, membrane proteins which do not (any longer) interact with the aggregate of regulated secretory proteins are sorted into the ISG-derived vesicle via a clathrin-involving process. These membrane proteins may include receptors such as the mannose-6-phosphate receptor or a putative receptor recognizing the S-M signal on regulated secretory proteins.

should be considered that lysosomal enzymes picked up by mannose-6-phosphate receptors in the TGN exit from this compartment in ISGs, from which they are removed and delivered to endosomes via the ISG-derived vesicle formed in the course of granule maturation (see discussion of clathrin on ISGs below).

In the case of the mannose-6-phosphate receptor, the acceptor membrane to which the ISG-derived vesicle is targeted would presumably be an endosomal membrane, possibly the early endosome. The same may be true for an ISG-derived vesicle involved in the recycling of the putative S-M signal receptor to the TGN, as there exists a recycling route from the endosome to the TGN. Alternatively, an ISG-derived vesicle may be targeted directly to the TGN. Whatever turns out to be the case, it is clear that the removal of membrane from the ISG in the course of granule maturation turns the ISG into a post-TGN sorting compartment specific to regulated secretory cells.

3.2.3 Molecular machinery and regulation

A characteristic feature which distinguishes the ISG from the MSG is the presence of a patchy clathrin coat (13, 63–65). Although direct evidence is lacking, it is likely that the presence of clathrin is related to the removal of transmembrane proteins from the maturing ISG, such as the above-mentioned S-M signal receptor or the mannose-6-phosphate receptor (Fig. 7). If correct, this would imply that the composition of the membrane of the ISG-derived vesicle is the result of an active sorting process, whereas that of its content is not (see above). Little is known about the clathrin adaptors on ISGs and other cytoplasmic proteins involved in the conversion of ISGs to MSGs. The same is true for any machinery regulating this conversion. That such a machinery exists, however, is suggested by the observation that the conversion of ISGs into either small MSGs (one MSG derived from one ISG) or large MSGs (one MSG originating from the self-fusion of several ISGs) may vary in the same cell (R. Bauerfeind and W. B. Huttner, unpublished observations).

4. Biogenesis of synaptic vesicles

4.1 Synaptic-like microvesicles

Synaptic-like microvesicles (SLMVs) of endocrine cells are very similar, if not identical, to the synaptic vesicles of mature neurones in terms of content, membrane composition, and membrane dynamics, and can be considered as their endocrine counterpart (2).

4.1.1 Origin

In contrast to secretory granules, SLMVs do not originate from the TGN. Rather, the biogenesis of SLMVs occurs from early endosomes (Fig. 1). This is indicated by two lines of evidence. First, newly synthesized synaptophysin, a membrane protein characteristic of synaptic vesicles and SLMVs (66), travels in CSVs from the TGN to the plasma membrane, cycles several times between the plasma membrane and early endosomes, and is then packaged into SLMVs (67). Second, the fluid-phase marker horseradish peroxidase, pre-internalized into early endosomes, is chased into SLMVs together with newly synthesized synaptophysin (60). Thus, the *de novo* generation of SLMVs occurs from the same compartment from which these vesicles are thought to be regenerated after exocytosis and endocytosis of the SLMV membrane, i.e. the early endosome (68, 69).

Is the traffic route of newly synthesized synaptophysin to SLMVs representative for that of the other membrane proteins characteristic of synaptic vesicles and SLMVs? An answer to this question has come from a study (60) on the intracellular localization of vesicular neurotransmitter uptake systems in the neuroendocrine cell line PC12. SLMVs were found to exhibit uptake and storage of acetylcholine. The same was the case for early endosomes, but not for secretory granules. Thus, the vesicular acetylcholine transporter does not travel to early endosomes and SLMVs via secretory granules. Rather, these data are consistent with this neurotransmitter transporter taking the same biosynthetic route as synaptophysin.

4.1.2 Assembly

As is the case for synaptic vesicles of neurones and in line with their biogenesis from early endosomes, SLMVs are not known to contain secretory proteins. Thus, their biogenesis is essentially a question of the proper assembly of the SLMV membrane proteins in the early endosome. The molecular machinery mediating this assembly, and the relevant structural features on the SLMV membrane proteins, are presently unknown.

4.2 Synaptic vesicles

Our understanding of the biogenesis of synaptic vesicles in neurones is even less complete than that of SLMVs in neuroendocrine cells. It is clear that synaptic vesicles, too, do not originate from the TGN. The TGN in neurones is confined to the perikaryon and excluded from the axon, yet no accumulation of synaptic vesicles is observed proximal to the site of an axonal transport block (70). Presumably, synaptic vesicles in analogy to SLMVs originate from axonal early endosomes (71), but direct evidence supporting this remains to be reported. As to the question of whether the synaptic vesicle membrane after exocytosis recycles through an axonal early endosome or not ('kiss and run'), the reader is referred to a recent discussion of this issue (72).

4.3 Small dense-core vesicles

The study of the intracellular localization of vesicular neurotransmitter uptake systems in the neuroendocrine cell line PC12 (60) provided other important information relevant to the biogenesis of neurosecretory vesicles. Uptake and storage of catecholamines was found in secretory granules, but not SLMVs, showing that the vesicular amine transporter expressed in these cells does not travel to the latter type of neurosecretory vesicle. This is remarkable considering the fact that monoamine uptake activity was detected in early endosomes, the compartment from which SLMVs originate (60). (The presence of the vesicular amine transporter in early endosomes presumably reflected the internalization of this secretory granule membrane component from the plasma membrane after exocytosis.) Thus, the early endosomes of PC12 cells do not sort this vesicular amine transporter, a 12 transmembrane domain protein (73), to SLMVs, although they are capable of performing similar sorting events since PC12 cell SLMVs contain SV2 (74), a 12 transmembrane domain protein (75), and the vesicular acetylcholine transporter (60), which is likely to belong to this family of membrane proteins (76).

These data require an explanation in light of the existence of the small dense-core vesicles (SDCVs) of sympathetic neurones. In these cells, SDCVs constitute one type of neurosecretory vesicle, and secretory granules (called large dense-core vesicles)

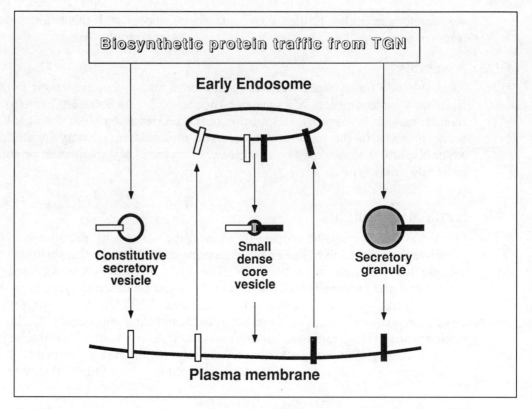

Fig. 8 The small dense-core vesicle is a chimeric vesicle composed of membrane proteins of both synaptic vesicles and secretory granules. Membrane proteins characteristic of synaptic vesicles such a synaptophysin (white rectangles) exit from the TGN in constitutive secretory vesicles (left). Membrane proteins characteristic of secretory granules such as dopamine β-hydroxylase and cytochrome b_{561} (black rectangles) exit from the TGN in secretory granules (right). It is postulated that upon fusion of these two types of post-TGN secretory vesicles with the plasma membrane, both classes of membrane proteins are endocytosed and delivered to early endosomes, where they are assembled into a chimeric membrane vesicle, the small dense-core vesicle.

constitute the other (77). SDCVs, like secretory granules, store catecholamines (which give rise to the electron dense core after certain fixation conditions) but, in contrast to secretory granules, do not contain secretory proteins (which is why they lack electron dense cores after standard aldehyde fixation) (78). SDCVs are similar in size to synaptic vesicles. Because of these properties, SDCVs have been considered to be catecholamine-containing synaptic vesicles (79). What, then, is the reason why SLMVs lack, but SDCVs contain, a vesicular amine transporter?

The answer to this question may lie in the existence of two isoforms of vesicular monoamine transporters, called VMAT-1 (or previously CGAT) and VMAT-2 (or previously SVAT) (73). Both VMAT-1 and VMAT-2 are found in secretory granules (60, 73). However, only VMAT-2, but not VMAT-1, appears to be present in SLMVs, as suggested by the following observations. The vesicular amine transporter expressed in PC12 cells is VMAT-1 (73); SLMVs in these cells lack catecholamines

(60). In contrast, bovine adrenal chromaffin cells express VMAT-2, and not only the secretory granules, but also the SLMVs in these cells contain catecholamines (80). This leads us to propose (60) that VMAT-2, but not VMAT-1, contains sorting information to synaptic vesicles. This hypothesis is consistent with the observation that neurones express VMAT-2 (73).

SDCVs, however, turn out not to be (as previously assumed) synaptic vesicles that just contain VMAT-2 instead of uptake systems for classical neurotransmitters. Besides VMAT-2, SDCVs contain several other membrane proteins characteristic of secretory granules, including dopamine β-hydroxylase and cytochrome b_{561} (81, 82). On the other hand, SDCVs also contain membrane proteins characteristic of synaptic vesicles and SLMVs, namely synaptophysin and synaptoporin (84). Thus, the SDCV membrane is neither formed solely from the membrane of secretory granules after their exocytosis, as proposed by some investigators (77), nor composed only of synaptic vesicle membrane proteins. Rather, SDCVs turn out to be hybrid vesicles generated by assembling the membrane proteins common to all synaptic vesicles together with selected components of the secretory granule membrane re-utilized after exocytosis. This assembly is likely to take place in early endosomes (Fig. 8).

Acknowledgements

We thank Dr Matthew Hannah for critically reading the manuscript. Work in the authors' laboratory was supported by grants from the Deutsche Forschungsgemeinschaft to W.B.H. (SFB 317 and SFB 352).

References

1. Burgess, T. L. and Kelly, R. B. (1987) Constitutive and regulated secretion of proteins. *Annu. Rev. Cell Biol.*, **3**, 243.
2. De Camilli, P. and Jahn, R. (1990) Pathways to regulated exocytosis in neurons. *Annu. Rev. Physiol.*, **52**, 625.
3. Griffiths, G. and Simons, K. (1986) The *trans* Golgi network: sorting at the exit site of the Golgi complex. *Science*, **234**, 438.
4. Simons, K. and Wandinger-Ness, A. (1990) Polarized sorting in epithelia. *Cell*, **62**, 207.
5. Huttner, W. B. and Dotti, C. G. (1991) Exocytotic and endocytotic membrane traffic in neurons. *Curr. Opin. Neurobiol.*, **1**, 388.
6. Rodriguez-Boulan, E. and Powell, S. K. (1992) Polarity of epithelial and neuronal cells. *Annu. Rev. Cell Biol.*, **8**, 395.
7. Kornfeld, S. and Mellman, I. (1989) The biogenesis of lysosomes. *Annu. Rev. Cell Biol.*, **5**, 483.
8. von Figura, K. (1991) Molecular recognition and targeting of lysosomal proteins. *Curr. Opin. Cell Biol.*, **3**, 642.
9. Hoflack, B. and Lobel, P. (1993) Functions of the mannose 6-phosphate receptors. *Adv. Cell Mol. Biol. Membranes*, **1**, 51.
10. Pearse, B. M. F. and Robinson, M. S. (1990) Clathrin, adaptors, and sorting. *Annu. Rev. Cell Biol.*, **6**, 151.

11. Goldstein, J. L., Brown, M. S., Anderson, R. G. W., Russell, D. W., and Schneider, W. J. (1985) Receptor-mediated endocytosis: concepts emerging from the LDL receptor system. *Annu. Rev. Cell Biol.*, **1**, 1.

12. Pfeffer, S. R. and Rothman, J. E. (1987) Biosynthetic protein transport and sorting by the endoplasmic reticulum and Golgi. *Annu. Rev. Biochem.*, **56**, 829.

13. Tooze, S. A., Chanat, E., Tooze, J., and Huttner, W. B. (1993) Secretory granule formation. In *Mechanisms of intracellular trafficking and processing of proproteins*. Peng Loh, Y. (ed.). CRC Press, Boca Raton, p. 157.

14. Wieland, F. T., Gleason, M. L., Serafini, T. A., and Rothman, J. E. (1987) The rate of bulk flow from the endoplasmic reticulum to the cell surface. *Cell*, **50**, 289.

15. Pelham, H. R. B. (1991) Recycling of proteins between the endoplasmic reticulum and Golgi complex. *Curr. Opin. Cell Biol.*, **3**, 585.

16. Fuller, R. S., Brake, A. J., and Thorner, J. (1989) Intracellular targeting and structural conservation of a prohormone-processing endoprotease. *Science*, **246**, 482.

17. Friederich, E., Fritz, H.-J., and Huttner, W. B. (1988) Inhibition of tyrosine sulfation in the *trans*-Golgi retards the transport of a constitutively secreted protein to the cell surface. *J. Cell Biol.*, **107**, 1655.

18. Balch, W. E., McCaffery, J. M., Plutner, H., and Farquhar, M. G. (1994) Vesicular stomatitis virus glycoprotein is sorted and concentrated during export from the endoplasmic reticulum. *Cell*, **76**, 841.

19. Karrenbauer, A., Jeckel, D., Just, W., Birk, R., Schmidt, R. R., Rothman, J. E., and Wieland, F. T. (1990) The rate of bulk flow from the Golgi to the plasma membrane. *Cell*, **63**, 259.

20. Bennett, M. K., Wandinger-Ness, A., and Simons, K. (1988) Release of putative exocytic transport vesicles from perforated MDCK cells. *EMBO J.*, **7**, 4075.

21. Tooze, S. A. and Huttner, W. B. (1990) Cell-free protein sorting to the regulated and constitutive secretory pathways. *Cell*, **60**, 837.

22. Salamero, J., Sztul, E. S., and Howell, K. E. (1990) Exocytic transport vesicles generated *in vitro* from the *trans*-Golgi network carry secretory and plasma membrane proteins. *Proc. Natl Acad. Sci. USA*, **87**, 7717.

23. Gravotta, D., Adesnik, M., and Sabatini, D. D. (1990) Transport of influenza HA from the *trans*-Golgi network to the apical surface of MDCK cells permeabilized in their basolateral plasma membranes: energy dependence and involvement of GTP-binding proteins. *J. Cell Biol.*, **111**, 2893.

24. Miller, S. G. and Moore, H.-P. H. (1991) Reconstitution of constitutive secretion using semi-intact cells: regulation by GTP but not calcium. *J. Cell Biol.*, **112**, 39.

25. Grimes, M. and Kelly, R. B. (1992) Intermediates in the constitutive and regulated secretory pathways released *in vitro* from semi-intact cells. *J. Cell Biol.*, **117**, 539.

26. Kahn, R. A., Kern, F. G., Clark, J., Gelmann, E. P., and Rulka, C. U. (1991) Human ADP-ribosylation factors. *J. Biol. Chem.*, **266**, 2606.

27. Moss, J. and Vaughan, M. (1991) Activation of cholera toxin and *Escherichia coli* heat-labile enterotoxins by ADP-ribosylation factors, a family of 20 kDa guanine nucleotide-binding proteins. *Mol. Microbiol.*, **5**, 2621.

28. Donaldson, J. G., Cassel, D., Kahn, R. A., and Klausner, R. D. (1992) ADP-ribosylation factor, a small GTP-binding protein, is required for binding of the coatomer protein β-COP to Golgi membranes. *Proc. Natl Acad. Sci. USA*, **89**, 6408.

29. Walker, M. W., Bobak, D. A., Tsai, S.-C., Moss, J., and Vaughan, M. (1992) GTP but not GDP analogues promote association of ADP-ribosylation factors, 20 kDa protein activators of cholera toxin, with phospholipids and PC12 cell membranes. *J. Biol. Chem.*, **267**, 3230.

30. Stamnes, M. A. and Rothman, J. E. (1993) The binding of AP-1 clathrin adaptor particles to Golgi membranes requires ADP-ribosylation factor, a small GTP-binding protein. *Cell*, **73**, 999.

31. Traub, L. M., Ostrom, J. A., and Kornfeld, S. (1993) Biochemical dissection of AP-1 recruitment onto Golgi membranes. *J. Cell Biol.*, **123**, 561.

32. Palmer, D. J., Helms, J. B., Beckers, C. J. M., Orci, L., and Rothman, J. E. (1993) Binding of coatomer to Golgi membranes requires ADP-ribosylation factor. *J. Biol. Chem.*, **268**, 12083.

33. Bauerfeind, R. and Huttner, W. B. (1993) Biogenesis of constitutive secretory vesicles, secretory granules and synaptic vesicles. *Curr. Opin. Cell Biol.*, **5**, 628.

34. Miller, S. G., Carnell, L., and Moore, H.-P. H. (1992) Post-Golgi membrane traffic: brefeldin A inhibits export from distal Golgi compartment to the cell surface but not recycling. *J. Cell Biol.*, **118**, 267.

35. Rosa, P., Barr, F. A., Stinchcombe, J. C., Binacchi, C., and Huttner, W. B. (1992) Brefeldin A inhibits the formation of constitutive secretory vesicles and immature secretory granules from the *trans*-Golgi network. *Eur. J. Cell Biol.*, **59**, 265.

36. Donaldson, J. G., Finazzi, D., and Klausner, R. D. (1992) Brefeldin A inhibits Golgi membrane-catalysed exchange of guanine nucleotide onto ARF protein. *Nature*, **360**, 350.

37. Helms, J. B. and Rothman, J. E. (1992) Inhibition by brefeldin A of a Golgi membrane enzyme that catalyses exchange of guanine nucleotide bound to ARF. *Nature*, **360**, 352.

38. Waters, M. G., Serafini, T., and Rothman, J. E. (1991) Coatomer: a cytosolic protein complex containing subunits of non-clathrin-coated Golgi transport vesicles. *Nature*, **349**, 248.

39. Barr, F. A., Leyte, A., Mollner, S., Pfeuffer, T., Tooze, S. A., and Huttner, W. B. (1991) Trimeric G-proteins of the *trans*-Golgi network are involved in the formation of constitutive secretory vesicles and immature secretory granules. *FEBS Lett.*, **294**, 239.

40. Leyte, A., Barr, F. A., Kehlenbach, R. H., and Huttner, W. B. (1992) Multiple trimeric G-proteins on the *trans*-Golgi network exert stimulatory and inhibitory effects on secretory vesicle formation. *EMBO J.*, **11**, 4795.

41. Jones, S. M., Crosby, J. R., Salamero, J., and Howell, K. E. (1993) A cytosolic complex of p62 and rab6 associates with TGN 38/41 and is involved in budding of exocytic vesicles from the *trans*-Golgi network. *J. Cell Biol.*, **122**, 775.

42. Huttner, W. B., Gerdes, H.-H., and Rosa, P. (1991) The granin (chromogranin/secretogranin) family. *Trends Biochem. Sci.*, **16**, 27.

43. Chanat, E. and Huttner, W. B. (1991) Milieu-induced, selective aggregation of regulated secretory proteins in the *trans*-Golgi network. *J. Cell Biol.*, **115**, 1505.

44. Chanat, E., Weiß, U., Huttner, W. B., and Tooze, S. A. (1993) Reduction of the disulfide bond of chromogranin B (secretogranin I) in the *trans*-Golgi network causes its missorting to the constitutive secretory pathway. *EMBO J.*, **12**, 2159.

45. Benedum, U. M., Lamouroux, A., Konecki, D. S., Rosa, P., Hille, A., Baeuerle, P. A., Frank, R., Lottspeich, F., Mallet, J., and Huttner, W. B. (1987) The primary structure of human secretogranin I (chromogranin B): comparison with chromogranin A reveals homologous terminal domains and a large intervening variable region. *EMBO J.*, **6**, 1203.

46. Gerdes, H.-H., Rosa, P., Phillips, E., Baeuerle, P. A., Frank, R., Argos, P., and Huttner, W. B. (1989) The primary structure of human secretogranin II, a widespread tyrosine-sulfated secretory granule protein that exhibits low pH- and calcium-induced aggregation. *J. Biol. Chem.*, **264**, 12009.

47. Yoo, S. H. (1993) pH-dependent association of chromogranin A with secretory vesicle membrane and a putative membrane binding region of chromogranin A. *Biochemistry*, **32**, 8213.

48. Tam, W. W. H., Andreasson, K. I., and Peng Loh, Y. (1993) The amino-terminal sequence of pro-opiomelanocortin directs intracellular targeting to the regulated secretory pathway. *Eur. J. Cell Biol.*, **62,** 294.

49. Winkler, H. and Fischer-Colbrie, R. (1990) Common membrane proteins of chromaffin granules, endocrine and synaptic vesicles: properties, tissue distribution, membrane topography and regulation of synthesis. *Neurochem. Int.*, **17,** 245.

50. Disdier, M., Morrissey, J. H., Fugate, R. D., Bainton, D., and McEver, R. P. (1992) The cytoplasmic domain of P-selectin (CD62) contains the signal for sorting into the regulated secretory pathway. *Mol. Biol. Cell*, **3,** 309.

51. Koedam, J. A., Cramer, E. M., Briend, E., Furie, B., Furie, B. C., and Wagner, D. D. (1992) P-selectin, a granule membrane protein of platelets and endothelial cells, follows the regulated secretory pathway in AtT-20 cells. *J. Cell Biol.*, **116,** 617.

52. Eipper, B. A., Stoffers, D. A., and Mains, R. E. (1992) The biosynthesis of neuropeptides: peptide α-amidation. *Annu. Rev. Neurosci.*, **15,** 57.

53. Milgram, S. L., Mains, R. E., and Eipper, B. A. (1993) COOH-terminal signals mediate the trafficking of a peptide processing enzyme in endocrine cells. *J. Cell Biol.*, **121,** 23.

54. Milgram, S. L., Johnson, R. C., and Mains, R. E. (1992) Expression of individual forms of peptidylglycine α-amidating monooxygenase in AtT-20 cells: endoproteolytic processing and routing to secretory granules. *J. Cell Biol.*, **117,** 717.

55. Rindler, M. J. (1992) Biogenesis of storage granules and vesicles. *Curr. Opin. Cell Biol.*, **4,** 616.

56. Taljanidisz, J., Stewart, L., Smith, A. J., and Klinman, J. P. (1989) Structure of bovine dopamine β-monooxygenase, as deduced from cDNA and protein sequencing: Evidence that the membrane-bound form of the enzyme is anchored by an uncleaved signal peptide. *Biochemistry*, **28,** 10054.

57. Pimplikar, S. W. and Huttner, W. B. (1992) Chromogranin B (secretogranin I), a secretory protein of the regulated pathway, is also present in a tightly membrane-associated form in PC12 cells. *J. Biol. Chem.*, **267,** 4110.

58. Fricker, L. D., Das, B., and Angeletti, R. H. (1990) Identification of the pH-dependent membrane anchor of carboxypeptidase E (EC 3.4.17.10). *J. Biol. Chem.*, **265,** 2476.

59. Tooze, S., Flatmark, T., Tooze, J., and Huttner, W. B. (1991) Characterization of the immature secretory granule, an intermediate in granule biogenesis. *J. Cell Biol.*, **115,** 1491.

60. Bauerfeind, R., Régnier-Vigouroux, A., Flatmark, T., and Huttner, W. B. (1993) Selective storage of acetylcholine, but not catecholamines, in neuroendocrine synaptic-like microvesicles of early endosomal origin. *Neuron*, **11,** 105.

61. Kuliawat, R. and Arvan, P. (1992) Protein targeting via the 'constitutive-like' secretory pathway in isolated pancreatic islets: Passive sorting in the immature granule compartment. *J. Cell Biol.*, **118,** 521.

62. Alberts, B., Bray, D., Lewis, J., Raff, M., Roberts, K., and Watson, J. D. (1989) *Molecular biology of the cell*. Garland Publishing Inc., New York.

63. Orci, L., Halban, P., Amherdt, M., Ravazzola, M., Vassalli, J.-D., and Perrelet, A. (1984) A clathrin-coated, Golgi-related compartment of the insulin secreting cell accumulates proinsulin in the presence of monensin. *Cell*, **39,** 39.

64. Tooze, J. and Tooze, S. (1986) Clathrin-coated vesicular transport of secretory proteins during the formation of ACTH-containing secretory granules in AtT20 cells. *J. Cell Biol.*, **103,** 839.

65. Tooze, S. A. (1991) Biogenesis of secretory granules: implications arising from the immature secretory granule in the regulated pathway of secretion. *FEBS Lett.*, **285,** 220.

66. Südhof, T. C. and Jahn, R. (1991) Proteins of synaptic vesicles involved in exocytosis and membrane recycling. *Neuron*, **6**, 665.

67. Régnier-Vigouroux, A., Tooze, S. A., and Huttner, W. B. (1991) Newly synthesized synaptophysin is transported to synaptic-like microvesicles via constitutive secretory vesicles and the plasma membrane. *EMBO J.*, **10**, 3589.

68. Cameron, P. L., Südhof, T. C., Jahn, R., and De Camilli, P. (1991) Colocalization of synaptophysin with transferrin receptors: implications for synaptic vesicle biogenesis. *J. Cell Biol.*, **115**, 151.

69. Linstedt, A. D. and Kelly, R. B. (1991) Synaptophysin is sorted from endocytic markers in neuroendocrine PC12 cells but not transfected fibroblasts. *Neuron*, **7**, 309.

70. Tsukita, S. and Ishikawa, H. (1980) The movement of membraneous organelles in axons. Electron microscopic identification of anterogradely and retrogradely transported organelles. *J. Cell Biol.*, **84**, 513.

71. Régnier-Vigouroux, A. and Huttner, W. B. (1993) Biogenesis of small synaptic vesicles and synaptic-like microvesicles. *Neurochem. Res.*, **18**, 59.

72. Fesce, R., Grohovaz, F., Valtorta, F., and Meldolesi, J. (1994) Neurotransmitter release: fusion or 'kiss-and-run'? *Trends Cell Biol.*, **4**, 1.

73. Liu, Y., Peter, D., Roghani, A., Schuldiner, S., Privé, G. G., Eisenberg, D., Brecha, N., and Edwards, R. H. (1992) A cDNA that suppresses MPP+ toxicity encodes a vesicular amine transporter. *Cell*, **70**, 539.

74. Lowe, A. W., Maddedu, L., and Kelly, R. B. (1988) Endocrine secretory granules and neuronal synaptic vesicles have three integral membrane proteins in common. *J. Cell Biol.*, **106**, 51.

75. Feany, M. B., Lee, S., Edwards, R. H., and Buckley, K. M. (1992) The synaptic vesicle protein SV2 is a novel type of transmembrane transporter. *Cell*, **70**, 861.

76. Alfonso, A., Grundahl, K., Duerr, J. S., Han, H.-P., and Rand, J. B. (1993) The *Caenorhabditis elegans* unc-17 gene: a putative vesicular acetylcholine transporter. *Science*, **261**, 617.

77. Winkler, H., Sietzen, M., and Schober, M. (1987) The life cycle of catecholamine-storing vesicles. *Ann. NY Acad. Sci.*, **493**, 3.

78. Klein, R. L., Lagercrantz, H., and Zimmermann, H. (1982) *Neurotransmitter vesicles*. Academic Press, London.

79. De Camilli, P. and Navone, F. (1987) Regulated secretory pathways of neurons and their relation to the regulated secretory pathway of endocrine cells. *Ann. NY Acad. Sci.*, **493**, 461.

80. Annaert, W. G., Llona, I., Backer, A. C., Jacob, W. A., and De Potter, W. P. (1993) Catecholamines are present in a synaptic-like microvesicle-enriched fraction from bovine adrenal medulla. *J. Neurochem.*, **60**, 1746.

81. Neuman, B., Wiedermann, C. J., Fischer-Colbrie, R., Schober, M., Sperk, G., and Winkler, H. (1984) Biochemical and functional properties of large and small dense-core vesicles in sympathetic nerves of rat and ox vas deferens. *Neuroscience*, **13**, 921.

82. Schwarzenbrunner, U., Schmidle, T., Obendorf, D., Scherman, D., Hook, V., Fischer-Colbrie, R., and Winkler, H. (1990) Sympathetic axons and nerve terminals: the protein composition of small and large dense-core and of a third type of vesicles. *Neuroscience*, **37**, 819.

83. Ohashi, M., and Huttner, W. B. (1994) An elevation of cytosolic protein phosphorylation modulates trimeric G-protein regulation of secretory vesicle formation from the trans-Golgi network. *J. Biol. Chem.*, **269**, 24897.

84. Bauerfeind, R., Jelinek, R., Hellwig, A., and Huttner, W. B. (1995) Neurosecretory vesicles can be hybrids of synaptic vesicles and secretory granules. *Proc. Natl Acad. Sci.*, **92**, 7342.

7 | Protein sorting during endocytosis

NICHOLAS T. KTISTAKIS and MICHAEL G. ROTH

1. Introduction

Movement and sorting of proteins along the endocytic pathway is complex. At most branching points in the pathway movement can potentially be multidirectional. For example, a protein passing through the endosomal membrane system has access to transport routes that could take it 'back' to the cell surface, 'forward' to the lysosomes, or in a 'lateral' direction towards the *trans* Golgi network (TGN). Secondly, movement along the endocytic pathway involves multiple terminal destinations, most notably the cell surface (either apical or basolateral in polarized cells), the lysosomes, or the TGN. Thirdly, in some instances multiple pathways can be used to achieve the same final destination. This is most evident in the case of some lysosomal proteins where targeting to lysosomes can be either intracellular (directly from the TGN) or via the cell surface following internalization.

In this chapter we will first discuss recent developments in understanding the amino acid and structural requirements of the signals which direct proteins at the cell surface to clathrin-coated pits. We will then discuss the possible relationship of the internalization signal to those signals which direct proteins to the lysosomes, the TGN, or the basolateral surface of polarized cells. Finally, we will summarize what is known about the organelles which receive and process endocytic cargo from the cell surface and the signals which control this aspect of trafficking. It should be noted that other excellent reviews and opinions are available on these topics (1–6).

2. Internalization from clathrin-coated pits

An indispensable requirement for the survival of mammalian cells is the ability to internalize nutrients and other ligands efficiently from the cell exterior. This process—receptor-mediated endocytosis—was first described in detail for the low-density lipoprotein (LDL) receptor, and later shown to apply to most cell surface receptors (6). An essential feature of this process is the segregation of proteins to be internalized away from other cell surface molecules into specialized regions of the plasma membrane, the clathrin-coated pits (7, 8). Historically, concentration into clathrin-coated pits has been thought to afford speed and selectivity to the process of receptor-mediated endocytosis (9, 10). Speed is undoubtedly an attribute of this system: up to half of the cell surface transferrin receptors can be internalized in one minute, and similar rates can be measured for other receptors. The question of selectivity remains to be re-examined in view of the growing number of proteins now known to use incorporation into clathrin-coated pits not as a means of internalizing a ligand or nutrient but rather as a means of targeting to their final destination. We will come back to this point later.

Pathways of internalization independent of clathrin-coated pits also exist and seem to differ in relative importance among different cell types. One such pathway is uncovered following treatments that interfere with the normal clathrin assembly–disassembly cycle (e.g. hypotonic shock and cytosol acidification) but nothing is known about its regulation (11–15). In addition, small molecules can enter cells by specialized plasma membrane regions called caveolae in a process termed potocytosis (16). This process might not involve membrane carriers, but rather the opening and closing of 'pores' on the plasma membrane. We will concentrate here on endocytosis through clathrin-coated pits.

2.1 Tyrosine-containing signals

The search for signals which direct proteins into coated pits was launched as a result of two observations concerning the internalization of the LDL receptor (LDLR). Two naturally occurring mutations which reduced internalization mapped to truncations of the cytoplasmic domain of the receptor, and a third mutation which also reduced internalization was the product of a cysteine to tyrosine substitution at position 807 in the cytoplasmic domain (17, 18). These observations clearly demonstrated that the information for internalization was in the cytoplasmic domain and a tyrosine residue therein was involved (19). It is worth noting that these mutations did not eliminate the ability of the receptor to be internalized, but they reduced it to about 20% of the wild-type levels.

Tyrosines were implicated as important contributors to the internalization signal by subsequent work on two fronts. On the one hand other receptors showed reduced internalization rates once tyrosine residues in their cytoplasmic domain were artificially mutated (20–27), and on the other it was shown that introducing a tyrosine into the cytoplasmic domain of the influenza haemagglutinin (HA), a protein normally excluded from coated pits, accelerates the internalization of this protein to levels comparable to some cell-surface receptors (50- to 100-fold higher than the wild-type protein) (28). Inspection of the cytoplasmic sequences of several internalized proteins did not reveal any obvious amino acid conservation, yet it was clear that the tyrosine could only function in the appropriate context because not all tyrosine substitutions in HA could specify internalization (four were tried in the 10-amino acid cytoplasmic domain, with one being successful) (28, 29).

The nature of the amino acid context required to specify internalization was addressed by a number of laboratories for a number of different receptors. In some instances the work involved single amino acid substitutions (usually alanine) along the length of the cytoplasmic domain and then analysis of the internalization rates of the resulting mutant proteins (30, 31). This type of analysis reveals amino acids which by themselves contribute to the formation of the signal but can miss cases where multiple amino acids interact to form the internalization signal. In cases where no single amino acid other than tyrosine could be shown to have a large effect on internalization, simultaneous mutagenesis of more than one position revealed an amino acid context required for efficient internalization (29).

Several models for amino acid sequence motifs that specify endocytosis have been proposed based upon the results of site-directed mutagenesis (29–32). One proposal suggested that a tyrosine residue within an 8–10 amino acid stretch of residues that are polar/positively charged and have the statistical tendency to break secondary structure (non-α-helix, non-β-sheet) that form a turn in the protein's cytoplasmic tail formed a general internalization signal (29). This was tested by introducing a tyrosine residue into a position that fulfilled these criteria in the cytoplasmic domain of a non-endocytosed protein (human glycophorin A) and showing that the mutant protein with the tyrosine in the correct context is internalized. Since the role of the amino acids surrounding the tyrosine is to provide a secondary structure that exposed the tyrosine, this proposal predicts that mutation of individual amino acids in the signal other than the tyrosine would have a minor effect on the ability of the signal to function. Only when more than one amino acid is mutated simultaneously can major effects be seen. In addition, this proposal implies the existence of a turn as part of the signal but does not postulate the geometry of the turn.

Another proposal based on extensive mutagenesis of the transferrin receptor makes specific predictions in the exact geometry of the structure surrounding the tyrosine. It was proposed that the internalization signal of the transferrin receptor (TR) consists of the tetrapeptide YTRF (31). Work with the LDLR had also suggested that the tetrapeptide NPVY forms the internalization signal of this protein (30) and both YTRF and NPVY tetrapeptides were found in proteins of known crystal structure to adopt a type 1 β-turn conformation. It was therefore proposed that the internalization signal consists of a tetrapeptide with the general sequence aromatic–X–X–aromatic/large hydrophobic (where X represents any amino acid) which adopts a tight turn conformation. Subsequently, two-dimensional NMR spectroscopy of synthetic peptides derived from putative internalization sequences of the insulin receptor (IR), LDLR, and the lysosomally targeted protein lysosomal acid phosphatase (LAP) confirmed that a region of these peptides containing the critical tyrosine residue could adopt a reverse-turn conformation. The relevant sequence for LAP was the tetrapeptide -PPGY- (33), for LDLR it was the tetrapeptide -NPXY- (34), and for the two internalization sequences of the IR, -NPEY- and -GPLY- (35). For the LDLR it was further shown that the tendency of various altered tetrapeptides to adopt the reverse-turn conformation in solution correlated well with their ability to specify internalization *in vivo*.

Table 1 Cytoplasmic sequences demonstrated to be important for internalization through coated pits

signals with an important aromatic residue at position 3			N-terminal to C-terminal Sequence					Reference
			1	2	3	4	5	6	
ASGP H1			K	E	Y	Q	D	L	(24)
TfR (Y20)hum		L	S	Y	T	R	F		(31, 172)
M6P/IGF-IIR			Y	K	Y	S	K	V	(43)
Cd-M6PR (Y45)			A	A	Y	R	G	V	(153)
pIgR (Y734)			L	A	Y	S	A	F	(23)
LAMP-1			A	G	Y	Q	T	I	(150, 26)
LAP		P	P	G	Y	R	H	V	(169)
HA+8*		I	Y	D	Y	K	S	F	(53)
HA-Y543*		S	L	Q	Y	R	I	F	(48)
HA-Y543*		S	L	Q	Y	R	I	R	(48)
glyc-Y106*		P	S	D	Y	K	P	L	(29)
TGN-38		A	S	D	Y	Q	R	L	(173, 167)
Glut-4		P	S	G	F	Q	Q	I	(174)
EGFR (Y974)			N	F	Y	R	A	L	(94)
PDGFR (Y579)			H	E	Y	I	Y	V	(175)

signals with an important aromatic residue at position 6									
TfR (C20)ham			L	S	C	T	R	F	(22, 132)
pIgR (Y668)			V	S	I	G	S	Y	(25)
LDL (Y807)			F	D	N	P	V	Y	(30)
Cd-M6PR (F18)			F	P	H	L	A	F	(153)
IR(Y960)			S	S	N	P	E	Y	(35)
IR(Y953)			P	L	G	P	L	Y	(35)
mannose rec.			F	E	N	T	L	Y	(176)
TfR (Y34)*			V	D	G	D	N	Y	(42)
EGFR (F999)			P	Q	Q	G	F	F	(94)

di-leucine motifs									
CD M6PR	D	E	D	H	L	L	P	M	(57)
M6P/IGF-IIR(L162)	S	D	E	D	L	L	H	V	(56)
CD3 δ	E	V	Q	A	L	L	K	N	(58)
CD3 γ	D	K	Q	T	L	L	Q	N	(58)
Limp II	E	R	A	P	L	I	R	T	(177)
CD4	Q	I	K	R	L	L	S	E	(59)

Outlined amino acids and boxed sequences have been shown to be important by site-directed mutagenesis. Sequences in grey have been shown to adopt tight-turns. *denotes an engineered internalization sequence

Following these initial observations, the internalization signals of many cell-surface molecules were established using mutagenesis of the cytoplasmic domains (Table 1). It was found in most cases that the region involved in coated-pit localization was four or six amino acids long with the general sequence aromatic–X–X–aromatic/

large hydrophobic or aromatic–X–X–X–X–aromatic (1) (and see below). For either of these sequence patterns, the signal depends on the primary sequence and on the propensity of this sequence to adopt a tight turn conformation. For example, in cases where the aromatic residue must be a tyrosine and not a phenylalanine for proper internalization (e.g. LAP), it was found by two-dimensional NMR that a phenylalanine in place of the tyrosine reduced the ability of the corresponding peptide to adopt a reverse turn conformation by 25% (36). In contrast, where phenylalanine is tolerated as well as tyrosine *in vivo* (e.g. LDLR), the reverse-turn conformation is adopted in solution by peptides containing either tyrosine or phenylalanine (34).

It should be emphasized that structural data which directly demonstrate tight turns as important components of the internalization signal sequence come from NMR data of peptides derived from three proteins: LDLR, IR, and LAP. In each of these cases, care in interpreting the data is warranted. In the case of both the LDLR and IR, the dipeptides -NP- and -GP- (and especially the proline at the second position) are known from other studies to initiate β-turns (37, 38), so it is to be expected that the tetrapeptides -NPXY- or -GPXY- can form a tight turn in solution. In addition, although -NPXY- is sufficient to form a turn, when the signal is transplanted on to another protein it must have the sequence -FXNPXY- (39). In the case of LAP, the tetrapeptide -PPGY- was shown to adopt a well-ordered β-turn conformation in solution (again with a proline at the second position) (33), but later experiments implicated the hexapeptide -PGYRHV- as being the tyrosine-containing internalization signal (36). The only tyrosine-based tetrapeptide without a proline for which NMR data are reported is the signal for TGN 38/41. In this case, the -YQRL- motif is part of a nascent helix and not a β-turn (40). In the absence of more structural data a conservative formulation of the structure of the internalization signal is that it is composed of amino acids which *allow* the formation of a β-turn. Whether a β-turn is formed independently leading to binding by the signal receptor, or whether the receptor induces the signal into a β-turn in the binding pocket following recognition remains to be investigated when more structural data (and the identity of the receptor) are at hand.

Leaving aside the question of fine structure, how well do we understand the amino acid composition of the internalization signal? A complete understanding of the signal would fulfil the following requirements:

- It should be possible to transplant the signal autonomously on to a reporter protein and induce internalization of the reporter protein.
- By scanning the sequence of the cytoplasmic domain of a protein it should be possible to identify internalization signal(s).
- Proteins with internalization signals not fully conforming to the signal could be made to internalize faster with strategically placed amino acid substitutions.

We will summarize progress toward achieving these requirements and then we will discuss what is still missing.

2.2 Signal transplants

In its simplest form this was achieved for three different proteins with the introduction of a single tyrosine residue. In the case of HA one of four tyrosine substitutions in the 10 amino acid cytoplasmic domain led to a protein that was internalized. The rate of internalization of the mutant HA in permanent cell lines was 8–20% per minute which compares well with lysosomal proteins and some receptors (28, 41). In the case of the transferrin receptor, it was shown that substitution of a tyrosine for a serine at position 34 in a cytoplasmic domain lacking the internalization signal restored wild-type levels of internalization to that protein (42). The third protein for which introduction of a single tyrosine could induce internalization was human glycophorin A. Although there is no rate information for this protein, the steady-state distribution was 40% internal which also compares well with other endocytosed proteins (29). The relative ease by which internalization signals were created in these three proteins suggests that the amino acid requirements of the signal other than the presence of a tyrosine are quite lax. More recent experiments have used longer sequences to analyse whether an identified region is sufficient to specify internalization (39, 43). The ability to cut and paste internalization signal sequences (usually tetra- or hexa-peptides) seems to be related to their position on the recipient relative to the transmembrane domain (44). Transplants very close to the transmembrane region result in low internalization rates suggesting that a spacer separating the signal from the membrane is required (31). Although in only a few cases the transplanted sequences restored internalization to 100% of the control levels, in most cases internalization was sufficiently restored to indicate that the majority of the information which specifies the signal is indeed contained within those short amino acid stretches (1).

2.3 Signal prediction

Any signal which depends on a specific three-dimensional structure for activity will be vulnerable to the structure of the protein surrounding it. For example, a tetrapeptide with sequence similarity to an internalization signal but forming a part of a stable β-sheet structure will probably not specify internalization. With this caveat in mind it is perhaps not surprising that the current consensus sequences for internalization signals are not very useful for predicting the probability that a primary sequence containing a tyrosine will function for internalization. For example, the VSV G protein contains in its tail the tetrapeptide -YTDI- which fits an internalization signal quite well yet is internalized very poorly (45). Mutagenesis of this sequence to -YTDF- only marginally improved internalization (46). Similarly, the tyrosine in the tetrapeptide -YSLL- of the Fc receptor is not involved in internalization, although it fits a signal consensus quite well (47). For HA, the hexapeptide -LQYRIC- specifies internalization, but the hexapeptide -LQCIRY- does not (29); although five functional internalization signals have been identified, including

Table 2 Proteins can contain cytoplasmic sorting sequences

Protein	Sequence N-terminal to C-terminal									Sorting Event Clathrin-coated pits	Golgi to endosome	Basolateral targeting
LDL (Y807)	F	D	N	P	V	Y	Q	K	T	x		x
LDL (Y825)	Q	D	G	Y	S	Y	P	S	R			x
IR (Y960)	S	S	N	P	E	Y	L	S	A	x		
IR (Y953)	P	L	G	P	L	Y	A	S	S	x		
CD-MPR (Y45)	N	V	P	A	A	Y	R	G	V	x		
CD-MPR (F18)	F	P	H	L	A	F	W	Q	D	x		
CD-MPR	E	R	D	D	H	L	L	P	M		x	
pIgR (Y734)	E	A	D	L	A	Y	S	A	F	x		
pIgR (V668)	R	R	N	V	S	I	G	S	Y	x		x
CD3 δ	K	N	E	Q	L	Y	Q	P	L	x	x	
CD3 δ	A	E	V	Q	A	L	L	K	N	x	x	
CD3 γ	Q	N	E	Q	L	Y	Q	P	L			
CD3 γ	S	D	K	Q	T	L	L	Q	N	x	x	
M6P/IGF-IIR	N	V	S	Y	K	Y	S	K	V	x	x	
M6P/IGF-IIR	D	S	D	E	D	L	L	H	V		x	

Positions shown in outline have been demonstrated to be important for sorting. The sequences shown are from references listed in Table 1. Information on basolateral targeting of the LDLR can be found in ref. 178 and for the pIgR in ref. 179.

two in the insulin receptor and one in the EGF receptor (EGFR), that have a single aromatic residue at the last position (Table 1). Taken together, these results suggest that factors other than the sequence of the tetra- or hexapeptide itself play important roles in determining whether a potential signal can, in fact, specify internalization. Such factors can be thought of as the context of the signal and may be different for different proteins. Many proteins contain more than one internalization signal in the same polypeptide chain (Table 2), or if they are multimeric, contain signals on each subunit. This allows sequences that vary considerably in the affinity with which they are bound by coated pits to function quite well for endocytosis, and probably explains why the sequence motifs can be so degenerate. Thus, the importance of any one internalization motif in a protein will depend upon multiple factors, including:

- how many signals are present in the protein;
- how well they bind to elements of coated pits;
- how well the structure of the cytoplasmic domain exposes the signal;
- and perhaps where the signal is located in space relative to the plasma membrane.

2.4 Signal fine-tuning

The rates of internalization of different naturally occurring proteins can vary by a factor of 10, and this implies that not all signals are recognized with the same affinity by the components of the coated pits. In an effort to investigate this further, a series of mutations were produced in the cytoplasmic domain of an internalization-competent HA mutant to convert the signal of this protein (YRIC) to a signal resembling that of the TR (YTRF) (48). In agreement with the prediction that the consensus sequence for internalization is aromatic residue–X–X–large hydrophobic residue, it was found that changing the cysteine to a phenylalanine increased the rate of internalization of HA. However, an increase in internalization was also seen when the cysteine was changed to an arginine, something not predicted by the consensus sequence. Similar mutagenesis of other endocytosed proteins also suggests that many different classes of amino acids can be tolerated at the fourth position away from the aromatic residue, although another aromatic or hydrophobic is preferred. In addition, a number of naturally occurring sequences are known that do not fit well to any consensus primary sequence for an internalization motif (Table 1).

In general, internalization signals of the form discussed above do not appear to be all-or-none switches between incorporation into coated pits or complete exclusion. For most proteins, altering a putative internalization signal by mutagenesis results in 2- to 5-fold reduction in internalization rate, and this is true even when the entire cytoplasmic sequence is deleted. This is a rather undramatic reduction of activity when compared with other binding reactions between cellular proteins where mutations cause a reduction in rate of several orders of magnitude. However, the actual filtering capacity of coated pits is greater than the 5-fold selectivity observed between receptors that have internalization signals and those lacking cytoplasmic sequences. The increase in internalization of HA is 50- to 100-fold when a tyrosine is introduced in the cytoplasmic domain. Other proteins, like Thy-1, also seem to be 'excluded' from coated pits (49). Thus, there are three classes of proteins with regards to internalization through coated pits:

- Endocytic receptors cluster in coated pits and internalize at rates 5- to 50-fold faster than the bulk uptake of membrane into coated vesicles.

- Other proteins such as receptors lacking cytoplasmic sequences do not concentrate in coated pits and are internalized at roughly the constitutive rate of coated-pit internalization, which is 1–2% per minute.

- Proteins like Thy-1 and HA are excluded from pits and are internalized 10- to 20-fold more slowly than the bulk uptake of membrane.

The existence of these three classes of proteins cannot be explained by a simple two-choice sorting mechanism for incorporation of proteins into coated pits. In addition to the well-described positive signal that allows the concentration of receptors into coated pits, either HA and proteins like it are actively excluded from coated pits, or additional information specifying internalization resides in proteins that are internalized in the absence of a tyrosine (or di-leucine) internalization signal. The latter possibility is more likely since both HA (50) and Thy-1 (51) are freely mobile at the plasma membrane and it is difficult to imagine how they can be excluded from coated pits by any mechanisms other than by simply leaving no space for them. If additional information can specify internalization, where might it be? A logical location is the transmembrane domain, since in several cases it has been shown that residual endocytic activity exists when the entire cytoplasmic domain of a protein is deleted or replaced with 'nonsense' sequences (52). In addition, chimeric HA proteins with foreign transmembrane domains but containing the 'signal-less' HA tail can be internalized at approximately 1–2% per minute (53).

The available information concerning the internalization signal can be summarized as follows:

- The hierarchy of interactions of a protein with coated pits has three levels. A protein can either be completely excluded from coated pits, internalized at a small but measurable rate (< 3% per minute), or internalized at a variable but fast rate (5–50% per minute).

- The 'tyrosine-dependent' signal which specifies rapid internalization is four or six amino acids long and has the general form $^1\Phi\text{-}^2X\text{-}^3\Phi\text{-}^4X\text{-}^5X\text{-}^6\Phi$, where Φ indicates either a large hydrophobic or aromatic residue, and X can be any amino acid. Either position 3 or 6 must contain an aromatic residue, either tyrosine (most common) or phenylalanine; no signals are known that contain tyrosine in both these positions. Signals are known that function quite well, although they have only a single aromatic residue and lack hydrophobic amino acids at the other conserved positions.

- The variation in the rates of internalization of proteins carrying this signal suggests that, depending on its amino acid composition, this signal can specify residence in coated pits (e.g. the polymeric immunoglobulin receptor (pIgR) with internalization rate of 50% per minute) or rapid recruitment into coated pits (lysosomal proteins with rates of 10% per minute).

- The context of the signal is important (not all sequences with agreement to this signal can specify internalization) and it may be simply a region flexible or 'idle' enough to allow the signal to assume its two-dimensional structure. For most proteins such a region exists upstream of the signal.

- The amino acids which make up the signal can be a part of a type I β-turn. *In vivo* this tendency may be used to fit the signal to its receptor, or it may be part of the initial recognition process.

- In the absence of this signal, or parts thereof, endocytic receptors are not excluded from coated pits but their internalization rate drops 5-fold. It is not known whether this slow internalization is signal-mediated with the default being exclusion from coated pits or whether exclusion from coated pits is signal-mediated. Some evidence suggests that the transmembrane domain contains information relevant to this question.

2.5 Signals that do not contain tyrosine

Several proteins are known to be internalized in the absence of a signal containing tyrosine or other aromatic amino acids (Table 1). In some cases aromatic residues simply do not exist in the cytoplasmic domain, whereas in others the aromatic residues have been shown to be dispensable for internalization. The cytoplasmic domain of CD4 contains no aromatic residues, yet this protein is internalized rapidly upon binding to MHC II (54) or challenge of the cells with phorbol esters (55). The cytoplasmic domain of the Fc receptor FcRII-B2 contains two tyrosines, but elimination of either or both does not significantly affect internalization (47). These data demonstrate that signals other than the tyrosine-containing one can be recognized by the coated pits. One such signal is a di-leucine (or leucine–isoleucine) peptide which is found in several proteins (the large and small mannose-6-phosphate receptors (56, 57), the T-cell antigen receptor γ-chain (58) and CD4 (59)) near the C-terminus. It is interesting that most proteins with the di-leucine signal also contain a tyrosine-dependent internalization signal in their cytoplasmic domain, and both signals are functional. Whether this signal assumes a particular secondary structure remains to be reported.

3. Internalization, lysosomal, TGN, and basolateral signals: a common theme for a homotypic membrane system?

The following four tetrapetides are clearly related to each other: -YTRF-, -YTD(I,F)-, -YQTI-, and -YQRL-. The first is from the TR and mediates rapid internalization; the second is from the VSV G protein and mediates basolateral sorting but not internalization; the third is from Lgp/120 and mediates basolateral sorting, internalization, and delivery to lysosomes; and the fourth is from TGN38/41 and mediates basolateral sorting, internalization, and TGN targeting. Although some exceptions exist, there is now a growing list of proteins with short signals which depend on a tyrosine, are related but not identical, and specify sorting to four distinct membrane subdomains: the clathrin-coated pits, the TGN, the basolateral surface of polarized cells, and the lysosomes (Table 3). This unexpected development has forced us to rethink how the different organelles of the cell maintain their identity and what mechanisms are needed to ensure that a protein is delivered to its correct final destination.

Is there a characteristic common to these four organelles that could account for the similarity of the targeting signals? A possible answer is suggested by work using the drug brefeldin A (BFA) (60). Treatment of cells with BFA has effects on the exocytic as well as the endocytic pathway (61). In general, this effect can be described as loss of identity and alteration in morphology of the various cellular organelles. A useful generalization that has been proposed about the effects of BFA is that it causes the intermixing of homotypic membrane systems which are otherwise kept distinct by, as yet, unidentified (but obviously BFA-sensitive)

Table 3 Cytoplasmic sorting sequences function for more than one sorting event

Protein	Sequence N-terminal to C-terminal									Sorting Event I	II	III	IV	Reference
ASGP H1		M	T	K	E	Y	Q	D	L	x		x		(24)
HA+8	I	C	I	Y	D	Y	K	S	F	x		x		(53)
HA-Y543	N	G	S	L	Q	Y	R	I	C	x		x		(41)
NGFR (PS)	R	W	N	S	L	Y	S	S	L	x		x		(180)
FcγRII	E	N	T	I	T	Y	S	L	L	x		x		(168)
LDL (Y807)	F	D	N	P	V	Y	Q	K	T	x		x		(168)
LDL (Y825)	Q	D	G	Y	S	Y	P	S	R			x		(178)
VSV G	K	K	R	Q	I	Y	T	D	I			x		(46)
LAMP-1 (Lgp-A)	R	S	H	A	G	Y	Q	T	I	x	x	x		(168, 159)
LAMP-2(Lgp-B)	R	H	H	T	G	Y	E	Q	F	x	x	x		(181)
LAP	A	Q	P	P	G	Y	R	H	V	x		x		(169)
TGN-38	P	K	A	S	D	Y	Q	R	L	x			x	(167, 173)
M6P/IGF-IIR		S	D	E	D	L	L	H	V		x		x	(56)
CD3 δ		A	E	V	Q	A	L	L	K	N	x	x		(58)
CD3 γ		S	D	K	Q	T	L	L	Q	N	x	x		(58)

Amino acids outlined have been demonstrated by site-directed mutagenesis to be important for recognition during sorting. Sorting events: I, clathrin-coated pits. II, Golgi to endosome. III, basolateral targeting. IV, endosome to Golgi.

mechanisms, probably involving coat proteins (62). Traffic within, but not between, homotypic membrane system continues in the presence of BFA. In this view, organelles of the secretory pathway encompassing the endoplasmic reticulum, the intermediate compartment, and the Golgi complex (but not the TGN (63)) form one homotypic system which collapses and intermixes after BFA treatment. Proteins in this secretory homotypic system continue to cycle in the presence of BFA but cannot be secreted. Another homotypic system is the TGN–endosome–plasma membrane system which also collapses and its contents intermixed in the presence of BFA (62, 64–66). This can be called the endosomal homotypic system. Lysosomes probably belong to this system as enough evidence exists to suggest that, in at least some cell types, BFA has morphological effects on the lysosomes without inhibiting delivery of proteins to them (66). In the presence of BFA, the organelles of this homotypic system lose their distinct morphology, but traffic of proteins between them (e.g.

endocytosis, recycling, and delivery to lysosomes) continues. For polarized epithelial cells the situation is more complex because the plasma membrane is divided into two distinct domains, the apical and the basolateral. In the presence of BFA, apical (but not basolateral) secretion is inhibited, and apical (but not basolateral) endocytosis is stimulated (67–70). Therefore if response to BFA can be used as a means of designating homotypic systems, then the basolateral domain belongs to the endosomal homotypic system, whereas the apical domain is distinct. This designation is also supported by the fact that transcytosis (which is the transport from the basal to the apical domain and would therefore be a jump from one homotypic system to another) is inhibited by BFA (67), although apparently this finding is not true for all receptors (71). So for polarized cells, in addition to the secretory and the endosomal membrane system, a third homotypic system can be postulated consisting of organelles involved in apical trafficking. The significant fact is that BFA preserves the identity and function of homotypic membrane systems indicating that, even in situations where organelle identity is severely challenged, mechanisms exist which maintain a protein in its own membrane system and prevent it from crossing unchecked from one homotypic system to another.

We propose that the similarity of the signals which sort proteins to the coated pits, the TGN, the lysosomes, and the basolateral surface of polarized cells reflects the fact that these four organelles belong to the endosomal homotypic membrane system. The significance of this similarity can be explained in two ways. It is possible that the similarity is simply an evolutionary phenomenon, i.e. as this membrane system evolved from one 'primitive' endosomal organelle into a more specialized collection of related but distinct organelles, the signal also evolved from a common ancestor into a series of related but distinct signals. In this view, the similarity is relevant from an evolutionary standpoint but does not imply any present-day mechanistic similarities. Alternatively, it is possible that the similarity of the signals in these homotypic organelles is the reason why traffic among them continues in the presence of BFA. This would imply the existence of related but distinct signals specifying targeting to organelles of a homotypic membrane system. One signal could be generic and simply allow a protein to enter and traffic within a homotypic system, whereas a second more specific signal would specify steady-state residence in one of the homotypic organelles. Using tyrosine-based signals as an example, the generic signal could be an 'exposed tyrosine' and the specific signal for rapid internalization through coated pits could be 'exposed tyrosine in a β-turn'. We will return to the problem of related signals at the end of this chapter.

4. How the signal works

4.1 Components which recognize and bind the signal

The search for proteins which could bind the internalization signal at the cell surface has focused on molecules normally found in clathrin-coated pits or vesicles. In addition to the clathrin heavy and light chains which probably mediate coated-pit

architecture and uncoating respectively (4, 72), a class of proteins has been identified originally based on its property to enhance the assembly of clathrin cages *in
vitro* and later localized to clathrin-coated pits *in vivo* (73–76). These proteins are
termed adaptins, or clathrin associated proteins (APs), and we will use both terms
interchangeably. Reconstitution studies have shown that the adaptins are located
between the plasma membrane and the clathrin coat, a space which will also contain the cytoplasmic domains of endocytosed proteins (77). Adaptins are hetero-
tetrameric complexes, with one complex localized to the plasma membrane (AP-2
or HAII) and another found intracellularly in the region of the TGN (AP-1 or HAI)
(76). The plasma membrane AP-2 tetramer is composed of a 105–112 kDa subunit
(alpha adaptin), a 104 kDa subunit (β-adaptin), a 50 kDa subunit, and a 16 kDa subunit (78–80). The TGN AP-1 is composed of subunits with analogous molecular
masses with the two largest termed γ- and β-adaptin. Based on sequence analysis
(81–83) and electron microscopic studies of the AP-2 α- and β-adaptins (84), it has
been proposed that these two molecules share a similar architecture which can be
divided into two subdomains. The N-terminal domain shows little sequence variation and is imagined to have a brick-like structure. The C-terminal domain is variable in sequence and is imagined to form an ear-like appendage. The hinge region
separating the two domains is sensitive to elastase digestion and provides a convenient locus at which to separate the two domains physically *in vitro*. Overall, a
heterodimer of α- and β-adaptins will have a central core region made up of the N-
termini of both subunits and two 'ears' made up of the two different C-termini.
Based on its structure and sequence variation, it has been proposed that the 'ear'
domain interacts with the tails of internalized proteins, whereas the 'brick' domain
anchors the complex to the membrane and provides an assembly point for clathrin
(85). *In vitro* reconstitution of clathrin-coated pits on stripped plasma membranes
supports the idea that the 'brick' domain of AP-2 is necessary and sufficient for the
assembly and invagination of coated pits (86). Are the adaptins—and more specifically the 'ear' domain—responsible for binding the receptor tails?

Until recently, all data concerning this question came from *in vitro* binding studies.
Pearse and co-workers were the first to show that an endocytosed protein (the man-
nose-6-phosphate receptor) assembled *in vitro* with clathrin and adaptins to form a
cage lacking a membrane but having a structure similar to a clathrin-coated vesicle
(87). In addition, an affinity matrix of bacterial fusion proteins containing the LDLR
internalization signal could bind AP-2, and this binding, although weak, could be
specifically inhibited by peptides derived from other endocytosed proteins but not
from HA which is not internalized (75). Using the same experimental strategy,
binding was also reported to two non-overlapping sites of the mannose-6-
phosphate receptor by AP-1 and AP-2 (88). This dual binding is expected since the
receptor interacts with clathrin-coated membranes both at the cell surface and at the
TGN during its itinerary (89, 90). That the AP-2 complex can bind the internalization signal is also supported by work using peptides derived from the cytoplasmic
domain of a lysosomal protein (LAP) which is targeted to lysosomes via the cell
surface and the endocytic pathway. It was shown that AP-2, but not AP-1, binds to

peptides containing the internalization signal of LAP and this binding (as did the internalization activity) depended on the presence of a tyrosine in the cytoplasmic domain (91).

Recent work focusing on the region of the adaptins which actually binds the internalization signal both *in vitro* and *in vivo* has provided us with a surprise. In one experiment, intact or partially proteolysed clathrin-coated vesicle proteins were resolved by SDS–PAGE, transferred to nitrocellulose, and probed with pure asialo-glycoprotein receptor (92). Apart from the amazing result that any binding to the APs was seen, the binding could be competed by the cytoplasmic domain of other endocytosed proteins and it was — contrary to expectations — to the N-terminal (head) region of the adaptins. The only data that provides evidence that the APs bind the internalization signal *in vivo* has come from work with the EGFR (93). It was recently shown in a variety of cell lines that when the EGFR is induced to internalize by EGF, it co-immunoprecipitates with α-adaptins, which are presumably part of the entire AP-2 complex. This association depends on the temperature, reaches a maximum after 12 minutes of EGF induction and involves at least 50% of the α-adaptins in the cells. Limited proteolysis showed that the N-terminal head of adaptin binds to the EGFR, thus confirming the data from *in vitro* binding. Since the internalization signals of this receptor have recently been identified (94), this will be the best experimental system to ask whether binding of adaptins to the signal is all that is required to initiate internalization and whether signal activity correlates with the strength of binding to the adaptins.

4.2 Formation and invagination of coated pits

Three steps are required to initiate a round of endocytosis at the plasma membrane:

- Proteins bearing internalization signals are recognized and concentrated.
- Clathrin molecules polymerize into polygonal lattices in the region that contains the signal-bearing proteins giving rise to clathrin-coated pits.
- The coated pits curve into a spherical structure (the clathrin-coated vesicle) which subsequently pinches off the membrane.

Although we are far from a complete description of any of these steps, some of the key components are beginning to be identified. Evidence from fluorescence photo-bleaching recovery techniques using the wild-type HA or internalized HA mutants suggests that the interaction of internalized proteins with coated pits is dynamic, so that a protein at the cell surface may bind reversibly and repeatedly with coated pits during the 1–2 min for which the pit exists on the plasma membrane (50). Also, it has been reported that overexpression of endocytosed proteins leads to an increase in the number of clathrin lattices at the cell surface, implying a dynamic interaction of pits and receptors that can change the equilibrium between soluble and membrane-bound clathrin (95). But what signals the formation of a pit? Undoubtedly the last step is the AP-dependent binding of clathrin to plasma membrane assembly sites (96), mediated by the N-terminal portion of AP-2 α- and β-subunits (86). What is less clear is the mechanism which recruits APs to the plasma membrane. One possibility is that APs are recruited to the plasma membrane by the signal-bearing cytoplasmic tails of endocytosed proteins. In this model, binding of isolated APs to isolated tails would eventually lead to aggregation driven by the self-associating properties of APs. Binding of clathrin to isolated AP–tail complexes could be prevented by somehow ensuring that APs will not bind clathrin unless they themselves were polymerized. The evidence which supports this model is that APs can bind the cytoplasmic domains of endocytosed proteins directly (see above). However, all the *in vitro* binding experiments used matrices of packed peptides which probably approximate the aggregated state of these proteins at the cell surface, and the *in vivo* binding experiment was successful with a receptor (EGFR) known to aggregate once it is challenged with its ligand (97). It is therefore likely that all of these experiments were biased towards detection of low-affinity interactions and might have bypassed a requirement for a prior interaction necessary for clustering receptors. An alternative model proposes that endocytosed proteins are recognized and concentrated at the cell surface by APs that have associated with the plasma membrane by binding to their own receptor. A receptor for APs has been postulated based on the observation that limited proteolysis of plasma membrane sheets abolishes AP-2 binding and releases an activity which binds AP-2 in solution and prevents it from binding to membranes (98). Recently it has been shown that the non-hydrolysable analogue of GTP, GTP-γS, (99) or treatment with cationic amphiphilic drugs (100) can induce APs to bind to endosomes, suggesting that the AP binding site normally cycles through the endocytic system and is turned on and off. Currently, it is easier to understand these recent results using the model that says that the APs have a separate, high-affinity and regulated binding site that is turned on at the plasma membrane and off after internalization. If so, APs would first bind to the plasma membrane and then bind to internalization signals on receptors. The stoichiometry of the high-affinity binding site with APs in a clathrin coat might not necessarily be one to one, since interactions between APs and clathrin triskelions could be sufficient to stabilize a growing clathrin lattice that was also anchored by weaker interactions with receptor cytosolic tails.

There is currently disagreement about the origin of curved, clathrin lattices (101, 102). Clathrin triskelions can form flat, hexagonal arrays, but must undergo multiple rearrangements to form the pentagons necessary to form a curved surface containing hexagons (85). If flat lattices do give rise to clathrin-coated pits, clathrin must be both de-polymerized and re-polymerized, probably with the addition of fresh triskelions from the cytoplasm. Alternatively, there might be more than one function for clathrin lattices in the cell, with flat lattices serving some structural role and only curved lattices serving for endocytosis.

Following its formation, the coated vesicle must pinch off the plasma membrane. A protein apparently involved in the step is the 100 kDa GTPase dynamin, first identified in *Drosophila* as the product of the shibire gene (103–105). Cells with temperature-dependent defects in shibire show a block in the budding of coated vesicles at the non-permissive temperature. This block must be at late stages of the pathway since the coated vesicles have a normal morphology but remain attached to the plasma membrane. The budding of coated vesicles has been reconstituted using cell-free systems (106, 107). It is similar to other vesicle budding events in that it is a multistep process which requires energy, Ca^{2+}, GTP, and cytosolic factors.

5. Sorting following internalization

The cargo of a coated vesicle has multiple terminal destinations. Some receptors release their ligand and recycle back to the cell surface, others travel along a series of endosomal compartments of increasing density and acidity before branching off to the TGN, others undergo transcytosis and are targeted to the opposite surface of polarized cells, while another class of proteins travel along endosomal compartments and are ultimately delivered to lysosomes. To complicate matters further, several models exist regarding the number and the pathways of biogenesis of the various endosomal compartments. These models range from a description of the endosomal compartment as a continuous reticulum which receives and processes a variety of cargo (108), to a more static view which postulates either distinct endosomal organelles (early and late endosomes) which communicate through vesicular carriers (2) or gradual maturation of early endosomal compartments into late endosomes/prelysosomes (109, 110). For the purpose of describing sorting mechanisms, knowledge of the exact structure and biogenesis of these organelles is not required as long as the endocytic pathway can be divided into discrete reversible steps (Fig. 1). In this model, clathrin-coated vesicles originating from the cell surface rapidly lose their clathrin coats and fuse with each other to form early endosomes in a process that depends on AP-2 (111), GTP-binding proteins (112–116), phospholipase A (117), time of vesicle formation (118), temperature, and ATP (119). Early endosomes are tubulovesicular structures which label with endocytic material within 5 min of endocytosis (hence early) (120, 121) and are the site where material to be recycled back to the cell surface is sorted away from material that continues in the pathway (122–125). The early endosome is also probably the site where material to be transcytosed is selected. What is not recycled or transcytosed is delivered to late endosomes through an intermediate, sometimes termed an 'endosomal carrier vesicle (ECV)' (2). The formation of ECVs requires the activity of a vacuolar-type H^+-ATPase, and it is sensitive to the drug bafilomycin (126) which inhibits this class of ATPase specifically (127). The delivery of CVs to late endosomes is enhanced by intact microtubules and the cytoplasmic motor protein dynein, and is sensitive to microtubule-depolymerizing drugs (113, 128, 129). Late endosomes contain four different kinds of molecules:

- lysosomal transmembrane proteins (Lgps) en route to the lysosomes;
- soluble lysosomal proteins delivered by the CI-M6PR also en route to the lysosomes;

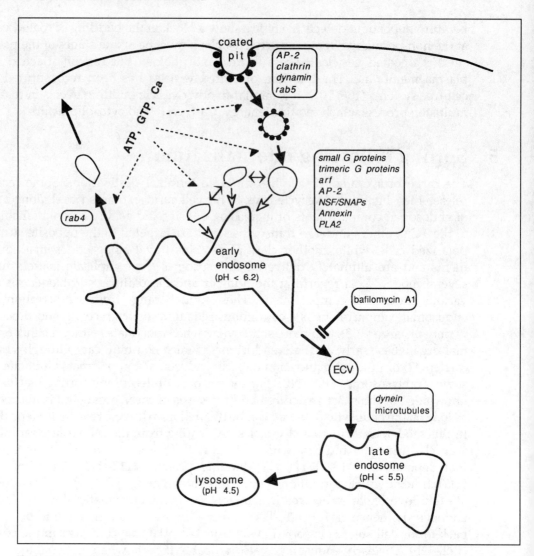

Fig 1. Movement in the endocytic pathway. The figure shows the proteins (italics) and other components known to be involved for each step of the endocytic pathway: formation and fission of clathrin-coated vesicles, uncoating, fusion with early endosomes, recycling or incorporation into endosomal carrier vesicles, delivery to late endosomes.

- proteins not destined for recycling but for degradation in the lysosomes (EGFR);
- and receptors which did not recycle through the early endosome that are destined to recycle through the TGN to the plasma membrane.

The final stop in the endosomal pathway is the lysosome. Although it is not clear whether lysosomes are formed from the maturation of late endosomes or are preexisting organelles receiving cargo, it appears that they are a terminal destination with no recycling routes (5). In the following sections we will summarize what is known about the sorting signals which operate at the various branch points of the endocytic pathway.

5.1 Recycling is by default

The strongest evidence that recycling of transmembrane proteins is not dependent on a signal sequence comes from studies using fluorescent lipid analogues (130, 131). When Chinese hamster ovary (CHO) cells were labelled with N-(N-[7-nitro-2,1,3-benzoxadiazol-4-y1]-ε-aminohexanoyl)-sphingosylphosphorylcholine (C6-NBD-SM) and subsequently were chased in the absence of the label, C6-NBD-SM was endocytosed with kinetics similar to bulk membrane internalization and recycled very rapidly with kinetics identical to fluorescent-labelled transferrin. Since the fluorescent lipid is unlikely to carry recycling information, the fact that it recycled with kinetics identical to those for the transferrin receptor suggests that no signal is required for recycling the latter. In addition, mutants of the transferrin receptor recycle with the same kinetics as the wild-type protein, even if they are severely impaired in the rate of internalization because they lack the internalization signal (42, 132). Mutant proteins such as an HA containing an artificial internalization signal also recycle efficiently (28, 29), although HA normally is excluded from the endocytic pathway and would be unlikely to have a specific recycling signal. However, it should be pointed out that in a mutant cell line with an acidification defect, recycling of membrane and of TR can occur at different rates (133). This can be interpreted as indicating either that recycling in some cases may require a signal, or that other sorting mechanisms in the early endosome are sensitive to endosomal pH and can influence recycling rates.

Although the mechanism of recycling is far from clear (134), it is expected that any recycling defect will interfere with the balance of membrane flow which must be maintained for proper functioning of the cellular organelles. This is evident in a mutant CHO cell line (TFT1.11) with a conditional defect in recycling (135, 136). At the restrictive temperature, these cells show pleiotropic defects in recycling, fluid-phase endocytosis, and lysosome biogenesis. Ultimately these cells die due to lack of iron because the transferrin receptor is trapped intracellularly. It is interesting that revertants of these cells still maintain the defects in recycling and lysosome biogenesis but have apparently developed a novel method for obtaining iron from the medium (137). These results indicate that cells can tolerate a defect in the function of endosomal organelles, and thus in the steady-state distribution of membranes, as long as they have a compensatory mechanism of obtaining nutrients.

A protein which is involved in the control of recycling has been shown to be the small GTPase rab4 (138). Overexpression of rab4 in cell lines was found to cause a block in the delivery of transferrin to acidic early endosomes and a redistribution of the transferrin receptor from early endosomes to the cell surface. Since the initial rates of internalization in these cells were unaffected, the defect was at the level of recycling. Similar to the TFT1.11 mutant, this defect in recycling was accompanied by a three-fold reduction in fluid-phase uptake but, unlike for TFT1.11, no effects on lysosome function were reported.

In terms of trafficking signals, internalization and recycling can be considered as a continuous pathway relying only on an internalization signal for its operation. A second signal is required to allow a protein to exit this pathway for transcytosis or delivery to the late endosome.

5.2 Transcytosis requires receptor phosphorylation or occupancy

Following internalization in polarized cells, a subset of proteins is delivered to the opposite cell surface from where endocytosis originated. This process is termed transcytosis. A model system used extensively to study this process is that of Madin–Darby canine kidney (MDCK) cells transfected with the poly immunoglobulin receptor (pIgR) (139). The pIgR is first delivered to the basolateral surface of MDCK cells where it binds IgA from the medium and, following internalization and transcytosis, delivers the IgA to the apical surface. At the apical surface it is internalized and recycles with its ligand bound until proteolysis releases the external domain of the receptor, still bound to IgA, into the medium (140). As expected from its itinerary, this receptor contains a multitude of sorting signals in its cytoplasmic domain (141). Among them, it has been shown that transcytosis of an unoccupied receptor depends on the presence of a phosphorylated serine residue at position 664 (142). Mutation of this serine to alanine reduces transcytosis which is restored by mutation to aspartate, presumably because the negatively charged aspartate can substitute for the phosphorylated serine. Subsequently, it was shown that if the receptor is bound to its physiological ligand (dimeric IgA), transcytosis can take place in the absence of a phosphorylated serine (143). These results suggest that transcytosis is normally signalled by binding of the ligand to its receptor, resulting in a change of the receptor cytoplasmic domain which can be mimicked by phosphorylation of a serine. Although the mechanism is not known, recent data indicate that it may involve activation of protein kinase C and/or heterotrimeric GTPases (144–146). It will be interesting to ask whether transcytosis depends on a single positive signal, or whether it requires the internalization signal to be switched off while at the same time presenting the signal for transcytosis to the sorting machinery.

Following their designation for transcytosis, vesicles and cargo derived from basolateral endosomes could fuse directly with the apical membrane or they could first fuse with the endosomal membrane system of the apical side and then be delivered apically. A recent report using the pIgR favours the second model, based on the observation that IgA endocytosed from the basal surface has access to compartments containing apically recycling material before reaching the apical cell surface (147, 148). If this model is correct it might imply that the transcytosis signal also serves as a signal for targeting to apical endosomes. Alternatively, the authors favour the interpretation that the endosomal systems from the apical and basolateral surface are a continuum. In this case, recycling to the basolateral surface may be by default, whereas recycling to the apical surface would require a transcytotic signal.

5.3 Sorting in the late endosome: proteins arriving by an intracellular route

Two classes of proteins are known to arrive at the late endosome/prelysosome compartment without first appearing at the cell surface: the two mannose-6-phosphate receptors (46 kDa and 215 kDa M6PR) transporting soluble lysosomal enzymes such as acid hydrolases (5), and, in some cell types, the transmembrane lysosomal proteins such as Lgp-A (149, 150). Although it is still debated whether these proteins intersect the endocytic pathway at the level of the early or late endosome (151, 152), it is clear that they must contain signals recognized by the endosomal sorting machinery directly, without first being sorted at the cell surface. The signals which mediate sorting of these proteins are at their cytoplasmic domain. The M6PRs contain two different signals: a tyrosine- (or phenylalanine-) based one proximal to the transmembrane domain, and a di-leucine signal distal to the first and very close to the C-terminus of these proteins (20, 56, 153, 154). For the 215 kDa M6PR, both signals contribute to direct intracellular sorting (56), whereas for the 46 kDa M6PR the C-terminal di-leucine signal is apparently sufficient for sorting of lysosomal proteins from the TGN (57), although microinjection of antibodies against the tyrosine-containing epitope also influenced the intracellular itinerary of this protein (155). Both the tyrosine-based and di-leucine motifs also function as internalization signals of high efficiency at the cell surface and are used to direct these receptors back to the endosomal system. For these proteins a common signal (or sets of signals) can potentially be recognized by multiple-sorting machineries (Fig. 2). This leads to a troubling question: if the signal for internalization can be recognized intracellularly for a pathway other than that leading to the cell surface, how do receptors carrying an internalization signal manage to 'escape' being trapped inside the cells following their biosynthesis? Either the internalization signal of receptors is masked during transport to the cell surface (there is no evidence for this), or the signal for direct delivery to endosomes carries additional information to distinguish it from an internalization signal. The 215 kDa M6PR is phosphorylated *in vivo* on two serines in its cytoplasmic domain (156), and it has been shown that this phosphorylation takes place only during exit of the protein from the TGN and not during its internalization from the cell surface (157). Similarly, the 46 kDa M6PR is phosphorylated *in vivo* on serine 56 in its cytoplasmic domain (158). These serines were mutated to test the hypothesis that, for these receptors, the internalization signal is recognized intracellularly and directly diverted to endosomal compartments only when the tail is also phosphorylated (57, 158). Unfortunately, as is often the case when site-directed mutagenesis is used to test an attractive hypothesis, mutagenesis of the phosphorylated serines in the cytoplasmic domains of both receptors had no effect on their function or subcellular distribution, leaving us ignorant of

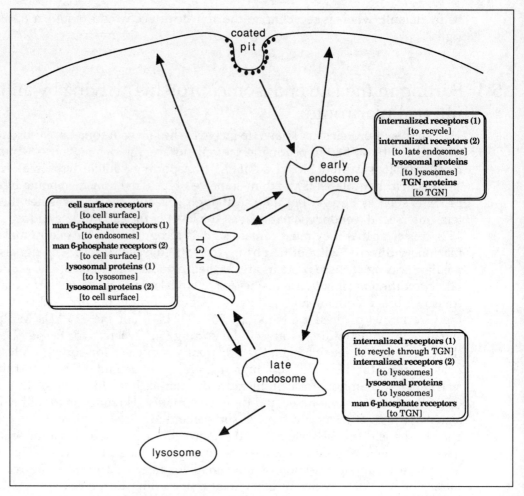

Fig 2. The existence of multiple routes with multiple terminal destinations probably accounts for the complexity and degeneracy of the signals involved in the endocytic pathway. For three of the organelles involved (early endosome, TGN, late endosome) we indicate the kinds and the destination of the molecules that are continously being sorted.

both the nature of the recognition event for intracellular delivery to lysosomes and the function of phosphorylation of these receptors.

Another group of proteins for which the problem of intracellular recognition versus recognition by plasma membrane coated pits must be addressed are the lysosomal membrane glycoproteins. It has been reported for Lgp120 (Lgp-A) that at low levels of expression this protein can reach the lysosomes directly, whereas if the expression levels are increased targeting to lysosomes involves passage from the cell surface and internalization (150). This protein contains in its cytoplasmic domain a tyrosine-based internalization signal which is necessary and sufficient for correct targeting to the lysosomes (26, 159). Although this signal involves an additional glycine residue not predicted to be involved in coated-pit internalization (150), the presence of this glycine alone cannot explain how the signal can work for targeting in two different pathways, because the lysosomal acid phosphatase, which also contains a glycine and a tyrosine-based signal, can only reach the lysosomes travelling through the cell surface (27, 160). We will return to the question of similar signals mediating distinct targeting events at the end of this chapter.

5.4 Sorting in the late endosome: proteins arriving from the cell surface

Another group of proteins arriving at the late endosome would have started their journey from the cell surface. These proteins would then travel to the lysosomes as a terminal destination or recycle to the TGN (Fig. 2). Among the first group would be either receptors which have internalized ligand and are delivering it to lysosomes (e.g. EGFR) (161), plasma membrane proteins destined to be degraded, or lysosomal transmembrane proteins targeted to the lysosomes through the cell surface (e.g. LAP) (160). Also present would be the M6PRs returning to the TGN to pick up newly synthesized lysosomal hydrolases (5) or receptors such as the LDLR which escaped the recycling endosomal pathway and are rerouted to the cell surface through the TGN (162). There are several different ways that proteins might be sorted in late endosomes towards these two destinations:

- Transport could be signal-mediated in both directions.
- Alternatively, one destination could require a signal and the other could occur by default.
- Since sorting is rarely observed to be completely efficient, transport to one or both destinations might require signals, but there might also be additional space available in the vesicles carrying proteins with specific signals to their destinations. Protein without signals might use this extra space to reach their destinations.

In this last case, the destination of proteins lacking signals will be determined by the amount and type of competition that they encounter in the form of proteins containing signals. This competition could vary in different cell types and even, perhaps, in the same cell type under different growth conditions.

Although it is commonly thought that delivery to lysosomes is the default destination for internalized proteins, the little direct evidence that exists argues against this. LDL receptors and transferrin receptors that escape the recycling pathway from the early endosome appear to recycle to the plasma membrane through the TGN or *trans* Golgi, rather than travelling to lysosomes (162), suggesting that recycling is by default and transport to lysosomes requires some positive recognition event. This is certainly the case for the endothelial cell adhesion molecule P-selectin, which contains a cytoplasmic determinant required for efficient delivery to lysosomes (163).

If transport of proteins from the late endosomes to the lysosomes is signal-mediated, what is the signal that directs this traffic? We will discuss the proteins destined for degradation separately from those which are delivered to the lysosomes in order to carry out their function, but it is important to remember that this pathway must accommodate both types of proteins. Several observations suggest that the state of aggregation of cell-surface receptors correlates with their ultimate delivery for degradation in lysosomes. This aggregation can be brought about by antibodies applied extracellularly (e.g. the macrophage Fc receptor) (164) or by ligand binding and subsequent structural changes (e.g. the EGFR). In both cases, aggregation leads to lysosomal delivery and degradation.

Additional evidence that the state of oligomerization of a protein influences its delivery to lysosomes comes from work with HA. A frameshift mutation at the C-terminal domain of HA results in a protein with an additional eight amino acids in its cytoplasmic domain (HA+8) (53). This protein is internalized at a rate of 50% per minute due to the presence in its cytoplasmic domain of a TR-like tyrosine based signal, -YKSF-. However, unlike the TR this protein recycles poorly following internalization and is degraded instead. When the biosynthesis, structure, and transport of HA+8 were compared with that of the wild-type HA, the only significant difference between them was that the former entered into higher-order oligomers and wild-type HA did not. Oligomer formation was not correlated with the rapid kinetics of internalization of HA+8, because second-site mutants of this protein internalized at the same rate but did not oligomerize to the same extent (165). In addition, other mutants of HA were identified that formed higher-order complexes but, having very poor internalization signals, were internalized slowly. However, these latter mutants had a larger internal population at steady-state than would have been predicted by their slow rate of internalization, suggesting that they were preferentially sorted towards lysosomes rather than recycling to the cell surface (29). Thus, there are several lines of evidence that a distinguishing feature of proteins destined for lysosomal degradation is their aggregated state. Whether this state is sensed by the sorting machinery through interactions of the external domains of these proteins, or through their the transmembrane and/or cytoplasmic domains is not known.

Another group of proteins delivered from the cell surface to the late endosomes en route to the lysosomes is best represented by LAP. This protein is apparently first targeted to the cell surface, enters the endocytic pathway by a tyrosine-based internalization signal, and recycles as many as 30 times back to the cell surface before being subsequently delivered to the lysosomes (160). How is delivery to the lysosomes determined? A recent observation concerning targeting of another lysosomal protein LAMP-1 may help answer this question. Mutagenesis of LAMP-1 suggested that the sequence -YQTI- at the carboxyl C-terminus is necessary and sufficient for lysosomal targeting, but when purified LAMP-1 protein was sequenced it was found to lack the last two amino acids (44). Perhaps some sort of modification of the cytosolic domains occurs to proteins like LAP during their recycling phase. These proteins would recycle to the cell surface as long as their cytoplasmic domain remained intact, but would be diverted to lysosomes following limited proteolysis or some other modification (166). It is interesting that a region that is required for proper targeting of LAP falls outside of the designated internalization signal, very near the C-terminus of the cytoplasmic domain (36). Is it possible that this region contains the additional information required for delivery of this protein to lysosomes?

6. A model for sorting in the endosomal homotypic membrane system

In Section 3, we discussed the possibility that the four sorting signals (for the (i) TGN, the (ii) clathrin-coated pits, the (iii) lysosomes, and the (iv) basolateral surface of MDCK (and similar domains in other polarized cells) are functionally related. That is, they are recognized by distinct adaptor proteins and sorted at different locations by a mechanism essentially similar to that operating in clathrin-coated pits. This hypothesis explains two puzzling observations, the fact that a very short segment of polypeptide can contain overlapping sorting signals (Table 3) and that proteins containing signals with quite similar primary sequences can be sorted into different pathways to reach the same final destination. An example of the latter behaviour is the lysosomal membrane protein LAMP-1, which has the signal -GYQTI- and is preferentially sorted in the TGN directly to late endosomes. The lysosomal enzyme precursor, LAP, has the signal -GYRHV- and is preferentially sorted in the TGN to the plasma membrane, where it enters the recycling pathway for a while before eventually joining LAMP-1 in lysosomes. To explain how such seemingly similar sequences might be recognized differently, we propose a simple model for sorting in the homotypic endosomal membrane system which relies on a minimum set of sorting proteins (Fig. 3). Central to this sorting model would be a dynamic membrane network composed of parts of the TGN, the recycling endosome and the late endosome (*endosomal network*, Fig. 3, shaded areas) into which proteins are sorted following their synthesis but before their arrival at the cell surface. Two points of entry into this network would be the TGN and the plasma membrane endocytic pits, whereas a point of exit without the possibility of re-entry would be the site in late endosomes where material is sorted to lysosomes. The default pathway for proteins entering this system would be delivery to the plasma membrane. We propose that all proteins carrying a certain common signal would enter this endosomal network following their post-translational modification at the Golgi complex and that targeting of these proteins to their correct final destination would require an additional signal element distinct for each case. The advantage of having one general signal that allows access to a whole membrane network and a specific signal that signifies residence in an organelle is that it allows proteins the dual capability of travelling between organelles and of achieving correct steady-state localization. In addition, this arrangement provides a convenient mechanism for retrieving proteins which have escaped correct initial targeting. For example, a molecule such as TGN38/41 would use the TGN-specific targeting signal for correct

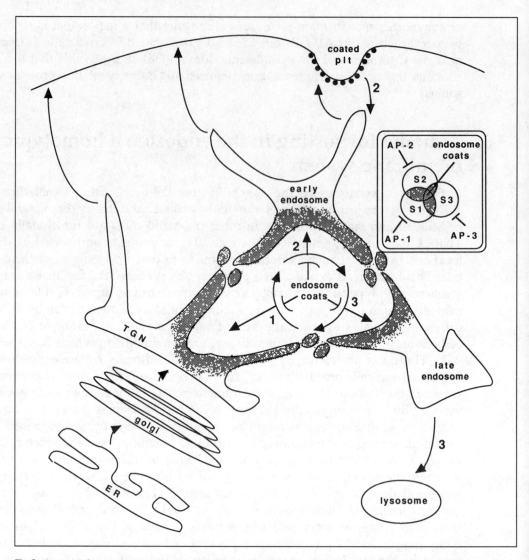

Fig 3. A model for sorting in the endosomal homotypic membrane system. Following their post-translational modifications in the Golgi complex, proteins bearing a common signal enter a dynamic BFA-sensitive membrane network composed of parts of the TGN and the early and late endosomes (shaded areas). Movement in this network is facilitated by a set of coat proteins (endosome coats) analogous to the ones involved in ER to Golgi transport, and which requires the presence of the common endosomal signal. Sorting from this network into the TGN (1) or the plasma membrane coated pits (2) and exit to the lysosomes (3) are controlled by three additional distinct but related signals (insert: S1, S2, and S3) recognized by three different sets of adaptors (insert: AP-1, AP-2, and AP-3). The common endosomal signal is not necessarily a distinct epitope—although this may be true for some proteins—but could be the minimum set of determinants common to the three signals (insert, shaded area). For polarized cells, the common endosomal signal is used for basolateral delivery and recycling.

localization in the TGN but the general endosomal membrane signal for retrieval from the cell surface. Similarly, a protein such as the M6PR would use the endosomal membrane signal for travelling from the TGN to late endosomes and a second specific signal for localization in the late endosome/prelysosome compartment. For a protein such as the TR, the coated pit-specific targeting signal would be used for internalization and the endosomal membrane signal for retrieval to the cell surface of molecules that escaped into late endocytic compartments. We have indicated the existence of two separate signals—one general and one specific—in order to describe this model better. In reality, if we imagine the signals for the TGN, the coated pits, and the late endosomes as overlapping but distinct structural elements, the common endosomal membrane signal would be the structural elements common to these three signals (Fig. 3, insert). Since both di-leucine and tyrosine-containing sequences can specify sorting within the endosomal system, perhaps the common structural elements are merely an exposed large hydrophobic residue in a flexible stretch of polypeptide chain that is located within a certain required distance from the cytosolic face of a membrane. Signals meeting this minimal requirement would interact with the general endocytic adaptor and at low affinity with other adaptors. Signals that specify residence in an organelle would have additional structural elements, for example 'tyrosine in a β-turn' for coated pits, and would bind with much higher affinity to the adaptor specialized for a certain organelle. From this viewpoint, it should be possible to distinguish signals of similar sequence (but with distinct targeting properties) by mutagenesis, and recent work on the signals for TGN38/41, LAP, and Lgp120 confirms this (167–169). For example, the TGN localization signal for TGN38/41 is the tetrapeptide -YQRL-, but mutagenesis to produce the tetrapeptide -YQDL- resulted in molecules that internalized but were not delivered to the TGN. From these data, we would argue that part of the discriminator which makes this signal specific for the TGN is the arginine at the third position. Loss of this residue decreases the affinity of this signal for the TGN adaptor so that the mutant protein now circulates within the endosomal system by virtue of the remaining structural elements common to the endosomal signal.

We suggest that recognition of the signals in the endosomal membrane system takes place at the two entry (TGN, coated pits) and one exit (lysosomes) sites by three sets of adaptors with distinct but related binding affinities (Fig. 3, insert). One set would recognize and bind motifs for trafficking between the coated pits and early endosomes and these would be the already known AP-2s. Another set would be specific for the TGN, whereas a third set would be adaptors for trafficking between the late endosomes/prelysosomes. Based on the available data, it would seem that the AP-1s would be specific for the TGN, which indicates that the late endosome adaptors remain unidentified. In addition, it is likely that a fourth set of proteins would maintain the structure of the organelles of the endosomal membrane system and mediate traffic between them (170), in analogy to the coat proteins which are involved in ER to Golgi transport (171). These proteins would recognize the general endosome membrane signal and be sensitive to BFA, and their dispersal in the presence of the drug would cause collapse and intermixing of the endosomal membrane system.

If these adaptors have binding affinities to related sets of signals how is mis-sorting prevented? The simplest way to accomplish this would be to spatially constrain the adaptors to specific membrane subdomains, and to ensure that their binding affinities to the signals be sufficiently distinct. For example, the LDLR cycling through the endosomal membrane system would be likely to encounter both early and late endosome adaptors, but if its sorting signal were recognized with a 10-fold higher affinity by the former than by the latter then two rounds of sorting would result in 99% correct localization.

A special case of this model is the situation in polarized epithelial cells. Sorting has been most studied in MDCK cells and we will confine our discussion to this example, but, in principle, the same mechanism could be employed in other types of cells for specialized domains functionally analogous to the basal membrane of MDCK cells. We propose, for MDCK cells, that any protein bearing the endosomal sorting signal would be delivered to the basal side and recycle back to this side following internalization. In these cells transcytosis would be accomplished either by a positive signal, or by the loss of the endosomal sorting signal. Proteins that do not carry the endosomal sorting signal would be sorted to the apical side in MDCK cells.

In summary, we propose that sorting in the endocytic pathway takes place in two steps by two classes of signals. Initially, proteins bearing a common signal are allowed access to a dynamic, BFA-sensitive membrane network composed of parts of the TGN, the early and the late endosomes. In this endosomal membrane system, trafficking is facilitated by a set of coat proteins (with properties similar to their counterparts in ER to Golgi transport) which recognize the general endosomal signal. In the absence of additional signals, the default pathway in this system would be the plasma membrane. Depending on their steady-state localization, proteins in this system then use three specific signals for targeting to coated pits, residency in the TGN, or delivery to late endosomes–prelysosomes. These three signals are recognized by three sets of adaptors, two of which are known (AP-1 and AP-2) and the third remains unidentified. In polarized cells, plasma membrane proteins will be sorted according to whether or not they are recognized by the general adaptor of the endosomal membrane system. In MDCK cells, proteins binding this adaptor will be delivered to the basal plasma membrane, although we do not currently have enough information to know whether this occurs by a route passing through the recycling endosome or directly to the plasma membrane.

References

1. Trowbridge, I. S., Collawn, J. F., and Hopkins, C. R. (1993) Signal-dependent membrane protein trafficking in the endocytic pathway. *Annu. Rev. Cell Biol.*, **9**, 129.
2. Gruenberg, K. and Howell, K. E. (1989) Membrane traffic in endocytosis: insights from cell-free assays. *Annu. Rev. Cell Biol.*, **5**, 453.

3. Smythe, E. and Warren, G. (1991) The mechanism of receptor-mediated endocytosis. *Eur. J. Biochem.*, **202,** 689.

4. Pearse, B. M. F. and Robinson, M. S. (1990) Clathrin, adaptors and sorting. *Annu. Rev. Cell Biol.*, **6,** 151.

5. Kornfeld, S. and Mellman, I. (1989). The biogenesis of lysosomes. *Annu. Rev. Cell Biol.*, **5,** 483.

6. Goldstein, J. L., Brown, M. S., Anderson, A. G. W., Russell, D. W., and Schneider, W. J. (1985) Receptor-mediated endocytosis: concepts emerging from the LDL receptor system. *Annu. Rev. Cell Biol.*, **1,** 1.

7. Pearse, B. M. F. (1975) Coated vesicles from pig brain: purification and biochemical characterization. *J. Mol. Biol.*, **97,** 93.

8. Anderson, R. G. W., Brown, M. S., and Goldstein, J. L. (1977) Role of the coated endocytic vesicle in the uptake of receptor-bound low density lipoprotein in human fibroblasts. *Cell,* **10,** 351.

9. Bretscher, M. S., Thomson, J. N., and Pearse, B. M. F. (1980) Coated pits act as molecular filters. *J. Cell Biol.*, **77,** 4156.

10. Bretscher, M. S. and Pearse, B. M. F. (1984) Coated pits in action. *Cell,* **38,** 3.

11. Daukus, G. and Zigmond, S. H. (1985) Inhibition of receptor-mediated but not fluid-phase endocytosis in polymorphonuclear leucocytes. *J. Cell Biol.*, **101,** 1673.

12. Sandvig, K., Olsnes, S., Peterson, O. W., and van Deurs, B. (1987) Acidification of the cytosol inhibits endocytosis from coated pits. *J. Cell Biol.*, **105,** 679.

13. Carpentier, J.-L., Sawano F., Geiger, D., Gorden, P., Perrelet, A., and Orci, L. (1989) Potassium depletion and hypertonic medium reduce 'non-coated' and clathrin-coated pit formation, as well as endocytosis through these two gates. *J. Cell. Physiol.*, **138,** 519.

14. Moya, M., Dautry-Varsat, A., Goud, B., Louvard, D., and Boquet, P. (1985) Inhibition of coated pit formation in Hep2 cells blocks the cytotoxicity of diphtheria toxin but not that of ricin. *J. Cell Biol.*, **101,** 548.

15. Hansen, S. H., Sandvig, K., and van Deurs, B. (1993) Clathrin and HA2 adaptors: effects of potassium depletion, hypertonic medium and cytosol acidification. *J. Cell Biol.*, **121,** 61.

16. Anderson, R. G., Kamen, B. A., Rothberg, K. G., and Lacey, S. W. (1992) Potocytosis: sequestration and transport of small molecules by caveolae. *Science,* **255,** 410.

17. Lehrman, M. A., Goldstein, J. L., Brown, M. S., Russell, D. W., and Schneider, W. J. (1985) Internalization-defective LDL receptors produced by genes with nonsense and frameshift mutations that truncate the cytoplasmic domain. *Cell,* **41,** 737.

18. Davis, C. G., Lehrman, M. A., Russell, D. W., Anderson, R. G. W., Brown, M. S., and Goldstein, J. L. (1986) The J.D. mutation in familial hypercholesterolemia amino acid substitution in cytoplasmic domain impedes internalization of LDL receptors. *Cell,* **45,** 15.

19. Davis, C. G., van Driel, I. R., Russell, D. W., Brown, M. S., and Goldstein, J. L. (1987) The low density lipoprotein receptor; identification of amino acids in cytoplasmic domain required for rapid endocytosis. *J. Biol. Chem.*, **262,** 4075.

20. Lobel, P., Fujimoto, K., Ye, R. D., Griffiths, G., and Kornfeld, S. (1989) Mutations in the cytoplasmic domain of the 275 kd mannose 6-phosphate receptor differentially alter lysosomal enzyme sorting and endocytosis. *Cell,* **57,** 787.

21. Jing, S., Spencer, T., Miller, K., Hopkins, C., and Trowbridge, I. S. (1990) Role of the human transferrin receptor cytoplasmic domain in endocytosis: localization of a specific signal sequence for internalization. *J. Cell Biol.*, **110,** 283.

22. Alvarez, E., Girones, N., and Davis, R. J. (1990) A point mutation in the cytoplasmic domain of the transferrin receptor inhibits endocytosis. *Biochem. J.*, **267**, 31.

23. Breitfeld, P. P., Casanova, J. E., McKinnon, W. C., and Mostov, K. E. (1990) Deletions in the cytoplasmic domain of the polymeric immunoglobulin receptor differentially affect endocytotic rate and postendocytotic traffic. *J. Cell Biol.*, **265**, 13750.

24. Fuhrer, C., Geffen, I., and Spiess, M. (1991) Endocytosis of the ASGP receptor H1 is reduced by mutation of tyrosine-5 but still occurs via coated pits. *J. Cell Biol.*, **114**, 423.

25. Okamoto, C. T., Shia, S. P., Bird, C., Mostov, K. E., and Roth, M. G. (1992) The cytoplasmic domain of the polymeric immunoglobulin receptor contains two internalization signals that are distinct from its basolateral sorting signal *J. Biol. Chem.*, **267**, 9925.

26. Williams, M. A. and Fukuda, M. (1990) Accumulation of membrane glycoproteins in lysosomes requires a tyrosine residue at a particular position in the cytoplasmic tail. *J. Cell Biol.*, **111**, 955.

27. Peters, C., Braun, M., Weber, B., Wendland, M., Schmidt, B., Pohlmann, R., Waheed, A., and von Figura, K. (1990) Targeting of a lysosomal membrane protein: a tyrosine-containing endocytosis signal in the cytoplasmic tail of lysosomal acid phosphatase is necessary and sufficient for targeting to lysosomes. *EMBO J.*, **9**, 3497.

28. Lazarovits, J. and Roth, M. (1988) A single amino acid change in the cytoplasmic domain allows the influenza virus hemagglutinin to be endocytosed through coated pits. *Cell*, **53**, 743.

29. Ktistakis, N. T., Thomas, D., and Roth, M. G. (1990) Characteristics of the tyrosine recognition signal for internalization of transmembrane surface glycoproteins. *J. Cell Biol.*, **111**, 1393.

30. Chen, W. J., Goldstein, J. L., and Brown, M. S. (1990) NPXY, a sequence often found in cytoplasmic tails, is required for coated pit-mediated internalization of the low density lipoprotein receptor. *J. Biol. Chem.*, **265**, 3116.

31. Collawn, J. F., Stangel, M., Kuhn, L. A., Esekogwu, V., Jing, S. Q., Trowbridge, I. S., and Tainer, J. A. (1990) Transferrin receptor internalization sequence YXRF implicates a tight turn as the structural recognition motif for endocytosis. *Cell*, **63**, 1061.

32. Vega, M. A. and Strominger, J. L. (1989) Constitutive endocytosis of HLA class I antigens requires a specific portion of the intracytoplasmic tail that shares structural features with other endocytosed molecules. *Proc. Natl Acad. Sci. USA*, **86**, 2688.

33. Eberle, W., Sander, C., Klaus, W., Schmidt, B., von Figura, K., and Peters, C. (1991) The essential tyrosine of the internalization signal in lysosomal acid phosphatase is part of a beta turn. *Cell*, **67**, 1203.

34. Bansal, A. and Gierasch, L. M. (1991). The NPXY internalization signal of the LDL receptor adopts a reverse-turn conformation. *Cell*, **67**, 1195.

35. Backer, J. M., Shoelson, S. E., Weiss, M. A., Hua, Q. X., Cheatham, R. B., Haring, E., Cahill, D. C., and White, M. F. (1992) The insulin receptor juxtamembrane region contains two independent tyrosine/beta-turn internalization signals. *J. Cell Biol.*, **118**, 831.

36. Lehmann, L. E., Eberle, W., Krull, S., Prill, V., Schmidt, B., Sander, C., von Figura, K., and Peters, C. (1992) The internalization signal in the cytoplasmic tail of lysosomal acid phosphatase consists of the hexapeptide PGYRHV. *EMBO J.*, **11**, 4391.

37. Chou, P. Y. and Fasman, G. D. (1977) Beta-turns in proteins. *J. Mol. Biol.*, **115**, 135.

38. Wilmot, C. M. and Thornton, J. M. (1988) Analysis and prediction of the different types of beta-turn in proteins. *J. Mol. Biol.*, **203**, 221.

39. Collawn, J. F., Kuhn, L. A., Liu, L. F., Tainer, J. A., and Trowbridge, I. S. (1991) Trans-

planted LDL and mannose-6-phosphate receptor internalization signals promote high-efficiency endocytosis of the transferrin receptor. *EMBO J.*, **10**, 3247.

40. Wilde, A., Dempsey, C., and Banting, G. (1994) The tyrosine-containing internalization motif in the cytoplasmic domain of TGN38/41 lies within a nascent helix. *J. Biol. Chem.*, **269**, 7131.

41. Brewer, C. B. and Roth, M. G. (1991) A single amino acid change in the cytoplasmic domain alters the polarized delivery of influenza virus hemagglutinin. *J. Cell Biol.*, **114**, 413.

42. McGraw, T. E., Pytowski, B., Arzt, J., and Ferrone, C. (1991) Mutagenesis of the human transferrin receptor: two cytoplasmic phenylalanines are required for efficient internalization and a second-site mutation is capable of reverting an internalization-defective phenotype. *J. Cell Biol.*, **112**, 853.

43. Jadot, M., Canfield, W. M., Gregory, W., and Kornfeld, S. (1992). Characterization of the signal for rapid internalization of the bovine mannose 6-phosphate/insulin-like growth factor-II receptor. *J. Biol. Chem.*, **267**, 11069.

44. Guarnieri, F. G., Arterburn, L. M., Penno, M. B., Cha, Y., and August, J. T. (1993) The motif tyr–x–x–hydrophobic residue mediates lysosomal membrane targeting of lysosome-associated membrane protein 1. *J. Biol. Chem.*, **268**, 1941.

45. Thomas, D. C., Brewer, C. B., and Roth, M. G. (1993) Vesicular stomatitis virus glycoprotein contains a dominant cytoplasmic basolateral sorting signal critically dependent upon a tyrosine. *J. Biol. Chem.*, **268**, 3313.

46. Thomas, D. C. and Roth, M. G. (1994) The basolateral targeting signal of the cytoplasmic domain of VSVG protein resembles a variety of intracellular targeting motifs related by primary sequence but having diverse targeting activities. *J. Biol. Chem.*, **269**, 15732.

47. Miettinen, H. M., Matter, K., Hunziker, W., Rose, J. K., and Mellman, I. (1992) Fc receptor endocytosis is controlled by a cytoplasmic domain determinant that actively prevents coated pit localization. *J. Cell Biol.*, **116**, 875.

48. Naim, H. Y. and Roth, M. G. (1994) Characteristics of the internalization signal in the Y543 influenza virus hemagglutinin suggest a model for recognition of internalization signals containing tyrosine. *J. Biol. Chem.*, **269**, 3928.

49. Lemansky, P., Fatemi, S. H., Gorican, B., Meyale, S., Rossero, R., and Tartakoff, A. M. (1990) Dynamics and longevity of the glycolipid-anchored membrane protein, Thy-1. *J. Cell Biol.*, **110**, 1525.

50. Fire, E., Zwart, D. E., Roth, M. G., and Henis, Y. I. (1991) Evidence from lateral mobility studies for dynamic interactions of a mutant influenza hemagglutinin with coated pits. *J. Cell Biol.*, **115**, 1585.

51. Ishihara, A., Hou, Y., and Jacobson, K. (1987) The Thy-1 antigen exhibits rapid lateral diffusion in the plasma membrane of rodent lymphoid cells and fibroblasts. *Proc. Natl Acad. Sci. USA*, **84**, 1290.

52. Verrey, F., Gilbert, T., Mellow, T., Proulx, G., and Drickamer, K. (1990) Endocytosis via coated pits mediated by glycoprotein receptor in which the cytoplasmic tail is replaced by unrelated sequences. *Cell Regul.*, **1**, 471.

53. Roth, M. G., Henis, Y. I., Brewer, C. B., Ktistakis, N. T., Shia, S., Lazarovits, J., Fire, E., Thomas, D. C., and Zwart, D. E. (1993) Sorting of membrane proteins in the endocytic and exocytic pathways. In *Cell biology and biotechnology*. Oka, M. S. and Rupp, R. G. (ed.). Springer-Verlag, New York, p. 137.

54. Weyand, C. M., Goronzy J., and Fathman, C. G. (1987) Modulation of CD4 by antigenic activation. *J. Immunol.*, **138**, 1351.

55. Pelchen-Mathews, A., Armes, J. E., and Marsh, M. (1989) Internalization and recycling of CD4 transfected into HeLa and NIH3T3 cells. *EMBO J.*, **8**, 3641.

56. Johnson, K. F. and Kornfeld, S. (1992) The cytoplasmic tail of the mannose 6-phosphate/insulin-like growth factor-II receptor has two signals for lysosomal enzyme sorting in the Golgi. *J. Cell Biol.*, **119**, 149.

57. Johnson, K. F. and Kornfeld, S. (1992) A His–Leu–Leu sequence near the carboxyl terminus of the cytoplasmic domain of the cation-dependent mannose 6-phosphate receptor is necessary for the lysosomal enzyme sorting function. *J. Biol. Chem.*, **267**, 17110.

58. Letourneur, F. and Klausner, R. D. (1992) A novel di-leucine motif and a tyrosine-based motif independently mediate lysosomal targeting and endocytosis of CD3 chains. *Cell*, **69**, 1143.

59. Aiken, C., Konner, J., Landau, N. R., Lenburg, M. E., and Trono, D. (1994) Nef induces CD4 endocytosis: requirement for a critical dileucine motif in the membrane-proximal CD4 cytoplasmic domain. *Cell*, **76**, 853.

60. Klausner, R. D., Donaldson, J. G., and Lippincott-Schwartz (1992) Brefeldin A: insights into the control of membrane traffic and organelle structure. *J. Cell Biol.*, **166**, 1071.

61. Pelham, H. R. (1991) Multiple targets for brefeldin A. *Cell*, **67**, 449.

62. Lippincott-Schwartz, J., Yuan, L., Tipper, C., Amherdt, M., Orci, L., and Klausner, R. D. (1991) Brefeldin A's effects on endosomes, lysosomes, and the TGN suggest a general mechanism for regulating organelle structure and membrane traffic. *Cell*, **67**, 601.

63. Chege, N. W. and Pfeffer, S. R. (1990) Compartmentation of the Golgi complex: brefeldin-A distinguishes *trans*-Golgi cisternae from the *trans*-Golgi network. *J. Cell Biol.*, **111**, 893.

64. Reaves, B. and Banting, G. (1992) Perturbation of the morphology of the *trans*-Golgi network following Brefeldin A treatment: redistribution of a TGN-specific integral membrane protein, TGN38. *J. Cell Biol.*, **116**, 85.

65. Wood, S. A., Park, J. E., and Brown, W. J. (1991) Brefeldin A causes a microtubule-mediated fusion of the *trans*-Golgi network and early endosomes. *Cell*, **67**, 591.

66. Wood, S. A. and Brown, W. J. (1992) The morphology but not the function of endosomes and lysosomes is altered by brefeldin. A. *J. Cell Biol.*, **119**, 273.

67. Hunziker, W., Whitney, J. A., and Mellman, I. (1991) Selective inhibition of transcytosis by brefeldin A in MDCK cells *Cell*, **67**, 617.

68. Sandvig, K., Prydz, K., Hansen, S. H., and van Deurs, B. (1991). Ricin transport in brefeldin A-treated cells: correlation between Golgi structure and toxic effect. *J. Cell Biol.*, **115**, 971.

69. Low, S. H., Wong, S. H., Tang, B. L., Tan, P., Subramaniam, V. N., and Hong, W. (1991) Inhibition by brefeldin A of protein secretion from the apical cell surface of Madin–Darby canine kidney cells. *J. Biol. Chem.*, **266**, 17729.

70. Prydz, K., Hansen, S. H., Sandvig, K., and van Deurs, B. (1992) Effects of brefeldin A on endocytosis, transcytosis and transport to the Golgi complex in polarized MDCK cells. *J. Cell Biol.*, **119**, 259.

71. Wan, J., Taub, M. E., Shah, D., and Shen, W. C. (1992) Brefeldin A enhances receptor-mediated transcytosis of transferrin in filter-grown Madin–Darby canine kidney cells. *J. Biol. Chem.*, **267**, 13446.

72. Brodsky, F. M., Hill, B. L., Acton, S. L., Nathke, I., Wong, D. H., Ponnambalam, S., and Parham, P. (1991) Clathrin light chains: arrays of protein motifs that regulate coated-vesicle dynamics. *Trends Biochem. Sci.*, **16**, 208.

73. Zaremba, S. and Keen, J. H. (1983). Assembly polypeptides from coated vesicles mediate assembly of unique clathrin coats. *J. Cell Biol.*, **97**, 1339.

74. Unanue, E. R., Ungewickell, E., and Branton, D. (1981) The binding of clathrin triskelions to membranes from coated vesicles. *Cell*, **26**, 439.

75. Pearse, B. M. F. (1988) Receptors compete for adaptors found in plasma membrane coated pits. *EMBO J.*, **7**, 3331.

76. Robinson, M. S. (1987) Coated vesicles and protein sorting. *J. Cell Biol.*, **87**, 203.

77. Vigers, G. P., Crowther, R. A., and Pearse, B. M. (1986) Three-dimensional structure of clathrin cages in ice. *EMBO J.*, **5**, 529.

78. Pearse, B. M. F. and Robinson, M. S. (1984) Purification and properties of 100 kd proteins from coated vesicles and their reconstitution with clathrin. *EMBO J.*, **3**, 1951.

79. Ahle, S., Mann, A., Eichelsbacher, U., and Ungewickell, E. (1988) Structural relationship between clathrin assembly proteins from the Golgi and the plasma membrane. *EMBO J.*, **7**, 919.

80. Virshup, D. M. and Bennett, V. (1988) Clathrin-coated vesicle assembly polypeptides: physical properties and reconstitution studies with brain membranes. *J. Cell Biol.*, **106**, 39.

81. Robinson, M. S. (1989) Cloning of cDNAs encoding two related 100-kD coated vesicle proteins (alpha-adaptins). *J. Cell Biol.*, **108**, 833.

82. Kirchhausen, T., Nathanson, K. L., Matsui, W., Vaisberg, A., Chow, E. P., Burne, C., Keen, J. H., and Davis, A. E. (1989) Structural and functional division into two domains of the large (100- to 115-kDa) chains of the clathrin-associated complex. *Proc. Natl Acad. Sci. USA*, **86**, 2612.

83. Ponnambalam, S., Robinson, M. S., Jackson, A. P., Peiperl, L., and Parham, P. (1990) Conservation and diversity in families of coated vesicle adaptins. *J. Biol. Chem.*, **265**, 4814.

84. Heuser, J. E. and Keen, J. (1988) Deep-etch visualization of proteins involved in clathrin assembly. *J. Cell Biol.*, **107**, 877.

85. Kirchhausen, T. (1993) Coated pits and coated vesicles — sorting it all out. *Curr. Opin. Struct. Biol.*, **3**, 182.

86. Peeler, J. S., Donzell, W. C., and Anderson, R. G. (1993) The appendage domain of the AP-2 subunit is not required for assembly or invagination of clathrin-coated pits. *J. Cell Biol.*, **120**, 47.

87. Pearse, B. M. F. (1985) Assembly of the mannose 6-phosphate receptor into reconstituted clathrin coats. *EMBO J.*, **4**, 2457.

88. Glickman, J. N., Conibear, E., and Pearse, B. M. F. (1989) Specificity of binding of clathrin adaptors to signals on the mannose-6-phosphate/insulin-like growth factor II receptor. *EMBO J.*, **8**, 1041.

89. Brown, W. J., Goodhouse, J., and Farhquar, M. G. (1986) Mannose-6-phosphate receptors for lysosomal enzymes cycle between the Golgi complex and endosomes. *J. Cell Biol.*, **103**, 1235.

90. Duncan, J. R. and Kornfeld, S. (1988) Intracellular movement of two mannose 6-phosphate receptors: return to the Golgi apparatus. *J. Cell Biol.*, **106**, 617.

91. Sosa, M. A., Schmidt, B., von Figura, K., and Hille-Rehfeld, A. (1993) *In vitro* binding of plasma membrane-coated vesicle adaptors to the cytoplasmic domain of lysosomal acid phosphatase. *J. Biol. Chem.*, **268**, 12537.

92. Beltzer, J. P. and Spiess, M. (1991) *In vitro* binding of the asialoglycoprotein receptor to the B adaptin of plasma membrane coated vesicles. *EMBO J.*, **10**, 3735.

93. Sorkin, A. and Carpenter, G. (1993) Interaction of activated EGF receptors with coated pit adaptins. *Science*, **261**, 612.

94. Chang, C. P., Lazar, C. S., Walsh, B. J., Komuro, M., Collawn, J. F., Kuhn, L. A., Tainer, J. A., Trowbridge, I. S., Farquhar, M. G., Rosenfeld, M. G., Wiley, H. S. and Gill, G. N. (1993) Ligand-induced internalization of the epidermal growth factor receptor is mediated by multiple endocytic codes analogous to the tyrosine motif found in constitutively internalized receptors. *J. Biol. Chem.*, **268**, 19312.

95. Miller, K., Shipman, M., Trowbridge, I. S., and Hopkins, C. R. (1991) Transferrin receptors promote the formation of clathrin lattices. *Cell*, **65**, 621.

96. Moore, M. S., Mahaffey, D. T., Brodsky, F. M., and Anderson, R. G. W. (1989) Assembly of clathrin coated pits onto purified plasma membranes. *Science*, **236**, 558.

97. Prywes, R., Linveh, E., Ullrich, A., and Schlessinger, J. (1986) Mutations in the cytoplasmic domain of EGF receptor affect EGF binding and receptor internalization. *EMBO J.*, **5**, 2179.

98. Mahaffey, D. T., Peeler, J. S., Brodsky, F. M., and Anderson, R. G. (1990) Clathrin-coated pits contain an integral membrane protein that binds the AP-2 subunit with high affinity. *J. Biol. Chem.*, **265**, 16514.

99. Seaman, M. N. J., Ball, C. L., and Robinson, M. S. (1993) Targeting and mistargeting of plasma membrane adaptors *in vitro*. *J. Cell Biol.*, **123**, 1093.

100. Wang, L. H., Rothberg, K. G., and Anderson, R. G. (1993) Mis-assembly of clathrin lattices on endosomes reveals a regulatory switch for coated pit formation. *J. Cell Biol.*, **123**, 1107.

101. Heuser, J. (1980) Three dimensional visualization of coated vesicle formation in fibroblasts. *J. Cell Biol.*, **84**, 560.

102. Pearse, B. M. and Crowther, R. A. (1987) Structure and assembly of coated vesicles. *Annu. Rev. Biophys. Biophys. Chem.*, **16**, 49.

103. Kosaka, T. and Ikeda, K. (1983) Reversible blockage of membrane retrieval and endocytosis in the garland cell of the temperature-sensitive mutant of *Drosophila melanogaster* shibire. *J. Cell Biol.*, **97**, 499.

104. vanderBliek, A. M. and Meyerowitz, E. M. (1991) Dynamin-like protein encoded by *Drosophila* shibire gene associated with vesicular traffic. *Nature*, **351**, 411.

105. Chen, M. S., Obar, R. A., Scroeder, C. C., Austin, T. W., Poodry, C. A., Wadsworth, S. C., and Vallee, R. B. (1991) Multiple forms of dynamin are encoded by shibire, a *Drosophila* gene involved in endocytosis. *Nature*, **351**, 583.

106. Schmid, S. (1993) Coated-vesicle formation *in vitro*: conflicting results using different assays. *Trends Cell Biol.*, **3**, 145.

107. Carter, L. L., Redelmeier, T., E., Woolenweber, L. A., and Schmid, S. L. (1993) Multiple GTP-binding proteins participate in clathrin-coated vesicle-mediated endocytosis. *J. Cell Biol.*, **120**, 37.

108. Hopkins, C. R., Gibson, A., Shipman, M., and Miller, K. (1990) Movement of internalized ligand–receptor complexes along a continuous endosomal reticulum. *Nature*, **346**, 335.

109. Stoorvogel, W., Strous, G. J., Geuze, H. J., Oorschot, V., and Schwartz, A. L. (1991) Late endosomes derive from early endosomes by maturation. *Cell*, **65**, 417.

110. Dunn, K. W. and Maxfield, F. R. (1992) Delivery of ligands from sorting endosomes to late endosomes occurs by maturation of sorting endosomes. *J. Cell Biol.*, **117**, 301.

111. Beck, K. A., Chang, M., Brodsky, F. M., and Keen, J. H. (1992) Clathrin assembly protein AP-2 induces aggregation of membrane vesicles: a possible role for AP-2 in endosome formation. *J. Cell Biol.*, **119**, 787.

112. Colombo, M. I., Mayorga, L. S., Casey, P. J., and Stahl, P. D. (1992) Evidence of a role for heterotrimeric GPT-binding proteins in endosome fusion. *Science*, **255**, 1695.

113. Gorvel, J. P., Chavrier, P., Zerial, M., and Gruenberg, J. (1991) rab5 controls early endosome fusion *in vitro*. *Cell*, **64**, 915.

114. Lenhard, J. M., Kahn, R. A., and Stahl, P. D. (1992) Evidence for ADO-ribosylation factor (ARF) as a regulator of *in vitro* endosome–endosome fusion. *J. Biol. Chem.*, **267**, 13047.

115. Bucci, C., Parton, R. G., Mather, I. H., Stunnenberg, H., Simons, K., Hoflack, B., and Zerial, M. (1992) The small GTPase rab5 functions as a regulatory factor in the early endocytic pathway. *Cell*, **70**, 715.

116. Stenmark, H., Valencia, A., Martinez, O., Ullrich, O., Goud, B., and Zerial, M. (1994) Distinct structural elements of rab5 define its functional specificity. *EMBO J.*, **13**, 575.

117. Mayorga, L. S., Colombo, M. I., Lennartz, M., Brown, E. J., Rahman, K. H., Weiss, R., Lennon, P. J., and Stahl, P. D. (1993) Inhibition of endosome fusion by phospholipase A2 (PLA2) inhibitors points to a role for PLA2 in endocytosis. *Proc. Natl Acad. Sci. USA*, **90**, 10255.

118. Gruenberg, J. E. and Howell, K. E. (1986) Reconstitution of vesicle fusions occurring in endocytosis with a cell-free system. *EMBO J.*, **5**, 3091.

119. Diaz, R., Mayorga, L., and Stahl, P. (1988) *In vitro* fusion of endosomes following receptor-mediated endocytosis. *J. Biol. Chem.*, **263**, 6093.

120. Griffiths, G., Back, R., and Marsh, M. (1989) A quantitative analysis of the endocytic pathway in baby hamster kidney cells. *J. Cell Biol.*, **109**, 2703.

121. Tooze, J. and Hollinshead, M. (1991) Tubular early endosomal networks in AtT20 and other cells. *J. Cell Biol.*, **115**, 635.

122. Wileman, T., Harding, C., and Stahl, P. (1985) Receptor-mediated endocytosis. [Review]. *Biochem. J.*, **232**, 1.

123. Steinman, R. M., Mellman, I. S., Muller, W. A., and Cohn, Z. A. (1982) Endocytosis and the recycling of plasma membrane. *J. Cell Biol.*, **96**, 1.

124. Stoorvogel, W., Geuze, H. J., and Strous, G. J. (1987) Sorting of endocytosed transferrin and asialoglycoprotein occurs immediately after internalization in HepG2 cells. *J. Cell Biol.*, **104**, 1261.

125. Mueller, S. C. and Hubbard, A. (1986) Receptor-mediated endocytosis of asialoglycoproteins by rat hepatocytes: receptor-positive and receptor-negative endosomes. *J. Cell Biol.*, **102**, 932.

126. Clague, M. J., Urbe, S., Aniento, F. and Gruenberg, J. (1994) Vacuolar ATPase activity is required for endosomal carrier vesicle formation. *J. Biol. Chem.*, **269**, 21.

127. Bowman, E. J., Siebers, A. and Altendorf, K. H. (1988) Bafilomycins: a class of inhibitors of membrane ATPases from microorganisms, animal cells, and plants. *Proc. Natl Acad. Sci. USA*, **85**, 7972.

128. Bomsel, M., Parton, R., Kuznetsov, S. A., Schroer, T. A., and Gruenberg, J. (1990) Microtubule- and motor-dependent fusion *in vitro* between apical and basolateral endocytic vesicles from MDCK cells. *Cell*, **62**, 719.

129. Aniendo, F., Emans, N., Griffiths, G. and Gruenberg, J. (1993) Cytoplasmic dynein-dependent vesicular transport from early to late endosomes. *J. Cell Biol.*, **123**, 1373.

130. Kok, J. W., Babia, T., and Hoekstra, D. (1991) Sorting of sphingolipids in the endocytic pathway of HT29 cells. *J. Cell Biol.*, **114**, 231.

131. Mayor, S., Presley, J. F. and Maxfield, F. R. (1993) Sorting of membrane components from endosomes and subsequent recycling to the cell surface occurs by a bulk flow process. *J. Cell Biol.*, **121**, 125.

132. Johnson, L. S., Dunn, K. W., Pytowski, B. and Mcgraw, T. E. (1993) Endosome acidification and receptor trafficking — bafilomycin A slows receptor externalization by a mechanism involving the receptors internalization motif. *Mol. Biol. Cell*, **4**, 1251.

133. Presley, J. F., Mayor, S., Dunn, K. W., Johnson, L. S. and McGraw, T. E. (1993) The End2 mutation in CHO cells slows the exit of transferrin receptors from the recycling compartment but bulk membrane recycling is unaffected. *J. Cell Biol.*, **122**, 1231.

134. Dunn, K. W., McGraw, T. E. and Maxfield, F. R. (1989) Iterative fractionation of recycling receptors from lysosomally destined ligands in an early sorting endosome. *J. Cell Biol.*, **106**, 3303.

135. Cain, C. C., Wilson, R. B. and Murphy, R. F. (1991) Isolation by fluorescence activated cell sorting of Chinese hamster ovary cell lines with pleiotropic, temperature-conditional defects in receptor recycling. *J. Biol. Chem.*, **266**, 11746.

136. Wilson, R. B., Mastick, C. C. and Murphy, R. F. (1993) A Chinese hamster ovary cell line with a temperature-conditional defect in receptor recycling is pleiotropically defective in lysosome biogenesis. *J. Biol. Chem.*, **268**, 25357.

137. Bucci, M. and Murphy, R. F. (1993) Isolation and characterization of revertants of the end6 mutant TFT1.11 reveals a genetic link between receptor recycling and lysosome biogenesis. *Mol. Biol. Cell*, **4**, 1884.

138. vanderSluijs, P., Hull, M., Webster, P., Male, P., Goud, B. and Mellman, I. (1992) The small GTP-binding protein rab4 controls an early sorting event on the endocytic pathway. *Cell*, **70**, 729.

139. Mostov, K. E. and Deitcher, D. L. (1986) Polymeric immunoglobulin receptor expressed in MDCK cells transcytoses IgA. *Cell*, **46**, 613.

140. Mostov, K. (1991) The polymeric immunoglobulin receptor. *Semin. Cell Biol.*, **2**, 411.

141. Mostov, K., Apodaca, G., Aroeti, B. and Okamoto, C. (1992) Plasma membrane protein sorting in polarized epithelial cells. *J. Cell Biol.*, **116**, 577.

142. Casanova, J. E., Breitfeld, P. P., Ross, S. A. and Mostov, K. E. (1990) Phosphorylation of the polymeric immunoglobulin receptor required for its efficient transcytosis. *Science*, **248**, 742.

143. Hirt, R. P., Hughes, G. J., Frutinger, S., Michetti, P., Perregaux, C., Poulain-Godefroy, O., Jeanguenat, N., Neutra, M. R., and Kraehenbuhl, J.-P. (1993) Transcytosis of the polymeric Ig receptor requires phosphorylation of serine 664 in the absence but not the presence of dimeric IgA. *Cell*, **74**, 245.

144. Bomsel, M. and Mostov, K. E. (1993) Possible role of both the alpha and beta gamma subunits of the heterotrimeric G protein, G_s, in transcytosis of the polymeric immunoglobulin receptor. *J. Biol. Chem.*, **268**, 25824.

145. Hansen, S. H. and Casanova, J. E. (1993) Stimulation of transcytosis by elevation of cellular cAMP in MDCK cells. *Mol. Biol. Cell*, **4**, 565.

146. Cardone, M. H., Smith, B. L., Mochly-Rosen, D., and Mostov, K. E. (1993) Phorbol ester mediated stimulation of transcytosis of the polymeric immunoglobulin receptor in MDCK cells involves protein kinase-C alpha translocation. *Mol. Biol. Cell*, **4**, 524.

147. Barroso, M. and Sztul, E. S. (1994) Basolateral to apical transcytosis in polarized cells is indirect and involves BFA and trimeric G protein sensitive passage through the apical endosome. *J. Cell Biol.*, **124**, 83.

148. Apodaca, G., Katz, L. A., and Mostov, K. E. (1994) Receptor-mediated transcytosis of IgA in MDCK cells is via apical recycling endosomes. *J. Cell Biol.*, **125**, 67.

149. Carlsson, S. R. and Fukuda, M. (1992) The lysosomal membrane glycoprotein lamp-1 is transported to lysosomes by two alternative pathways. *Arch. Biochem. Biophys.*, **296**, 630.

150. Harter, C. and Mellman, I. (1992) Transport of the lysosomal membrane glycoprotein lgp120 (lgp-A) to lysosomes does not require appearance on the plasma membrane. *J. Cell Biol.*, **117,** 311.

151. Griffiths, G., Hoflack, B., Simons, K., Mellman, I., and Kornfeld, S. (1988) The mannose 6-phosphate receptor and the biogenesis of lysosomes. *Cell*, **52,** 329.

152. Ludwig, T., Griffiths, G., and Hoflack, B. (1991) Distribution of newly synthesized lysosomal enzymes in the endocytic pathway of normal rat kidney cells. *J. Cell Biol.*, **115,** 1561.

153. Johnson, K. F., Chan, W., and Kornfeld, S. (1990) Cation-dependent mannose 6-phosphate receptor contains two internalization signals in its cytoplasmic domain. *Proc. Natl Acad. Sci. USA*, **87,** 10010.

154. Canfield, W. M., Johnson, K. F., Ye, R. D., Gregory, W., and Kornfeld, S. (1991) Localization of the signal for rapid internalization of the bovine cation-independent mannose 6-phosphate/insulin-like growth factor-II receptor to amino acids 24–29 of the cytoplasmic tail. *J. Biol. Chem.*, **266,** 5682.

155. Schulze-Garg, C., Boker, C., Nadimpalli, S. K., von Figura, K., and Hille-Rehfeld, A. (1993) Tail-specific antibodies that block return of 46,000 M(r) mannose 6-phosphate receptor to the *trans*-Golgi network. *J. Cell Biol.*, **122,** 541.

156. Meresse, S., Ludwig, T., Frank, R., and Hoflack, B. (1990) Phosphorylation of the cytoplasmic domain of the bovine cation-independent mannose 6-phosphate receptor. Serines 2421 and 2492 are the targets of a casein kinase II associated to the Golgi derived HAI adaptor complex. *J. Biol. Chem.*, **265,** 18833.

157. Meresse, S. and Hoflack, B. (1993) Phosphorylation of the cation-independent mannose-6-phosphate receptor is closely associated with its exit from the *trans*-Golgi network. *J. Cell Biol.*, **120,** 167.

158. Hemer, F., Korner, C. and Braulke, T. (1993) Phosphorylation of the human 46-kDa mannose 6-phosphate receptor in the cytoplasmic domain at serine 56. *J. Biol. Chem.*, **268,** 17108.

159. Mathews, P. M., Martinie, J. B., and Fambrough, D. M. (1992) The pathway and targeting signal for delivery of the integral membrane glycoprotein LEP100 to lysosomes. *J. Cell Biol.*, **118,** 1027.

160. Braun, M., Waheed, A., and von Figura, K. (1989) Lysosomal acid phosphatase is transported to lysosomes via the cell surface. *EMBO J.*, **8,** 3633.

161. Hanover, J. A., Willingham, M. C., and Pastan, I. H. (1984) Kinetics of transit of transferrin and epidermal growth factor through clathrin-coated membranes. *Cell*, **39,** 283.

162. Green, S. A. and Kelly, R. B. (1992) Low density lipoprotein receptor and cation-independent mannose 6-phosphate receptor are transported from the cell surface to the Golgi apparatus at equal rates in PC12 cells. *J. Cell Biol.*, **117,** 47.

163. Green, S. A., Setiadi, H., Mcever, R. P., and Kelly, R. B. (1994) The cytoplasmic domain of p-selectin contains a sorting determinant that mediates rapid degradation in lysosomes. *J. Cell Biol.*, **124,** 435.

164. Mellman, I. and Plutner, H. (1984) Internalization and degradation of macrophage Fc receptors bound to polyvalent immune complexes. *J. Cell Biol.*, **98,** 1170.

165. Fire, E., Gutman, O., Roth, M. G., and Henis, Y. I. (1995) Dynamic or stable interactions of influenza hemagglutinin mutants with coated pits. *J. Biol. Chem.*, **270,** 21075.

166. Luzio, J. P. and Banting, G. (1993) Eukaryotic membrane traffic: retrieval and retention mechanisms to achieve organelle residence. *Trends Biochem. Sci.*, **18,** 395.

167. Humphrey, J. S., Peters, P. J., Yuan, L. C., and Bonifacino, J. S. (1993) Localization of

TGN38 to the *trans*-Golgi network: involvement of a cytoplasmic tyrosine-containing sequence. *J. Cell Biol.*, **120**, 1123.

168. Hunziker, W., Harter, C., Matter, K., and Mellman, I. (1991) Basolateral sorting in MDCK cells requires a distinct cytoplasmic domain determinant. *Cell*, **66**, 907.

169. Prill, V., Lehmann, L., von Figura, K., and Peters, C. (1993) The cytoplasmic tail of lysosomal acid phosphatase contains overlapping but distinct signals for basolateral sorting and rapid internalization in polarized MDCK cells. *EMBO J.*, **12**, 2181.

170. Hunziker, W., Whitney, J. A., and Mellman, I. (1992) Brefeldin A and the endocytic pathway. Possible implications for membrane traffic and sorting. *FEBS Lett.*, **307**, 93.

171. Rothman, J. E. and Orci, L. (1992) Molecular dissection of the secretory pathway. *Nature*, **355**, 409.

172. Girones, N., Alvarez, E., Seth, A., Lin, I. M., Latour, D. A., and Davis, R. J. (1991) Mutational analysis of the cytoplasmic tail of the human transferrin receptor. Identification of a sub-domain that is required for rapid endocytosis. *J. Biol. Chem.*, **266**, 19006.

173. Bos, K., Wraight, C., and Stanley, K. K. (1993) TGN38 is maintained in the *trans*-Golgi network by a tyrosine-containing motif in the cytoplasmic domain. *EMBO J.*, **12**, 2219.

174. Piper, R. C., Tai, C., Kulesza, P., Pang, S., Warnock, D., Baenziger, J., Slot, J. W., Geuze, H. J., Puri, C., and James, D. E. (1993) GLUT-4 NH_2 terminus contains a phenylalanine-based targeting motif that regulates intracellular sequestration. *J. Cell Biol.*, **121**, 1221.

175. Mori, S., Ronnstrand, L., Cleason-Welsh, L., and Heldin, C. (1994) A tyrosine residue in the juxtamembrane segment of the platelet derived growth factor b-receptor is critical for ligand-mediated endocytosis. *J. Biol. Chem.*, **269**, 4917.

176. Kruskal, B. A., Sastry, K., Warner, A. B., Mathieu, C. E., and Ezekowitz, A. B. (1992) Phagocytic chimeric receptors require both transmembrane and cytoplasmic domains from the mannose receptor. *J. Exp. Med.*, **176**, 1673.

177. Ogata, S. and Fukuda, M. (1994) Lysosomal targeting of Limp II membrane glycoprotein requires a novel Leu–Ile motif at a particular position in its cytoplasmic tail. *J. Biol. Chem.*, **269**, 5210.

178. Matter, K., Hunziker, W., and Mellman, I. (1992) Basolateral sorting of LDL receptor in MDCK cells: the cytoplasmic domain contains two tyrosine-dependent targeting determinants. *Cell*, **71**, 741.

179. Aroeti, B., Kosen, P. A., Kuntz, I. D., Cohen, F. E., and Mostov, K. E. (1993). Mutational and secondary structural analysis of the basolateral sorting signal of the polymeric immunoglobulin receptor. *J. Cell Biol.*, **123**, 1149.

180. LeBivic, A., Sambuy, Y., Patzak, A., Patil, N., Chao, M., and Rodriguez-Boulan, E. (1991) An internal deletion in the cytoplasmic tail reverses the apical localization of human NGF receptor in transfected MDCK cells. *J. Cell Biol.*, **115**, 607.

181. Nabi, I. R. and Rodriguez-Boulan, E. (1993) Increased LAMP-2 polylactosamine glycosylation is associated with its slower Golgi transit during establishment of a polarized MDCK epithelial monolayer. *Mol. Biol. Cell*, **4**, 627.

Note added in proof. For a recent review of this field see for example: Robinson, M. R., Watts, C., and Zerial, M. (1996) Membrane dynamics in endocytosis. *Cell*, **84**, 13.

Index